UNIVERSITY OF STRATHCLYDE
30125 00413976

Natural Food Colorants

Natural Food Colorants

Edited by

G.A.F. HENDRY

Honorary Lecturer
NERC Unit of Comparative Plant Ecology
University of Sheffield

and

J.D. HOUGHTON

Lecturer
Department of Biochemistry and Molecular Biology
University of Leeds

Blackie
Glasgow and London

Published in the USA by
avi, an imprint of
Van Nostrand Reinhold
New York

Blackie and Son Ltd
Bishopbriggs, Glasgow G64 2NZ
and
7 Leicester Place, London WC2H 7BP

Published in the United States of America by
AVI, an imprint of
Van Nostrand Reinhold
115 Fifth Avenue
New York, New York 10003

Distributed in Canada by
Nelson Canada
1120 Birchmount Road
Scarborough, Ontario M1K 5G4, Canada

16 15 14 13 12 11 10 9 8 7 6 5 4 3 2 1

First published 1992

British Library Cataloguing in Publication Data

Natural food colorants.
I. Hendry, G.
664

ISBN 0-216-93146-0

Library of Congress Cataloging-in-Publication Data

Natural food colorants / edited by George Alexander Forbes Hendry.
p. cm.
ISBN 0-442-31464-7
1. Coloring matter in food. I. Hendry, George Alexander Forbes.
TP456.C65N37 1991
664′.06—dc20 91-12549
CIP

Phototypesetting by BPCC-AUP Glasgow Ltd
Printed in Great Britain by Thomson Litho Limited, East Kilbride, Scotland

Preface

Natural food colorants have been widely used in the preparation of foods and beverages for centuries and, alongside artificial colorants, they continue to make a significant contribution to the preparation and processing of food. Recent years have seen an increased interest in natural colorants; in some cases the natural product can be priced competitively compared with the artificial alternative and, in others, the consumer has preferred not to use the artificial colorant. As yet, however, up-to-date and readily accessible information on natural pigments is not available, and several workers have commented on the errors and omissions that have become enshrined in the scientific and technological literature on food colorants. It is our aim to improve this situation by bringing together a full consideration of the biochemistry, chemistry and biology of natural pigments since these are the core areas on which the food chemist or technologist bases his or her creative skills.

The book has been written by experts on natural food colorants and the biochemistry of natural pigments, who have been drawn from both the academic and industrial spheres. We have allowed a certain amount of overlap between different chapters written with a different emphasis. Carotenoids, for example, are discussed as pigments in biology (chapter 1), as colorants with industrial applications (chapter 2) and as colorants of the future (chapter 7), as well as in the principal chapter devoted to them (chapter 5).

One of the aims in launching this book is to extend the horizons of those using natural colorants. Each contributor has considered pigments which are widely used today, and those of potential use in the near future. Natural pigments which are rare, of limited distribution in biology, or expensive to acquire, may soon become commonplace in the hands of the biotechnologist and offer considerable potential for exploitation. This is supported by the large increase in the number of patent applications for food colorants, covering a large number of unusual natural pigments. In the following chapters, over 400 natural pigments are described, with data on structure, stability and natural occurrence. We hope that our enthusiasm for exploring nature's rich box of colours will become apparent.

G.A.F.H.
J.D.H.

Contributors

Dr G. Britton	Department of Biochemistry, University of Liverpool, PO Box 147, Liverpool L69 3BX
Professor F.J. Francis	Department of Food Science, College of Food and Natural Resources, University of Massachusetts, Amherst, MA 01003
Dr G.A.F. Hendry	NERC Unit of Comparative Plant Ecology, Department of Animal and Plant Sciences, University of Sheffield, Sheffield S10 2TN
Mr B.S. Henry	Phytone Limited, Unit 2B Boardman Industrial Estate, Hearthcote Road, Swadlincote, Burton-on-Trent, Staffordshire DE11 9DL
Dr J.D. Houghton	Department of Biochemistry and Molecular Biology, University of Leeds, Leeds LS2 9JT
Profesor R.L. Jackman	Department of Food Science, Ontario Agricultural College, University of Guelph, Guelph, Ontario N1G 2W1, Canada
Professor J.L. Smith	Department of Food Science, Ontario Agricultural College, University of Guelph, Guelph, Ontario N1G 2W1, Canada

Contents

1 Natural pigments in biology

G.A.F. HENDRY

1.1 Summary

This introductory chapter provides a working definition of 'natural' pigments. An outline is given of the way colours or hues are defined throughout the book, and of the areas of confusion surrounding colour description. The variation in colour perception by human eyes is also described. Chemical nomenclature and certain physical attributes common to pigmented compounds are provided, including a brief synopsis of the structural features needed to make an otherwise colourless molecule pigmented. The greater part of the chapter consists of a survey of pigmented compounds found in biology. Two systems of classification are adopted, one based on structural affinities, the second based on the natural occurrence of the pigment in biology. The latter system provides a resumé of almost all but the rarest pigments in higher and lower animals, plants including algae, fungi, lichens and bacteria. Where appropriate, the classifications are cross-referenced to the fuller treatment of certain pigments in the succeeding chapters.

1.2 Introduction

The immediate purpose of this chapter is to introduce the major and at least the most widespread of the minor pigments occurring naturally in plants, fungi, lichens, animals and bacteria. Whether or not these have previously been exploited as natural food colorants is not of immediate consideration. Neither is the availability of the pigment for commercial exploitation, particularly in view of the recent and promised advances in gene transfer and biotechnology. Theoretically, almost all reasonably stable biological pigments can now be considered for biotechnological production and, in future, may become available to the food and other industries as natural colorants.

1.3 Definitions

By definition, a natural pigment in biological systems is one that is synthesized and accumulated in, or excreted from living cells. In addition, certain pigments, particularly the more simple phenolic derivatives such as the anti-coagulant coumarins, may be formed by the dying cell. What is *not* a natural pigment in biology is less easy to define. For the purposes of this chapter, inorganic pigments based on, for example, iron or titanium are not considered natural if only because they do not appear to play a significant or even minor role in biology. Nor are the several organic-iron complexes formed synthetically from naturally occurring organic complexes. That 'work-horse' of the food colorist's palette, caramel, is also not considered in depth in this chapter since it is not a pigment of living cells, even if its constituent parts (sugars and amino acids and amides) are.

This working definition and the delimitation of natural pigments inevitably highlight certain 'grey' areas, and so common sense and a little pragmatism have been employed to keep within the spirit of the above definitions while remaining relevant to the particular interests of the food colorant user. For example, green chlorophylls are truly natural compounds: the grey-green phaeophytin pigments extracted from green plants are natural in that they may be formed from chlorophyll during the natural senescence of the dying leaf and incidentally during industrial extraction of chlorophylls from leaves. The bright green Cu-chlorophyllins (see chapter 3) are not found in nature but represent man's close attempt to restore the original colour of true chlorophyll from these phaeophytins. All of these green pigments are considered here as 'natural'. Indigo, a well-known plant dye, is present in living tissues but as the colourless glycoside. Only on extraction, hydrolysis and oxidation does this pigment acquire its blue colour. For this reason, an expanded definition of a natural colorant as a 'pigment formed in living or dead cells of plants, animals, fungi or micro-organisms, including organic compounds isolated from cells and structurally modified to alter stability, solubility or colour intensity' is offered. The definition must also allow for pigments whose synthesis is engineered in genetically transformed organisms. For pragmatic reasons, structural colours due to diffraction, interference or Tyndall scattering of light are not discussed in depth; nevertheless, many play a significant role in the coloration of animals.

1.4 Colour and colour description

Throughout this book and most of the literature colours are described by the wavelength (λ) of the maximum absorption (A_{max}) in the visible part of the electromagnetic spectrum, expressed in nanometres (nm). White light is seen as the simultaneous incidence of the full range of the visible spectrum (wavelengths between 380 and 730 nm approximately) at the *same relative intensities*. Any object that lessens the intensity of one part of the spectrum of white light by absorption (as in a monochromatic filter) will bring about the perception of colour in the residual transmitted light, as well as a reduction in irradiance (measured in Watts per m^2, Einsteins or moles of photons) or more simply a darkening. As more of the spectrum is filtered out the perception of colour will be increasingly monochromatic (shades of grey) darkening ultimately to black.

To those familiar with quantifying and qualifying light, a compound with an absorption maximum of, for example 600 nm, in the yellow part of the spectrum, will be thought of as deep blue in perceived colour. One absorbing light at say 470 nm, the blue-green part of the spectrum, will be predictably a yellow, leaving the unfamiliar reader thoroughly confused. The reason for the confusion is readily explained: light, or more specifically visible radiation, when perceived simultaneously over the complete range of the spectrum as in unfiltered sunlight, is seen as white light by the human eye. When separated by passing the light through a prism, the refracted beams of light are displayed as the sequence of colours familiar in a rainbow. The human eye 'senses' six hues:

1. Red light at around 700 nm;
2. Orange at 625 nm;
3. Yellow at about 600 nm;
4. Green at 525 nm;
5. Blue at around 450 nm; and
6. Violet at and below 400 nm.

Hence the casual laboratory description of a compound as having a 'strong red absorbance' instead of the more precise 'absorbance maximum at 680 nm.' Such a compound would probably appear to most human eyes to have a deep blue or green-blue colour! The confusion arises because the pigment in solution with say a strong red *absorbance* when placed in front of a beam of light, would allow only the passage, that is *transmission*, of the remaining non-red part of the spectrum—the broad area yellow-green-blue-violet. In white light, such a compound would appear to be blue (or green-blue). For a workable translation of description based on absorption maxima into hue perceived, see Table 1.1. The first line of Table 1.1 represents the colours absorbed by a range of pigments (dyes and colorants) that absorb light at

Table 1.1 Perception of colours according to the wavelength of light absorbed

Colour absorbed	Red	Orange	Yellow	Yellow-green	Green	Green-blue	Blue	Violet
Absorption wavelength (maximum) nm	675	600	585	570	540 525	490	460	410
Colour perceived	Blue-green	Blue	Violet	Mauve Red	Orange	Yellow	Yellow-green	

various wavelengths (second line) leaving the remaining colours (third line) to be transmitted and so perceived by the human eye as a distinct colour. Thus a pigment such as anthocyanin under white light will absorb the green portion (first line) of the visible light spectrum at around 520 to 550 nm (second line). Under white light the pigment will be perceived as mauve (third line). Green light has been 'removed' or absorbed by the anthocyanin leaving purple-blue, orange-yellow and red, which, as a mixture, appears as mauve.

In the succeeding chapters data will be presented with the absorption maxima of several hundred different natural pigments. By reference to Table 1.1 the portion of the spectrum of white light absorbed by the pigment can be predicted as well as the colour of the compound as seen by the human eye. Table 1.1 can also be used in the reverse order to predict the part of the spectrum that any coloured compound might absorb when viewed under white light. Thus from the previous example, grape juice anthocyanins that appear to be mauve (line 3) under white light are likely to absorb light at around 510 to 540 nm (line 2), that is the green part of the spectrum (line 1). There is then little point in observing the colour of a treasured glass of wine under the soft green lights sometimes favoured by restaurateurs for the wine will appear as black as ink!

1.5 Colour perception

Numerous definitions of colour are available, the most useful one in this context being that part of the electromagnetic spectrum visible to the human eye and generally regarded as lying between 380 and 730 nm. The key element in the special context of food colorants, however, is the hue *perceived* by the eye of the normal, healthy and average adult. Such an eye is less well-defined.

One problem with defining pigments by their apparent or perceived hue is that one man's mauve is another man's purple or red (and here male gender *is* meant). Precisely where red moves to the description mauve is less a matter of definition but more a subjective decision dependent on variations in colour reception and processing that exist in the eyes of different humans—particularly males. While the human eye can detect up to six major colours and many more intermediate blends, just three predominate—red, yellow and blue. These 'primary' colours (nearly) coincide with the three types of eye pigments.

The retina at the back of the human eye is made up of some 3 million colour-perceiving cone cells and about 100 million rod cells involved in providing monochromatic vision under low light. Each cell contains one of several types of visual pigment, which are structurally based on the β-carotene and vitamin A derivative 11,*cis*-retinol and bound to a particular class of proteins, the opsins. 11,*cis*-retinol itself, without modification, absorbs light maximally at around 500 nm, coinciding conveniently with the spectral maximum of sunlight. The cone-cell pigment proteins are sensitive to different ranges of predominantly blue, green and red light. Colour in this sense is the sum of excitation of red-sensitive, green-sensitive and blue-sensitive cones. The difference between the 'red' and 'green' protein lies in just 15 out of 348 amino-acid residues, and there is evidence from studies of higher mammals that the evolution of the 15-residue difference may have occurred relatively recently in humans. Perhaps because of this recent evolution, considerable variation appears to persist within the human genetic make-up in these 15 or so residues. Small variations between two individuals in the DNA coding for the protein can lead to quite distinct differences in distinguishing the colour perceived by the different viewers. Disagreements over the matching or separation of colours lie particularly at the junctions of what are described as blue/green, red/orange and mauve/violet hues. These differences are not classed in normal parlance as colour blindness since what is green, blue or blue/green is a matter of unspoken consensus. The degree of variation in colour perception and its prevalence is unknown but is believed to be considerable.

Clinically defined 'colour blindness' is, in most cases, the result of an X chromosome-linked recessive mutation occurring in approximately 1 in 12 males. The incidence in females is about 1 in 170. As a brief description of what is a complex and fascinating subject in its own right, the following is offered as an introductory guide. An individual classed as 'colour blind' and lacking say the 575 nm red-light cones will 'see' only green and blue, or if lacking the 540 nm green-light cones will 'see' green as red, the latter condition being the more common. Other individuals carry cone cells with a maximum absorbance lying between green and red, resembling the cone cells of the more primitive apes. In most cases of colour-blindness, the individual is well able to distinguish the colour between blue and yellow but distinctions between green and red are possible often only as differences in brightness. In the most common forms of colour blindness, a red and green mixture of light is seen as either too 'red' or too 'green' by the standards prevailing among the non colour-blind majority. Given the immense effort to achieve food colorants that are perceived by the food chemist as a reproducibly distinct green or red hue, it is worth noting that the endeavour will never be appreciated by one in twelve male consumers!

1.6 Colour and chemical name

One way of pinpointing coloured compounds from a list of organic chemicals is through the name of the substance. There has been a long tradition particularly among German chemists of the last century of giving coloured compounds a name derived from a Greek or Latin pigment. These names have persisted in the trivial (and familiar) names of many chemicals. A short list is given in Table 1.2. A more comprehensive treatment of the subject is provided by Stearn [1] and Flood [2].

Table 1.2 Chemical names implying colour[a]

Chemical prefix or suffix	Classical derivation or root	Implied colour	Chemical example
anil-	al-nil (Arab)	Deep blue	Aniline
arg-	argentum (L)	Silver	Arginine
aur-	aurum (L)	Gold	Aureomycin
-azur	azul (OSp)	Sky blue	Aplysioazurin
chloro-	chloros (Gk)	Yellow-green	Chlorophyll
chrom- or -chrome	chroma (Gk)	Coloured	Chromotropic acid
chryso-	chrysos (Gk)	Gold	Chrysin
citr-	citrus (L)	Lemon-coloured	Citruline
-cyanin	kyanos (Gk)	Blue	Anthocyanin
erythro-	erythros (Gk)	Red	Phycoerythrin
-flavin	flavus (L)	Pale yellow	Riboflavin
fulv-	fulvus (L)	Red-yellow	Fulvine
fusc-	fuscus (L)	Brown	Fuscin
haem- or heme-	haima (Gk)	Blood red	Haematin
indigo-	indikon (Gk)	Deep blue	Indigo
leuco-	leukos (L)	Colourless/white	Leucoptrein
lute-	luteus (Gk)	Yellow	Lutein
mela-	melas (Gk)	Black	Melanin
phaeo-	phaios (Gk)	Dark-coloured	Phaeophytin
porph(o)-	porphyros (Gk)	Purple	Porphyrin
purpur-	purpura (L)	Purple	Purpurin
-pyrrole	pyrros (Gk)	Fiery-red	Pyrrole
rhodo-	rhodon (Gk)	Rose-coloured	Rhodophyllin
rub-	ruber (L)	Red	Bilirubin
ruf-	rufus (L)	Red-brown	Anthrarufin
sepia-	sepia (L)	Brown-pink	Sepiapterin
stilb-	stilbein (Gk)	Glitter	Stilbene
verd- or virid-	viridis (L)	Bright green	Biliverdin
viol-	viola (L)	Purple-blue	Aplysioviolin
xanth-	xanthos (Gk)	Yellow	Xanthophyll

[a] Arab = Arabic; Gk = Ancient Greek; L = Latin; OSp = Old Spanish.

Caution is needed, however, in using these names as indicators of *particular* colours. For example, porphyra of the Greek aristocratic toga gives rise in translation to purple; such a toga seen today would be described by most observers as crimson. Similarly, not all melanins are black as implied by the Greek melas. Given these limitations, however, the terminology is a fast way to recognize a coloured chemical.

1.7 Electronic structure of pigments

Whether or not a biological molecule is coloured, is determined by its structure, particular electronic, size of the molecule, solubility and elemental composition. Fortunately most natural pigments have several common features that immediately distinguish them from the larger number of colourless compounds found in biological material. All biological material is composed of a limited range of elements within the periodic table. Within this narrow selection, pigmented organic compounds generally possess three features:

1. Almost all biological molecules are composed of no more than 17 elements within the periodic table. Of these, just four elements, H, C, N and O predominate.
2. Most pigmented compounds, particularly those other than yellow, contain either N or O, often both.
3. Most are relatively large molecules. Among the more common pigmented compounds, molecular weights range from about 200 (anthraquinones), 300 (anthocyanidins), 400 (betalaines), 500 (carotenoids) to 800 (chlorophylls).

Much greater molecular weights are of course found in pigment polymers such as melanins.

With this limited range of attributes, it is possible to restrict a discussion on the electronic structure of biological pigments to bare essentials (see Brown [3] for a more full explanation). The electrons in the outer shells over the surface of biological molecules, coloured or not, exist in distinct orbits. Each shell holds a finite number of electrons arranged to occupy the least energy-demanding orbits possible. Such electrons are described as being in a resting state. When irradiated with light of sufficient energy, the electrons in the outermost orbitals oscillate in resonance with light of sufficient energy, and are raised from the resting state to a higher excited state. After the excitation, as the electrons decay (relax) to the original resting state so the energy is released in a less energetic form, most commonly as heat. As biological molecules are composed of generally no more than four, perhaps five out of seventeen elements, almost all of the electronic orbitals participating in colour are from just three orbitals designated d, p and n. On excitation, one or more electrons within these orbitals may be raised to respectively d* and p* orbitals.

Transitions from d to d*, which are characteristic of saturated hydrocarbons (with no double bonds), are the most energy demanding, requiring more energy than that provided by the visible part of the spectrum and so appearing colourless. Transitions from p to p* as in unsaturated hydrocarbons (which contain double bonds) are somewhat less demanding energetically but are usually still beyond the energy levels provided by the visible part of the spectrum beyond the ultra-violet. The larger molecules at least may appear as pale yellow, a characteristic of many vegetable oils. The third transition n to d*, characteristic of saturated hydrocarbons particularly with nitrogen (N) or oxygen (O) substitutions may also just appear in the visible part of the spectrum as pale yellow compounds, as in many fungal pigments. However, the least energy demanding are the n to p* transitions found in compounds that are essentially unsaturated hydrocarbons with N or O substitutions (Table 1.3) and it is this group that provide the greatest range of pigmented organic molecules.

Table 1.3 Range of electro-magnetic spectrum absorbed by molecules and causing excitation of electrons from lower (resting state) to higher orbital state

Light causing excitation	Electronic transition	Molecular form	Expected colour
Visible	n → p*	Unsaturated hydrocarbon with N or O	Red, blue or green
Violet to blue	n → d*	Saturated hydrocarbon with N or O	Pale yellow
Near UV	p → p*	Unsaturated hydrocarbon	Probably colourless
Far UV	d → d*	Saturated hydrocarbon	Colourless

The simple picture presented for alkenes needs to be improved, however, to understand why some biological molecules appear to be coloured while most do not. Several important structural modifications cause the electrons of the modified molecule to absorb and to become excited by light at wavelengths considerably longer (and less energetic) than would be predicted from the unmodified alkene. This shift to less energetic wavelengths is called a bathochromic shift.

Six types of structural modification are associated with a bathochromic shift as follows:

1. An increase in chain length;
2. Branching of the carbon chain;
3. Arrangement in a single ring structure (cyclization);
4. Addition of N or O, particularly with 3;
5. Bonding of two or more rings; and
6. Addition of certain transition metals.

Ring structures, particularly in relatively large molecules, permit sharing or delocalization of particular electrons orbiting over the face of the molecule. Such delocalized electrons are more readily raised to an excited state on irradiance with visible light and contribute to most of the non-yellow pigments in biology. The prime biological example of one class of compounds with all of these modifications (that is a large number of conjugated double bonds, branched structures, addition of N and O, linked rings and addition of a transition metal) is the iron porphyrins such as the red haems.

1.8 Classification of biological pigments

It will be apparent that pigmented compounds in biology will tend to be large and complex molecules. Given the long time-scale over which these complex structures have evolved, it is no surprise to find that they are also highly varied. Among just one group, the anthocyanins, over two hundred and fifty structurally and spectrally distinct pigments have been recorded so far [4]. The classification of anything approaching the majority of pigmented compounds in biology into a single workable synopsis will inevitably result in grouping together compounds that have little or no biosynthetic, functional or structural or even colour relationship. Several bold attempts have been made by other workers and the account that follows is in no way the only means of classifying pigments. Good alternatives can be found in the works of Britton [5], Fox [6] and Fox and Vevers [7], also, in more detail, in Needham [8].

Two methods of classification are adopted here, which are linked through cross references and through the index. The first is based on structural affinities. This has the advantage of brevity but the disadvantage that it groups together compounds whose biosynthetic origins are quite unrelated. The second is based on the natural occurrence of pigments in biology. For convenience, the groupings used here cover the pigments of plants including algae, fungi, lichens, higher animals (vertebrates) and lower animals (invertebrates) and bacteria. Throughout, reference will be made to chapters in this book that specifically cover either the pigment itself or at least a group of structurally related pigments.

1.9 Classification based on structural affinities

Almost all biological pigments can be reduced to no more than six major structural classes: the tetrapyrroles, tetra-terpenoids, quinones, O-heterocyclic, N-heterocyclic and metallo-proteins. Given the wealth of variation thrown up by 3000 million years of evolution there will inevitably also be a group of miscellaneous pigments that defy simple classification. In Table 1.4 this latter category has been limited by excluding pigments that are rare or limited in occurrence.

1.9.1 Tetrapyrroles

This is a relatively small group of pigments that contribute the greatest range of colours as well as being the most abundant globally. All are based on the same structure, the tetrapyrrole either in its linear form (Figure 1.1) and

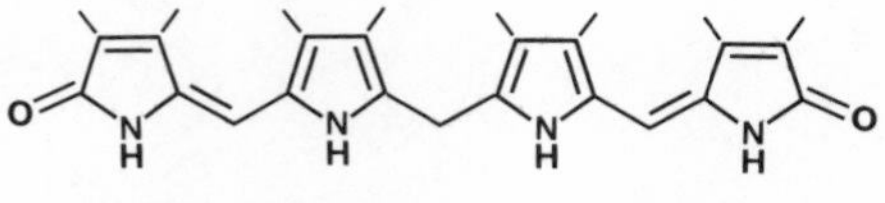

Figure 1.1 A linear tetrapyrrole.

Table 1.4 Naturally occurring pigments in biology

Group	Alternative or familiar name	Major example	Predominant colour	See chapter
Tetrapyrroles	Porphyrins and porphyrin derivatives	Chlorophylls	Green	3
		Haems (hemes)	Red	4
		Bilins (Bile pigments)	Blue-green-yellow-red	4
Tetraterpenoids	Carotenoids	Carotenes	Yellow-red	5
		Xanthophylls	Yellow	5
O-heterocyclic compounds	Flavonoids	Anthocyanins	Blue-red	6
		Flavonols	Yellow-white	
		Flavones	White-cream	
		Anthochlors	Yellow	
Quinones	Phenolic compounds	Naphthaquinones	Red-blue-green	7
		Anthraquinones	Red-purple	7
		Allomelanins	Yellow-brown	
		Tannins	Brown to red	
N-heterocyclic compounds	Indigoids and indole derivatives	Betalaines	Yellow-red	6
		Eumelanins	Black-brown	
		Phaeomelanins	Brown	
		Indigo	Blue-pink	
	Substituted pyrimidines	Pterins	White-yellow	7
		Purines	Opaque white	
		Flavins	Yellow	
		Phenoxazines	Yellow-red	
		Phenazines	Yellow-purple	
Metalloproteins		Cu-proteins	Blue-green	
		Haemerythrin	Red	
		Haemovanadins	Green	
		Adenochrome	Purple-red	
Miscellaneous		Lipofuscins	Brown-grey	
		Fungal pigments	Various but commonly yellow	

known as bilins or bile pigments or in the cyclic form (Figure 1.2) where it constitutes the chromophore of haem proteins (chelated with iron) or chlorophyll proteins (chelated with magnesium). In total, the number of naturally occurring cyclic tetrapyrroles (haems, chlorophylls and their pre-

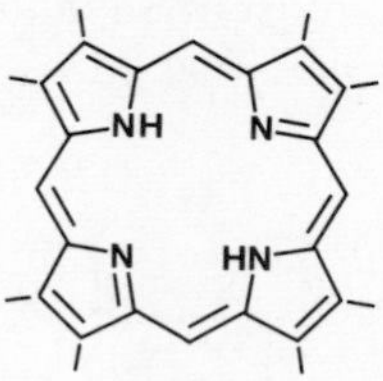

Figure 1.2 A cyclic tetrapyrrole, porphyrin.

cursors) that have been recorded is about 28, with a further 6 or so linear tetrapyrroles (see Hendry and Jones [9] for a full list).

The proteins to which all cyclic and most of the linear pigments are linked alter the solubility properties of the tetrapyrrole and under natural conditions probably determine the stability of the chromophore. In living, healthy cells, free haems and chlorophylls that have been detached from their proteins, probably do not exist or at least do not accumulate to detectable concentrations. The linear bilins are also bound either to a protein (as in phycobilins of certain algae) or to sugars or as amorphous polymers when formed as breakdown products of, typically, haemoglobin.

The colour of the particular tetrapyrrole is determined largely by the structure and substitutions to the tetrapyrrole molecule itself and relatively little by the metal. The colour range of this group of pigments is considerable and is summed up in Table 1.5.

Evolutionary evidence suggests that the tetrapyrroles must have been among one of the earliest of biological pigments functioning in photosynthesis (the various chlorophylls and phycobilins), electron transfer (the haem-based cytochromes), oxygen transport (haem-based haemo- and myoglobin) and in protection from reactive forms of oxygen (peroxidases). As all of these functions are of primary importance in cellular metabolism, tetrapyrroles are found in all biological organisms—probably without exception.

In living cells the tetrapyrroles are relatively stable with half-lives of many weeks. On isolation and particularly where ingested by animals, the tetrapyrroles are rapidly degraded. Chlorophylls ingested with the rest of the photosynthetic apparatus retain the potential to generate electrons on exposure to light. In transparent animals, this could lead to the formation of oxygen radicals and at best indigestion! There are examples of animals that retain ingested chlorophyll for an appreciable length of time (certain hydra, sea anemones, corals and sea slugs) but in most cases the chlorophyll is rapidly and perhaps selectively degraded.

Table 1.5 Colour range of naturally occurring tetrapyrroles

Colour	Tetrapyrrole	Occurrence
Blue	Phycocyanin	Blue-green algae
Blue-green	Chlorophyll *a*	Plants
Green	Chlorophyll *b*	Plants
Yellow-green	Phaeophytin	Senescing plants
Yellow-orange	Phycoerythrin	Red algae
Orange	Bilirubin	Higher animals
Red	Haem	All organisms

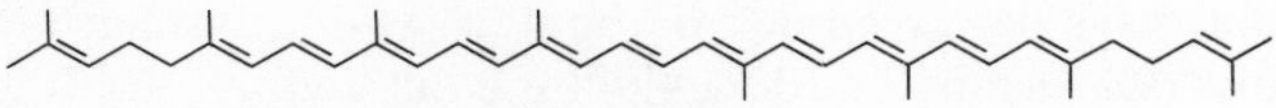

Figure 1.3 A tetraterpenoid, lycopene.

1.9.2 Tetraterpenoids

More familiar as carotenoids, this group of pigments is also present, in abundance globally, both in plants and in animals. In total over 600 carotenoids have been described making this one of the larger groups of natural pigments.

Most carotenoids are yellow to yellow-orange. Distinctly red and indistinctly yellow-green examples can also be found. The structures of all carotenoids are based on a common 40-carbon (40-C) linear terpene (Figure 1.3). In plants, carotenoids are principally associated with chlorophylls in the photosynthetic organelle, the chloroplast or its degenerate successor, the chromoplast. At this point, carotenoids enter the animal food chains by ingestion with chlorophylls. In many animals, the carotenoids are selectively assimilated for a wide range of purposes (see chapter 5), while in almost all animals the chlorophylls are excreted in a partially if not wholly degraded form. This perhaps emphasizes the potentially positive value of the carotenoids to animals in contrast to the possible photo-toxicity of chlorophylls should they be absorbed. Carotenoid-protein complexes are common in animals but in photosynthetic organisms (plants and bacteria) carotenoids are more usually present dissolved in lipid membranes. The more widespread carotenoids in nature are summarized in Table 1.6.

Table 1.6 The more widespread of the naturally occurring carotenoids in biology (unless otherwise indicated all are to be found in higher plants)

α-Carotene
β-Carotene (particularly widespread)
Lutein
Violaxanthin
Neoxanthin
β-Cryptoxanthin
Antheraxanthin
Fucoxanthin (brown algae)
Lycopene (and several related derivatives)
Astaxanthin (fish, crustacea)
Canthaxanthin (beetles)

1.9.3 O-heterocyclic compounds

An enormous and somewhat diverse group, these compounds are more familiar as flavonoids and related structures. They are pre-eminently compounds of higher (flowering) plants and appear to have evolved along with flower petals and pollinator attractants. A total of some 4100 flavonoids have been described so far in structural detail (see Harborne [4]), and almost one quarter within the past 6 years. Of the several classes of flavonoids, most are at best pale yellow in appearance, the majority being visible only under ultraviolet (UV) light. To that extent, most flavonoids are not pigments in the narrow sense of this book. The most truly colourful of the flavonoids are the anthocyanins, the pigments contributing to much of the colour of petals and ripening fruits of higher plants and to senescing leaves of trees in autumn.

Structurally, the flavonoids are based on a common tri-cyclic phenyl benzopyran (Figure 1.4) consisting of two benzene rings (designated A and B) joined with an oxygen-containing pyran group. Further modifications to the heterocyclic ring determine the class (and indeed colour) of the various flavonoids. In terms of visible-light pigments, much the most important group are the anthocyanidins (Figure 1.5). In biological systems, anthocyanidins are probably always bound to one or more sugars, most usually β-D-glucose, β-D-galactose and α-D-rhamnose with one, often more, hydroxyl groups on the benzene rings, these latter substitutions dictating the colour of the anthocyanidin. Although some 256 anthocyanins (the flavonoid plus sugar) have been documented (Harborne [4]), only 17 anthocyanidins (the aglycone structure) are involved. Omitting the uncommon or rare, Table 1.7 lists the six most widespread anthocyanidins with their structural attributes. The trivial names reflect, in most cases, the plant from which the pigment was

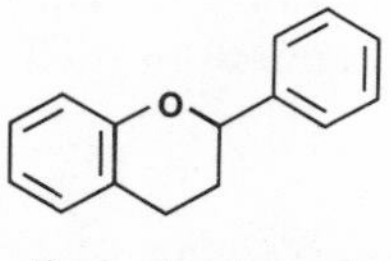

Figure 1.4 Basic structure of a flavonoid.

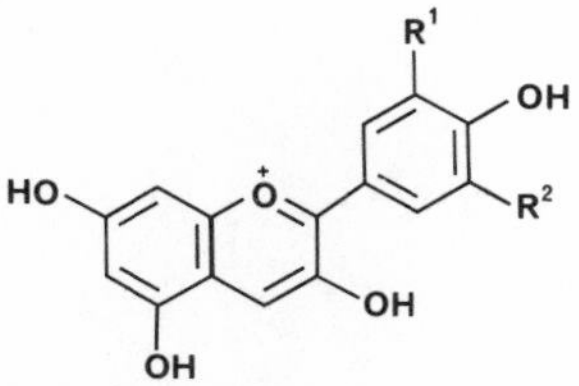

Figure 1.5 Basic structure of an anthocyanidin (where R^1 and R^2 are typically H, OH, OCH_3).

Table 1.7 The most widely occurring anthocyanidins

Name	Colour	λ-max in acidic methanol
Cyanidin	Blue-red	535
Delphinidin	Purple-blue	546
Malvidin	Purple	542
Pelargonidin	Scarlet-red	520
Peonidin	Blue-red	532
Petunidin	Purple-blue	543

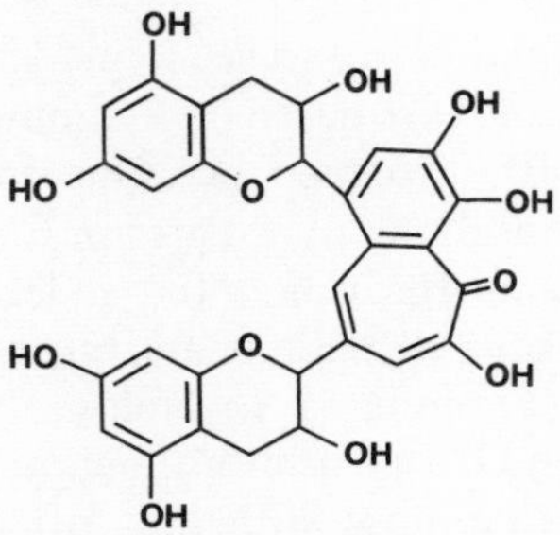

Figure 1.6 A theaflavin.

first isolated and disguise the very widespread presence of the six throughout the flowering plants.

In addition, interactions of the ring B with certain metals, usually aluminium or iron, and interactions elsewhere in the molecule with magnesium confer a blueing to the pigment. Interactions between anthocyanins and between flavonoids extend the range of colours. Flavonoid polymers are common. The reddish-brown pigment in the beverage tea theaflavin (Figure 1.6) is one example.

1.9.4 Quinoids

These phenolic compounds vary from the simple monomeric dihydroxyphenol derivatives such as 1,4-benzoquinone (Figure 1.7), to the reduced

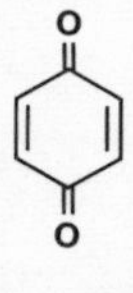

Figure 1.7 1,4-benzoquinone.

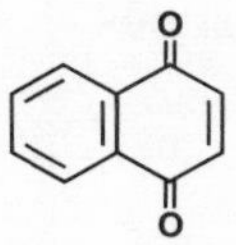

Figure 1.8 1,4-naphthaquinone.

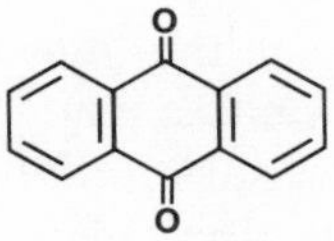

Figure 1.9 9,10-anthraquinone.

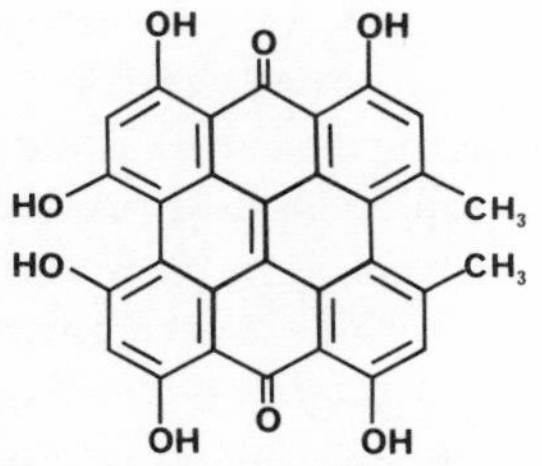

Figure 1.10 Hypericin.

dihydroquinone, to the dimer such as 1,4-naphthaquinone (Figure 1.8) to the trimer 9,10 anthraquinone (Figure 1.9) on to the more complex polymeric forms as in hypericin (Figure 1.10).

The simpler quinones may function as electron and proton carriers in primary metabolism and are to be found, if only in trace amounts, in all actively respiring organisms. In terms of contributing to natural pigmentation, observable coloured quinones are widely distributed in plants, particularly trees where they contribute to the pigmentation of the heartwood and therefore have considerable interest to the timber and furniture trades.

Most quinones and their derivatives are probably bitter to the taste and relatively toxic. Perhaps for that reason they figure in the literature as anti-herbivore defences, that is chemical defences formed in plants against challenges by would-be herbivores. There are problems with proving that this is the natural function of this class of pigment if only because it is difficult to show that animals avoid plants rich in say naphthaquinones because of the taste of the quinone or to something else. Certainly the plant genus *Hypericum*, which contains the toxin hypericin, has few natural enemies and is a notorious poisonous plant to livestock. It is an introduced weed species in Australia that is a serious problem.

Quinoids, as a group, probably make little contribution to the colours of the visible parts of plants. They do, however, contribute to some of the duller colours of certain fungi and lichens (typically yellow, orange or brown) but also to the brilliant red, purple and blues of sea urchins, crinoids (sea lilies), coccid insects and several species of visibly coloured aphids.

In general, the simpler quinones, such as benzoquinones, are colourless except in high concentration when they may have a pink hue. The naphthaquinones and anthraquinones are frequently present in plants with complex additions or substitutions. These substitutions provide colours ranging from almost black or deep purple to deep wine-reds to orange and through to yellow. Further colour changes are readily made *in vitro* by simple additions of hydroxyl groups under basic conditions.

Historically, certain quinone pigments have been the mainstay of the textile dyestuff trade. Quinoid-rich plants were formerly cultivated solely for their pigment. The deep-red pigment from the roots of *Rubia* species, such as madder (*R. tinctorum*) is a mixture of anthraquinones, while the yellow-red-brown naphthaquinones of *Lawsonia alba* produce the cosmetic dye henna. Two related anthraquinones have long been established as food colorants, the dyestuff kermes from the coccid insect *Kermococcus ilicis* and cochineal (carminic acid) from another insect *Dactylopius coccus* (see chapter 7).

A final group of quinoid pigments are the allomelanins, usually dark brown to black pigments of fungi and some plants, which are chemically unrelated to the true or eumelanins of animals. The precise biosynthetic origin of most of the allomelanins is unknown but they appear, structurally at least, to be related to polymers of simple phenols and their quinones.

1.9.5 N-heterocyclic compounds (other than tetrapyrroles)

From a range of largely unrelated N-heterocyclic compounds found in biology there are two groups that show at least some structural affinities as well as being pigmented; namely (i) indigoid and indole derivatives and (ii) substituted pyrimidines.

1.9.5.1 Indigoid and indole derivatives. The blue textile dye indigo (Figure 1.11) was until relatively recently an economically important product of the semi-tropical plant *Indigofera tinctoria* (and near relatives) and the European woad plant *Isatis tinctoria*. Indigo constitutes one of the oldest colorants known to man. The intact living plants, however, contain not indigo but the colourless glycoside of 3-hydroxyindole. Only when extracted, hydrolyzed, oxidized and dimerized is the blue dye, indigo, formed. Another indigoid, the ancient dye Tyrian purple, was derived by photochemical oxidation and extraction of a colourless indigoid from certain Mediterranean molluscs to

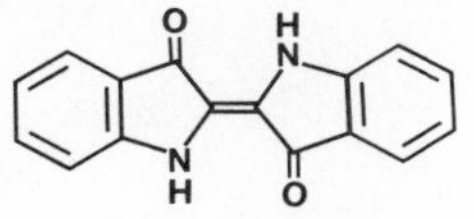

Figure 1.11 Indigo.

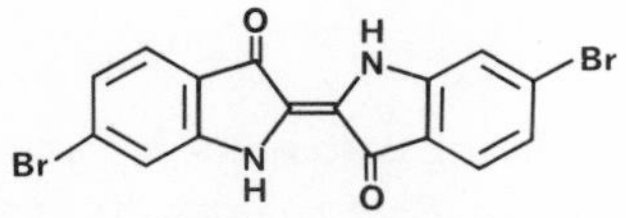

Figure 1.12 Tyrian purple, dibromoindigo.

yield 6,6′-dibromo indigo (Fig. 1.12). This illustrates the important point that several pigments considered to be 'natural' are actually colourless in the living cell. Only on isolation from the host organism, and following relatively complex chemical modifications, do they provide dyes or colorants. It is often their long history of use that endows them with the accolade 'natural'. For all their artificiality, these indigo dyes are vigorously defended as 'natural dyes' by the wool dyers and weavers in contrast to the identical indigo synthesized by Baeyer in 1878 and for many years an important product formed from distillation of coal tar (formed from fossilized land plants) and later from oil (formed from degraded marine plants). The boundaries between 'natural' and 'artificial' come no closer than in indigo chemistry.

Structurally related to indigo through the common indole group, an important group of plant pigments are the betalains. Unlike indigo, the betalains are true pigments in the living plant that give rise to the distinct deep-red pigmentation of beetroot and the crimson of *Amaranthus* flowers. As they are considered in detail in chapter 6, only an outline will be provided here. The betalains consist of two sub-groups, the betacyanins (Figure 1.13) and

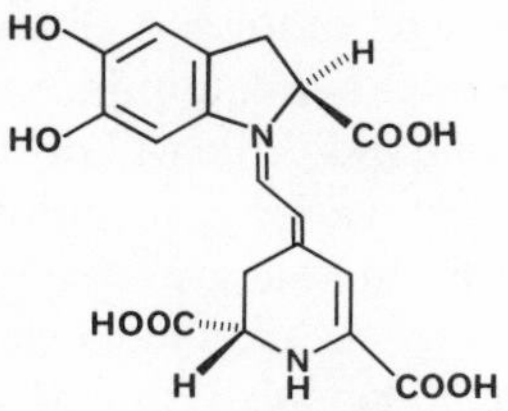

Figure 1.13 A betacyanin, betanidin.

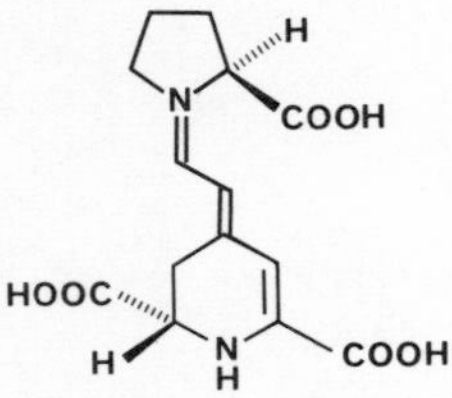

Figure 1.14 A betaxanthin, indicaxanthin.

betaxanthins (Figure 1.14). Being water-soluble and produced in abundance by, for example, the beet plant *Beta vulgaris*, the betacyanins are an established food colorant. Of interest to those studying the origin and evolution of higher plants, the betalains are something of a mystery. They are found in just one sub-group of extant flowering plants, the *Centrospermae*, and one extinct group of primitive trees, the *Bennetales*. The living *Centrospermae* have either lost, or perhaps never acquired, the genes for the synthesis of the other great family of plant pigments, the flavonoid anthocyanins. Genes for the synthesis of betalains appear again in a few otherwise unrelated fungi, most colourfully perhaps in fly agaric *Amanita muscaria* as violet and yellow pigments.

The betalains are, however, structurally and biosynthetically related to the widespread group of animal pigments the melanins, more correctly the eumelanins. In mammals, including man, the melanins are the major pigments familiar as the colorant of black and brown skin and hair. Although the natural function of eumelanins is probably the protection from ultraviolet light (UV-B radiation 280 to 320 nm) and, in association with the pineal gland, in light perception interest in these pigments has arisen largely because they may serve as indicators of malignant melanomas in man. In mammals, melanins are frequently linked to metals (Fe, Cu and Zn) and attached to proteins isolated in discrete packages called melanosomes, which are present in light-receptive areas of the body. Other related but simpler melanins have quite different functions; in the cuttle fish *Sepia officinalis*, for example, melanins are released as a defensive ink.

For many years the structure of the eumelanins was debated, the starting material for most analyses being an amorphous, insoluble and frequently impure protein-bound isolate. The simplified 'textbook' structure was that of a long-chain polymer of units of indole-5,6-quinone (Figure 1.15). In practice, the true structure of a eumelanin molecule will probably remain elusive given numerous and perhaps random variants in points of attachment and the presence of branches and cross-linkages. Whatever the eumelanins lack in structural consistency, they are all virtually insoluble in water and most organic solvents and, except under fairly drastic chemical conditions, they are highly stable. Eumelanins are essentially black in perceived colour.

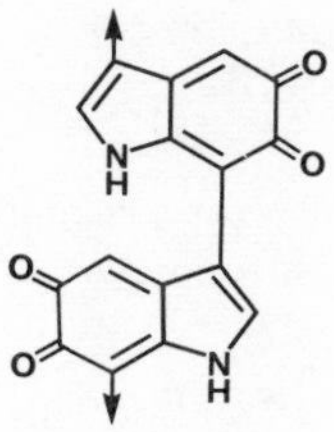

Figure 1.15 Eumelenin, part of a linked series of indole-5,6,quinone.

An unrelated class of pigments the phaeomelanins also contribute to mammalian colours as light brown, red, yellow or blond colours of hair. Similar pigments contribute to some of the colorants present in feathers of birds. At least the red phaeomelanins reveal an indole origin within the complex structures of these sulphur-containing pigments. Unlike the eumelanins, the phaeomelanins can be dissolved in dilute alkali but in other respects they are stable and relatively inert.

1.9.5.2 Substituted pyrimidines. Components of this diverse group show little biosynthetic relationship, their grouping here is entirely due to their superficial structural similarity and to some important physical characteristics in common.

The major groups are the purines (Figure 1.16), and the pterins (Figure 1.17) each with the 6C pyrimidine ring substituted with a 5C imidazole or 6C pyrazine ring respectively, bearing in all four nitrogens. In addition, three minor groups are also conveniently considered here, structurally related to pteridine by a further 6C substitution to form flavin (note spelling) (Figure

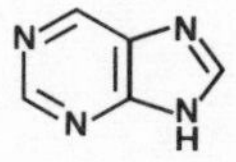

Figure 1.16 Basic structure of a purine.

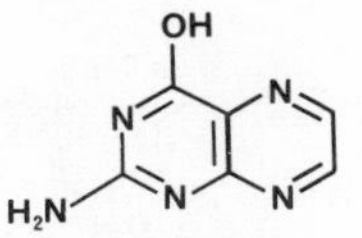

Figure 1.17 Basic structure of a pterin.

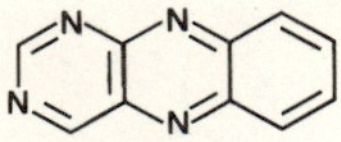

Figure 1.18 Isoalloxazine, basic structure of flavin.

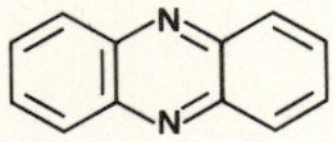

Figure 1.19 Basic structure of phenazine.

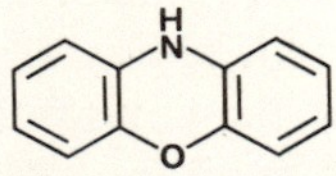

Figure 1.20 Basic structure of phenoxazine.

1.18) or without the nitrogens on the first ring to form the colourful phenazines (Figure 1.19) or with an O-substitution for a N-group and called phenoxazine (Figure 1.20). Given this somewhat tenuous link between these diverse structures, they possess several common features. While many are not at all common in biology, each group contains several individually important pigments.

A significant group of pigments in the animal world is formed from purines, particularly guanine, xanthine and uric acid. In the form of micro-crystals or granules they are the chemical basis for several structural colours, notably white, semi-transparent cream colours and silver. They are particularly common in the scales of fish.

More widespread perhaps are the pterins, contributing to the white, cream, yellow and reds of many insect groups. For example, the intricate patterns of colour in the wings of butterflies and moths are often formed from pterins including the mustard-yellow colours of leucopterin (Figure 1.21). The bright yellow of many wasps are due to xanthopterin (Figure 1.22) while other yellows, oranges and reds of crustaceans, fish, amphibians and reptiles are

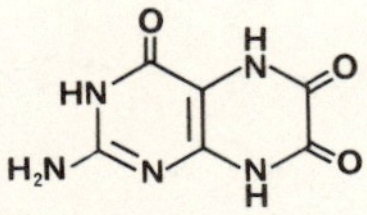

Figure 1.21 Leucopterin.

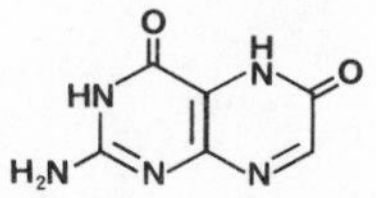

Figure 1.22 Xanthopterin.

due partly to several other pterins. Most of the pterins are soluble in dilute acid or alkali and in the dark are relatively stable. Unfortunately most are also photo-labile when isolated from their biological carrier.

Another widespread group of substituted pyrimidines are the flavins, and one in particular, riboflavin, is an important component of redox proteins throughout biological systems, particularly in respiratory enzymes. Its more familiar name is vitamin B_2. Despite its widespread occurrence, riboflavin seems rarely to function as a visible pigment in biology. Exceptionally, a small number of yellow-coloured marine invertebrates owe their pigmentation to riboflavin, and flavin-pigmented bacteria are occasionally found in nature, although more recently in deliberately genetically engineered vitamin B_2-forming microbes. The free riboflavin is highly soluble in water, appears bright yellow in concentrated solution and has an absorbance spectrum similar to the otherwise unrelated carotenoids.

The phenazines have, superficially, a structural resemblance to the flavins. Reported so far only from certain bacteria, most commonly the species of *Pseudomonas* and *Streptomyces* the phenazines are often brightly coloured—mostly yellow but exceptions include the deep-blue pyocyanine (Figure 1.23), and violet-blue iodinin (Figure 1.24). Occasionally, these and other phen-

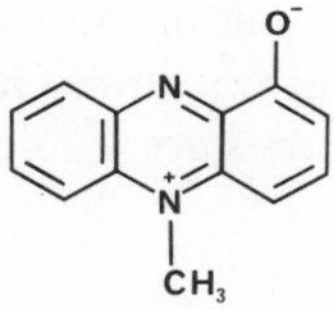

Figure 1.23 Pyocyanine.

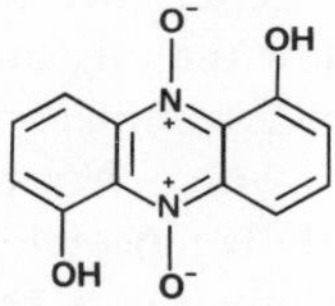

Figure 1.24 Iodinin.

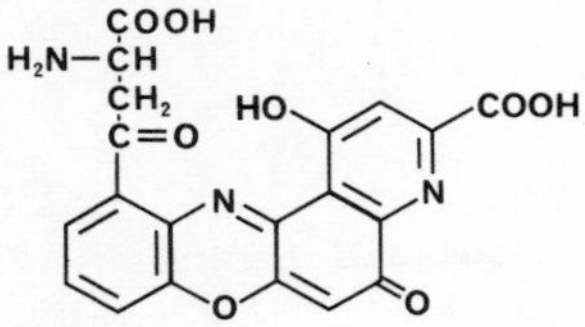

Figure 1.25 Xanthommatin.

azines are found in mammalian (including human) tissues associated with, for example, infected wounds, presumably from a bacterial invasion. Many of the phenazines are (or can be readily converted to become) water-soluble and appear to be relatively stable at least under laboratory conditions. Functionally, the phenazines may be formed as chemical protectors (antibiotics) against other bacteria. They have important applications as redox mediators in biochemistry and some are available commercially.

The final group of substituted pyrimidines, the phenoxazines, are structurally similar to the phenazines but with an oxygen (O–) substituted for a nitrogen (N–). They constitute a group of invertebrate pigments, the ommochromes, and enjoy a welter of names. Among the more common ommochromes is the yellow xanthommatin (Figure 1.25), most of these pigments being yellow, golden-yellow or yellow-brown, although darker browns are common too. Occasional variants give rise to red-browns and mauves. Many ommochromes may serve to screen out stray light in the eyes of some invertebrates. They are also widespread in higher invertebrates where they may function in camouflage. Phenoxazines are also reported from microbes, for example some *Streptomyces* spp. produce a pink-red phenoxazine, actinomycin, with antibiotic properties. Soluble in acids and in acidified alcohols, many phenoxazines, undergo a colour shift often acquiring a red hue. In alkali, they are more yellow and not infrequently fairly unstable. It seems likely that *in vivo* the ommochromes are stable in the reduced form.

1.9.6 Metalloproteins

An otherwise unrelated group of pigments, the metalloproteins form an important group of proteins in biological systems. One group, the metalloporphyrin haems and chlorophylls, has already been described above. A list of the more widespread (and a few that are uncommon) metalloproteins are given in Table 1.8. The list is incomplete however. Many proteins, on isolation, appear to be associated with one or more metals. Such associations may be artifacts of extraction. Other protein-metal complexes almost certainly exist *in vivo* but have perhaps rarely remained intact on isolation or purification. Many other metalloproteins are well-known but are not shown

Table 1.8 The more widespread metalloporphyrins with minor examples of interest

Metal co-factor	Chelator	Chromophore	Chromophore plus protein	Colour
Fe	Protoporphyrin	Haems (of several types)	Haemoglobin	Red
			Myoglobin	Red
			Chlorocruorin	Green
	Complexed directly to protein		Oxyhaemerythrin	Brick-red to scarlet
			Ferritin	Red
			Haemosiderin	Brick-red
Mg	Protoporphyrin	Chlorophylls (of several types)	Chlorophyll-proteins (several types)	Green
Cu	Complexed directly to protein		Oxyhaemocyanin	Blue
			Ceruloplasmin	Blue
			Erythrocuprein	Blue-green
			Oxyplastocyanin	Blue
	Uroporphyrin	Turacin	—	Purple
V	Complexed directly to protein		Haemovanadin	Apple-green

in Table 1.8 because they are not present in sufficient quantities to form a recognizable pigment. The cobalt-containing porphyrin-derivative chromophore of vitamin B_{12} that is naturally present in certain bacteria in trace amounts is therefore excluded from the list, as are the many haem-containing enzymes such as the cytochromes and peroxidases, the large number of other iron and iron-sulphur containing proteins, many of which have been purified and concentrated to the point where the protein shows a clear colour. Cytochrome *c* for example in the reduced state is bright scarlet. The Fe-S ferredoxin is yellow. Neither is present in any organism in such concentrations that they make any distinguishable contribution to the colour of the cell. It has to be conceded, however, that through biotechnology it might be possible to synthesize many of the metalloproteins in substantial amounts; therefore these should be considered as potential colorants of the future. However, to keep Table 1.8 to a reasonable size and within the bounds of the natural and present-day availability of the proteins, only those metalloproteins that function naturally as pigments or are present in quantities high enough to be visibly pigmented are included.

Iron-proteins are widespread in biology; they are usually pigmented and inevitably come in a wide variety of forms. In one important group, the iron is bound to its own non-protein chromophore, a porphyrin (see Figure 1.2), which, in turn, is associated in various ways with different proteins. This group includes the familiar haem (or heme) proteins that are, in the reduced form at least, usually bright red. One exceptional haem-protein is the oxygen-carrying pigment chlorocruorin (also known as chlorohaemoglobin, Spirographis haem, or haem *s* protein), the pigment in the green 'blood' of four families of polychaete worms. However, in most other major examples of the iron-porphyrin proteins, where there *is* any distinct colour, then it is red to purple.

Haems, in whatever form, are pigments composed of or derived from protoporphyrin chelates of iron (see chapter 3 for fuller descriptions). The word 'haem' therefore has a precise meaning. Unfortunately this was not always so. The use of the name haem for several red (and indeed blue) non-porphyrin pigments stems from the 19th century and reflects unfortunate attempts to name and classify chemicals by their characteristic spectral qualities. The legacy is a number of metalloproteins with haem as part of the name but which are not haems in the modern meaning of the word (Fe-protoporphyrin or its Fe-derivatives). One of these is the misnamed haemerythrin, a simple iron-protein, again functioning in oxygen transport, which although red in colour has no structural resemblance to haem proper. Present in a few otherwise unrelated marine invertebrates, the iron is directly bound to amino acids of the protein. Copper (Cu) metalloproteins, generally blue to blue-green, are present in a number of invertebrates. Exceptionally a Cu-containing porphyrin, turacin, is a bronze-purple colorant in one family of tropical birds. A vanadium-containing protein, haemovanidin, is apple-green in colour.

1.9.7 Miscellaneous

There are many pigments throughout biology, particularly within the fungi and invertebrates, that are often of very limited distribution. No single comprehensive available work attempts to cover this area. However, reviews from time to time do give some exposure to these often obscure pigments. Among the more useful sources both for a perspective and for reports of the more unusual structures the annual (or more frequent) reviews in *Progress in the Chemistry of Organic Natural Products* (*Fortschritte der Chemie Organischer Naturstoffe*) and *Natural Product Reports* and the journals *Liebigs Annalen der Chemie*, *Lloydia*, *Phytochemistry*. The pharmacological literature is also helpful.

1.10 Colourless compounds in perspective

Most biological molecules are not coloured, that is, they do not absorb light in the visible part of the spectrum. As a class, carbohydrates ranging from simple sugars to polymers of glycogen, starch and cellulose, are colourless (they appear white). Most naturally occurring amino acids are colourless as are many proteins (which tend to have a strong absorbance between 250 and

300 nm). Lipids, including oils and waxes, are at best pale yellow but generally colourless. Nucleic acids are also without colour. All of these widespread organic compounds function directly or indirectly in the primary metabolism of all organisms but such functions do not include the absorption of light.

Coloured compounds are then much less common than colourless ones and, in most cases, have quite specialized functions. As a consequence, their distribution is often restricted. For example, pigments in higher animals are frequently involved in camouflage (as in amphibians) or in mating attraction (as in male birds). Pigments in plants and in the fruiting bodies of some fungi may function as animal attractants in seed or spore dispersal, other pigments in lichens and algae may be synthesized in response to extreme desiccation although their mode of action thereafter may not be understood. All of these functions, however vital to the particular organism, generally reflect the highly specialized life led by that organism. Only in the cases of chlorophylls (chapter 3), haems and bilins (chapter 4), and some carotenoids (chapter 5) is there a more widespread occurrence reflecting directly the involvement of these few pigments in fundamental process of oxygen and electron transport, photosynthesis and as anti-oxidants.

1.11 Pigment classification by natural distribution

The second approach to summarizing the pigments found in biological systems is to classify them according to the type of organisms (animal, plant or bacteria) in which they occur. In this way, it is possible to pinpoint those pigments that are confined to one particular type, family or species. For example, many lichen pigments are, in terms of abundance, relatively uncommon. In a classification based on structure, they would inevitably be grouped as a 'hotch-pot' of miscellanea. By considering lichens separately as a source of pigments, attention is drawn to one otherwise obscure class of pigments known as depsides. These depsides and their derivatives form the basis of the long-practised art of textile dye. In more recent decades, depside derivatives such as litmus have formed the basis for several chemical indicators and cytological stains.

Despite the considerable advances in biotechnology and gene transfer across the whole field of biology, the gene products of certain types of organism lend themselves to exploitation more readily than others. In the context of food colorants, bacteria, single-celled and simple fungi together with single-celled algae and perhaps the simpler zooplankton are the most likely sources of new pigments because they have the potential of being exploited using existing culture techniques. The pigments of higher organisms, animal, plant and fungal, may be less accessible to exploitation because of the structural complexity of the pigment-bearing tissue or because the pigment is formed only at critical points of development within a complex life cycle.

For example, pigments that function as attractants in sexual reproduction may be formed only after completion of other aspects of the life cycle. They may not then be amenable to exploitation through genetic manipulation. Pigments that are present in low concentrations for much of the life cycle of the organism and which function in, for example, stress tolerance may perhaps be more readily exploited.

1.11.1 Plants including algae

While the colour of vegetation is visible from orbiting satellites and contributes to the overall colour of the planet, the range of pigments present in plants is surprisingly small. The predominant colours in land plants result from two types of chlorophyll, no more than four or five carotenoids with a seasonal contribution from three or so flavonoids. The oceans yield four common chlorophylls, perhaps six or seven widespread carotenoids and two forms of phycobilins. The contribution from other plant pigments, including the betalains, melanins, anthraquinones, naphthaquinones, the less-common carotenes, xanthophylls, and the several thousand flavonoids is relatively insignificant when considered on a global basis. This fact helps explain why a few plant pigments are produced, or at least can be produced, commercially in many countries; chlorophyll, β-carotene and grape anthocyanins being good examples. Other pigments, such as annatto (extracted from the fruits of one species of tree *Bixa orellana*), saffron (from the anthers of *Crocus sativa*), and turmeric (an extract from roots of *Curcuma longa*), are the products of one or two countries only. Table 1.9 shows the major classes of pigments present in plants, including algae, with a short list of the most commonly occurring forms. More detailed treatments of the occurrence of these pigments is given in chapters 3 to 7. The purpose of Table 1.9 is to highlight the most abundant plant pigments.

Several thousand plant pigments are excluded from Table 1.9 simply because they are uncommon (many carotenoids), or of limited natural distribution (*e.g.* turmeric pigments), or natural production occurs only under specific environmental conditions (*e.g.* certain anthocyanins of ripening fruits or senescing leaves) or simply because in overall quantity the pigment is small relative to other pigments of the type (*e.g.* bacteriochlorophylls).

1.11.2 Higher animals (vertebrates)

Fox [6] and Fox and Vevers [7] have provided excellent summaries of the distribution of pigments among all of the major taxa of animals. The account provided by Needham [8] is more comprehensive. Table 1.10 is then a digest of a much more extensive consideration but at least emphasizes the relatively limited distribution and range of pigments in the higher animals. All ver-

Table 1.9 The most abundant pigments in plants including algae

Pigment	Common examples	Natural occurrence and abundance
Chlorophylls	*a*	All photosynthetic eukaryotes
	b	All land plants, many algae
	c, *d*	Brown and other algae
Phycobilins	Phycocyanin	Blue-green and other algae
	Phycoerythrin	Red and other algae
Carotenoids	Lutein	Most abundant xanthophyll. Most photosynthetic organisms
	β-Carotene	Most abundant carotene. Most photosynthetic organisms
	Violaxanthins	Common in higher plants
	Neoxanthins	Common in higher plants
	Fucoxanthin	Brown and other algae
Anthocyanidins	Cyanidin	The most common anthocyanidin widely distributed in higher plants
	Pelargonidin, Delphinidin	Common in higher plants
Betalains	Betacyanin	Widely distributed but confined to one order of plants

tebrates (with a few rare exceptions) contain haemoglobin and myoglobin, which may also contribute to the pigmentation of the animal.

Of the higher animals, the more colourful classes are the birds, amphibians, the bony fish and some reptiles. Mammals are generally remarkably dull and considering that they constitute a substantial element of the human diet, they provide relatively little to the intake of natural pigments. More natural pigments are consumed through a diet of fish.

Table 1.10 The major pigments of the vertebrates

Class	Pigment
Mammals	Mainly melanins
Birds (including their eggs)	Melanins Carotenoids Tetrapyrroles
Reptiles and amphibians and bony fish (teleosts)	Melanins Carotenoids Pterins Riboflavin
Cartilaginous fish (elasmobranchs)	Melanins

1.11.3 Lower animals (invertebrates)

The distribution of pigments in the lower animals is considerably greater than in the vertebrates and rivals the higher plants in terms of variety. Rather than give an extensive table covering the many invertebrate classes, Table 1.11 lists only the major invertebrate groupings that are significantly pigmented. Needham [8] gives a more comprehensive account.

In the context of human diets, the crustacea and molluscs provide perhaps the most diverse source of pigments including several unusual carotenoids, bilins, ommochromes and, in many instances, pigments that have not been characterized. It is for this reason that no serious attempt can be made to establish a comprehensive list of 'natural' pigments appearing in the average human diet, particularly one that is based, at least in part, on an intake of sea food. Natural food colorant regulations that are based on the dozen or so pigments widely consumed in a diet of mammalian meat, fish, vegetable and fruits (from widely grown sources) must inevitably ignore the fact that

Table 1.11 The major pigments of the invertebrates and protozoa

Phyllum or major class	Pigment
Echinoderms (star fish, sea urchins, crinoids)	Generally highly pigmented, mainly carotenoids, quinones and melanins
Molluscs (snails, bivalves, squids)	Melanins, porphyrins (particularly in shells), ommochromes, many with haemocyanin as a visible pigmentation. The group also contain numerous and so far unidentified pigments and others that are rare but of potential economic interest
Insects	Often highly coloured and containing many carotenoids, flavonoids, wide range of melanins, ommochromes, pterins, flavins and occasional bilins and unusual polycyclic quinones
Malacostraca (centipedes, millipedes)	Melanins and carotenoids predominate but many examples of ommochromes and rare pigments. Haemocyanin may contribute to pigmentation
Crustacea (crabs, lobsters, shrimps)	Highly pigmented with carotenoids, bilins, melanins, pterins, flavins and some with haemocyanin as a visible pigment
Arachnida (spiders, scorpions and allies)	Varied pigments, including unusual carotenoids, pterins, quinones (especially spiders), some with haemocyanin and pterins, many with so far unidentified pigments
Annelids (segmented worms and leeches)	Varied in pigmentation, often due to bilins, some with haemoglobin or chlorohaemoglobin
Cnidaria (jellyfish, anemones and allies)	Often with unusual carotenoids, some with bilins, the group contains numerous poorly described pigments
Porifera (sponges)	Mainly carotenoids
Protozoa (plankton)	Some with photosynthetic pigment (chlorophylls and carotenoids)

significant proportions of the population ingest natural pigments of which we have little knowledge. The most ill-defined foods, from a description of their pigments, will be among the shrimps, crabs, krill, lobsters, crayfish and their relatives. On this point alone, most of us have no idea what we are eating! Apart from a couple of insects and molluscs (see above), the invertebrates have contributed little to commercialized colorants.

1.11.4 Fungi

Fungi, particularly the fruiting bodies, are often brightly pigmented and even where the colours are dull, exposure of the fungal sap to air (as occurs in wounding) often yields pigmented oxidation products. The number of different pigments so far described from all classes of fungi including slime moulds probably exceeds 1000. Gill and Steglich [10] give structural descriptions of over 500 in their review. Steglich [11] provides a broad summary of groups of pigments in the larger fungi while Shibata *et al.* [12] give detailed coverage to pigments from other orders.

The summary that follows here is a much reduced digest, highlighting the major classes of pigment at the expense of the minor or obscure. For the greater part, the review by Gill and Steglich [10] should be referred to for fuller details of structure and natural occurrence.

Perhaps the most remarkable feature of fungal pigments is their extraordinary diversity. If flavonoids are excluded, then fungi must rival plants in the variety of pigments they contain. This becomes even more significant when one considers the types of pigments *not* found in fungi. The chlorophylls, abundant in plants including algae, are entirely absent from fungi. Carotenoids are absent from the major families of the familiar toadstools (with a few exceptions) and confined to just four orders of the *Phragmobasidiomycetidae* spp. and one order of *Ascomycetes* (*Discomycetes*). Flavonoids have not been reported (reliably) from fungi. Instead, the fungal kingdom excels in other areas by accumulating pigmented compounds that are either unreported from other biological groups or if present then in only trace amounts. An example of the latter is riboflavin (vitamin B_2), which is present in most organisms where it functions in electron transport but in concentrations that do not contribute to any visible pigmentation. In the fungal genera *Russula* and *Lyophyllum*, riboflavin and its close derivatives are accumulated to such high concentrations in some species as to contribute a deep yellow colour to the organism. Almost the only pigments common to fungi and to plants or animals are the betalains, melanins, a relatively small number of carotenoids and certain anthraquinones. Many of the pigments found in fungi (or lichens) have not been recorded in any other biological organisms. Fungi, particularly the more simple single-celled organisms amenable to large-scale culture, offer the greatest potential for exploration in the field of natural pigments.

As with plants, fungi as a group have a well-developed terpenoid metabolism; however, remarkably little of this results in the accumulation of pigmented compounds. This presumably reflects the limited role of pigments in fungal biology. Sesquiterpenoids are widespread in many orders of higher plants as aromatic and volatile oils but rarely in concentrations that would impart visible pigmentation. Most unusually then, in one fungal genus *Lactarius*, sesquiterpenoids accumulate in the latex of the fruiting body to yield a range of red, orange, green and blue pigments. Tetraterpenoid carotenoids are present in all higher plants and widespread in the animal and bacterial kingdoms. They have, however, a limited distribution in fungi and are generally confined to β-carotene with smaller amounts of α- and γ-carotene and lycopene. Xanthophylls appear to be rare in fungi (Britton [5]). In the higher fungi (which include the familiar toadstools and bracket fungi) (class *Basidiomycetes*), carotenoids appear to be largely confined to three orders, the jelly fungi *Tremellales* and *Dacrymycetales*, and to the stinkhorn *Phallales* where odour and pigmentation function as insect attractants in spore distribution. Unexpectedly, at least on systematic grounds, carotenoids appear again in the familiar chantarelles (*Cantharellus* and related spp.). In the edible North American *C. cinnabarinus* the dominant orange-red pigment has been reported to be canthaxanthin (β,β-carotene 4,4′-dione) (although this has been disputed). If true, this would at least justify the claim that this pigment, familiar to the food colorant industry, is a 'natural' colorant. Older (and often still cited) reports of carotenoids in other groups of higher fungi, including the *Agaricales*, are to be treated with reservation (Gill and Steglich [10]) although, as with much else in biology, there are always the surprising exceptions.

Among the Ascomycetes, carotenoids are also of limited distribution and again largely confined to β-carotene and its close derivatives. Carotene pigmentation is characteristic of the orange-peel fungi (genus *Aleuria*), the related genera *Morchella* (morels) and *Helvella* (false morel). Other carotene-containing groups of discomycetes include the *Helotiales*. A more extensive list of fungal species from which carotenoids have been reported is provided by Gill and Steglich [10].

A biosynthetically unrelated group of malonate derivatives, the ketides, are found predominantly in the Ascomycetes and notably in the genera *Daldinia*, *Bulgaria*, *Hypoxylon*, *Chlorosplenium* or in the Basidiomycete order *Polyporales*. The simplest, the tetraketides, contribute to the yellow pigments of *Albatrellus* and *Phlebia* spp. and to the red-brown pigments of *Gloeophyllum* spp.

The dimeric oosporeine (Figure 1.26) that is widespread among the ascomycetes and deuteromycetes, illustrates the quinoid structure of many, but not all, of these and more elaborate ketides. Naphthalene and naphthaquinone structures (also known as pentaketides) are widespread as dark-red, brown or black pigments. 1,8-Dihydroxynaphthalene (colourless) on dimer-

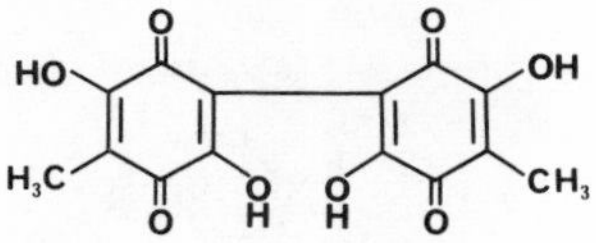

Figure 1.26 Oosporeine.

ization forms the brown and black perlene quinone of the fruiting bodies of *Daldinia* and *Bulgaria* spp. Several of these naphthaquinones are identical to those isolated from higher plants.

Other pentaketides are purple to violet and more elaborated forms (hexaketides) contribute to the pigmentation of *Penicillium* and the otherwise unrelated *Hypoxylon* spp., a feature emphasizing the impossibility of relating the distribution of many of these pigments to any recognized systematic arrangement. As many of these naphthalene-naphthaquinone derivatives absorb strongly in the ultraviolet part of the spectrum and particularly below 300 nm, it is possible that they play a part in both higher plants and fungi as UV-B-screening pigments. Another group of pigments with no established function, the octaketide anthraquinones, feature large in the Agaric genus *Dermocybe* but are also present in several otherwise unrelated higher fungi as yellow, orange, pink, red, purple and brown pigments. Many are also related, if not identical to, anthraquinones of higher plants. Ironically, anthraquinones (and naphthaquinones) of trees have been shown, in isolation, to have anti-fungal properties, which does nothing to explain the presence of similar anthraquinones in many fungi of woodlands. The dimeric anthraquinone xylindeine is the beautiful turquoise-green pigment of dead wood infested with *Chlorosplenium aeruginosum*. Other dimeric forms may have antiherbivore potential; one pink-red anthraquinone hypericin (Figure 1.10) present in the fungal genus *Dermocybe* is the probable toxic component of the chemically well-defended plant genus *Hypericum*. Among the often brilliantly pigmented slime moulds, of which the bright yellow *Metatrichia vesparium* is an example, octaketides are important yellow, orange and red colorants.

As with one group of higher plants, the *Centrospermae*, so several genera of higher fungi produce deep-red betalain pigments identical to the red pigments of beet. Unlike beet, most of these fungal betalains are not produced in large quantities although there are exceptions. The commercial mushroom *Agaricus bisporus* and others show on wounding, albeit faintly sometimes, a colour change from pale pink to grey-black, recording the oxidation of the betalain precursor dihydroxyphenylalanine (L-dopa) to melanin. The familiar halucinogenic toadstool *Amanita muscaria* and others of the genus produce a range of purple, red, orange and yellow betalain pigments (Figure 1.13) derived from L-dopa. Final elaboration of some of these involves con-

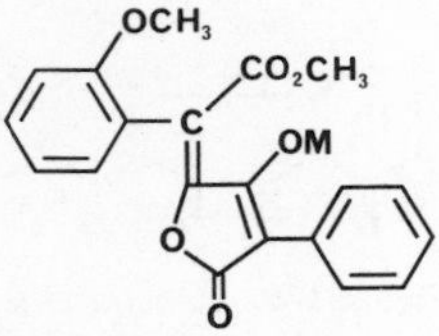

Figure 1.27 Pulvinic acid derivative, leprarinic acid.

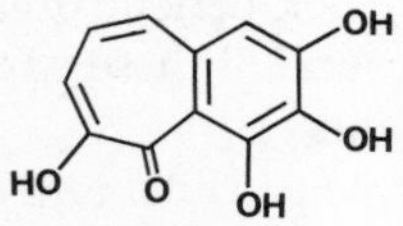

Figure 1.28 Purpurogallin.

densation of betalamic acid with unusual amino acids to form a group of yellow musca-flavin, orange musca-aurin and purple musca-purpurin pigments, several seemingly unique to *Amanita* and *Hygrocybe*. Some of these yellow pigments are structurally related to betaxanthins (Figure 1.14) of higher plants and, unlike the carotenoids, are water soluble.

One pre-eminently fungal pigment group is the terphenylquinones of which pulvinic acid and its modified forms are examples (Figure 1.27). In their many variants, these arylpyruvic acid derivatives produce intense colours across the complete spectrum, although sometimes unrevealed in the intact (uninjured) fungus. The blue pigment of boletes (*Boletales*), for example, appears rapidly on damage due to oxidation of variegatic (tetrahydropulvinic) acid. In the hydroxylated forms, pulvinic acids are responsible for the yellow and reds of most boletes. In other groups, terphenylquinones produce striking pigments particularly greens, violets and dark reds and, in isolation, will readily form other colours when titrated against alkali. In the dimeric form, the terphenylquinones contribute to the deep-chocolate colour of boletes.

Another group of fungal phenol derivatives, more properly phenylpropanoid and cinnamic acid derivatives, share some similarities, at least in biosynthesis, to higher plants. The wood-rotting *Fomes fomentarius* produces brown-red purpurogallin (Figure 1.28) derivatives which form striking blood-red pigments on treatment with alkali. Simpler hydroxybenzoic acid derivatives, often with terpene side-chains, contribute to the yellows and reds of several fungi including *Suillus*, *Polyporus*, *Gomphidius*, *Chroogomphus* and *Coprinus* spp. Many appear to have antibiotic properties.

1.11.5 Lichens

No single group of organisms has such an ancient and extensive use as dyes as the lichens. These fungal-algal and fungal-cyanobacterial associations provide pigments found nowhere else in biology and this may explain the relatively limited number of comprehensive (and readable) accounts on the subject. Ely *et al.* [13] have provided a major review of recent discoveries, Richardson [14], Hale [15] and Hawksworth and Hill [16] have published brief summaries.

As a class, many lichens are brightly pigmented. Some of these pigments may act as sunlight filters that protect the photosynthetic algal component; others appear to have antibiotic properties at least *in vitro*. For the majority, however, the functions of these pigments are unknown. Neither the light filtration nor antibiotic properties have promoted a significant amount of research. Inevitably, more is known about the pigments that have (or have had) a place in human economy, the lichen textile dyes, although these probably constitute a relatively minor part of the total complement of pigments in the group. Once of considerable commercial interest, formation of lichen textile dyes depends largely on the structural modification of compounds, many of which are colourless in their native state. Interest in these dyes waned considerably in the latter half of the 19th century, although their production has continued in northern Scotland, Norway and elsewhere in home-based natural dyeing. Unfortunately, their limited distribution and very slow rates of growth make their collection neither commercially worthwhile nor desirable from a conservation viewpoint. Small quantities of lichen dyes have also been produced for many years as reagents for biological stains and as pH indicators. Perhaps the most important group of lichen dyes has been those derived from the numerous depsides and depsidones characteristic of many lichens. The basis of depside dye formation is now well understood as shown:

Depsides
gyrophoric acid
evernic acid
lecanoric acid
erythrin → decomposition products → orcinol (orcin)
(mild alkaline hydrolysis) (brown)

$$\text{orcinol} + NH_4OH + O_2 \rightarrow \text{orcein (purple)} \qquad (1.1)$$

Orcein itself (Figure 1.29) is a mixture of oxy- and amino-phenoxazon or phenoxazin, the individual components varying from brown-red to blue-violet. Insoluble in water, most components are soluble in dilute acid or alkali. Some eight different pigmented components of orcein have been characterized with several others largely unidentified. One group of orcein-based dyes from *Rocella tinctoria, Ochrolechia tartarea* and other lichens gives rise to the purple to red-mauve pigments known as orchil, or to the Scottish variants

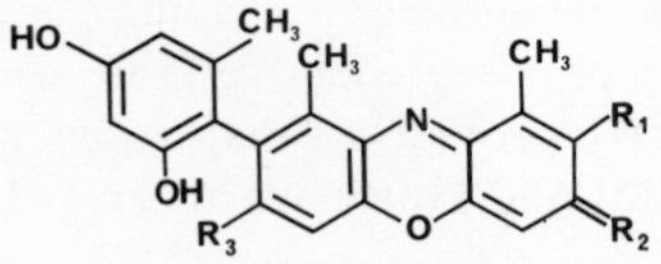

Figure 1.29 Orcein (orcinol).

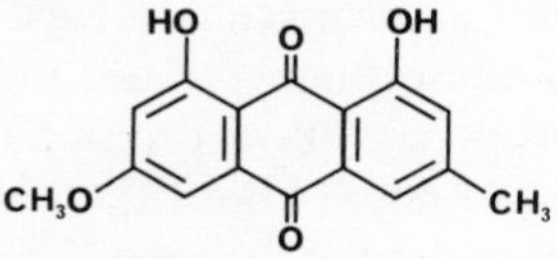

Figure 1.30 Parietin.

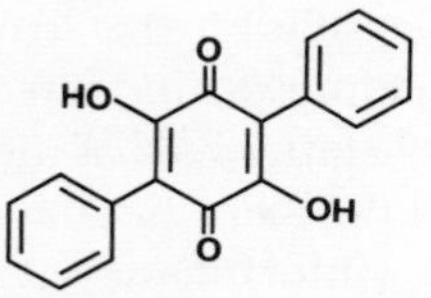

Figure 1.31 Polyporic acid.

corkir and cudbear. Related species have provided the source material for the pH indicator litmus. Litmus itself has also been used for colouring beverages (Hawksworth and Hill [15]).

Probably more widespread than despides are the phenolic pigments of lichens, although as a group they are less well-documented. Some 40 or so anthraquinones, ranging from yellow to red, have been described in some detail and are exemplified by the bright orange to orange-red colours found in the genus *Xanthoria*. The structure of one, parietin, typifies the group (Figure 1.30). Naphthaquinones contribute more to the red to blood-red colours found, for example, in the apothecia of *Cladonia* (rhodocladonic acid) and *Haematomma* spp. (haemoventosin). The structurally more varied terphenylquinones provide deep-red to purple colours of which polyporic acid (Figure 1.31) is one of a small number of pigments in this class. Biosynthetically related to the terphenylquinones, pulvinic acid and its derivatives represent a widespread class which provides a range of yellow to orange pigments. Some of the pulvinic acid derivatives may have a role as anti-herbivore defence chemicals. One, vulpinic acid from *Letharia vulpina*, appears to be one of the ingredients in a concoction designed to poison wolves in Lapland. Absorption if not digestion needs to be aided by the addition of ground glass in the bait (Hawksworth and Hill [16])! The more elaborate chromones, of which siphulin is an example, have a structural resemblance to flavonoids, including the anthocyanins of higher plants. As lichen pigments they are generally pale yellow.

Bright and dull yellow or yellow-orange pigments are particularly widespread among lichens. Apart from the phenyl derivatives discussed above lichen xanthones, norlichexanthone and lichexanthone (Figure 1.32), are relatively widespread pale yellow to yellow pigments. Chlorinated examples have also been recorded, although these are not apparently common. Some have larvicidal properties. Even more widespread is the class of pale yellow pigments based on usnic acid (Figure 1.33) and its several dibenzofurane derivatives. Relatively little is known of the function of this large class of pigments. In some species, usnic acid may function as a sunlight filter under conditions of high irradiance and desiccation but its antibiotic properties have given rise to a substantial literature. All lichens contain carotenoids (as well as chlorophylls) if only because part of the dual organism is photosynthetic. However, several xanthophylls have been found; these are synthesized within the fungal component. In addition, a few of the algal *Xanthophyta* and *Phaeophyta* form lichens so giving rise to marine xanthophylls (and chlorophyll *c*) in land plants.

Most reviews of lichen pigments lack a perspective either by concentrating on one class of pigment to the exclusion of others (*e.g.* dyes) or tend to be chemotaxonomic 'block-busters' with no indication of the obscurity or commonness of each group of pigments. Chemotaxonomy rarely demands quantification. The following summary is an attempt to redress this state. Chemically, the lichens are outstandingly rich in compounds derived from relatively simple phenols including quinones. It is doubtful if there are many genera that do not accumulate phenol derivatives to concentrations where they impart visible coloration to the organism. Many of these compounds,

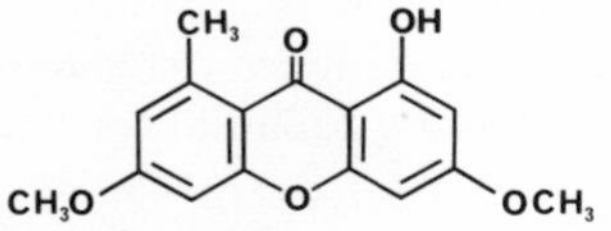

CH_3 O OH
CH_3O O OCH_3

Figure 1.32 Lichexanthone.

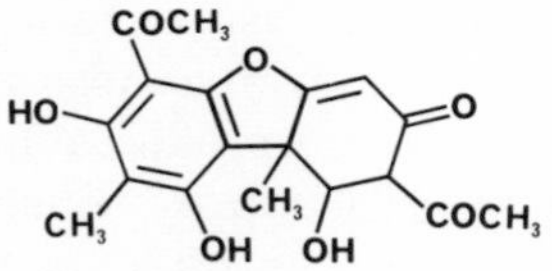

$COCH_3$ O
HO =O
CH_3 CH_3 $COCH_3$
OH OH

Figure 1.33 Usnic acid.

perhaps the majority, are unique to lichens. The most widespread groups probably include the anthraquinones, rivalling higher plants in diversity, pulvinic and usnic acid derivatives and perhaps the xanthones. Despite their apparent abundance throughout the lichens, many are poorly described (particularly structurally). From reviews such as that of Ely *et al.* [13] it is highly likely that many more of these compounds, particularly the terphenylquinoids, await description. Just as widespread are the carotenoids common to many algae (and higher plants). The carotenoids (trans)formed within the fungal partner are probably less common. At the other end of the scale of abundance, the depsides, particularly those of commercial interest, are not at all widespread, being confined to a few species. The abundant literature covering lichen depsides can give a false view of these uncommon compounds.

1.11.6 Bacteria

In general, bacteria contain many pigments that are similar, if not identical, to those of more complex organisms—particularly plants. Bacterial chlorophylls differ from plant chlorophylls in the reduction of one double bond. Although bacteria also possess their own complement of distinct carotenoids, these pigments are structurally and biosynthetically closely related to the carotenoids of plants and animals. Many bacteria, both photosynthetic and non-photosynthetic, also accumulate the more familiar carotenoids such as β- and γ-carotene.

In general, bacteria appear to be relatively poorly endowed with pigmented quinoids, with melanins and (almost) not at all with flavonoids. They do, however, make up for this in the production of some strikingly brilliant pigments apparently unique to bacteria and often to just one genus. One group of pigments apparently confined to bacteria are the phenazines based on a dibenzopyrazine skeleton. Among these often intensely coloured compounds, are the purple iodinin from species of *Chromobacterium* and the dark blue (in acid solution) pyocyanine (Figure 1.23) isolated from *Pseudomonas aeruginosa*. Many of the several dozen phenazines so far described (see Ingram and Blackwood [17]) have potential commercial interest particularly as antibiotics.

Several other intensely coloured compounds have been isolated from certain bacteria and which have little resemblance to pigments in other biological systems. Indigoidine or 'bacterial indigo', a dimeric pyridine structurally unrelated to the indigo of plants, is found in *Pseudomonas indigofera*. The highly pigmented genus *Chromobacterium* has also yielded (in acidic solution) the dark-red antibiotic prodigiosin with a most uncommon structure, a trimeric pyrrole. The same genus also produces dimeric indoles such as the purple violacein, although this has, at least, some resemblance to the indole derivatives of higher plants. Little is known of the function and properties of these unusual pigments.

1.12 Biological systems as sources of pigments for commercial exploitation

Increasingly, with the improvements in fermentation and other biotechnological techniques, bacteria, single-celled fungi and protozoa (including photosynthetic plankton) offer considerable scope for the commercial production of many pigments. The more future prospects using gene transformation techniques will allow, in theory, the biological production of almost any stable natural pigment. The limits, however, are not in the technologies, despite some major problems, but in our ignorance of what is in the biological world. At its simplest, despite centuries of interest in natural pigments, we have had a very myopic view and knowledge of the coloured natural pigments.

The chemistry, biochemistry and chemosystematics of higher plant pigments are relatively well-developed, often to considerable biophysical detail. In contrast, those of algae, particularly single-celled ones, are not. For example, of the five new chlorophylls described in the past 20 years, all are algal. The likelihood is that the next two or three chlorophylls will also be from algae (S.W. Jeffrey [18]). Yet algae lend themselves to biotechnological exploitation. Some progress is being made with a few organisms of which *Dunaliella* is one example (see later chapters). In contrast, bacterial pigments appear to be relatively well-documented and several pigments (including pigmented antibiotics) are already produced industrially from bacteria.

Much less is known of animal pigments, perhaps because of their diversity particularly among the invertebrates. Despite this plethora, animals have never contributed more than perhaps a dozen pigments of commercial value and it is not clear why this has occurred. The mounds of mussel shells from archaeological digs of dye works in the Mediterranean suggests that our forefathers had a different attitude to the value of animal bi-products.

Perhaps the richest source of pigment 'pickings' will be among the fungal kingdom. Many of these, including the myxomycetes (slime moulds), are highly pigmented. A significant proportion of fungal (including lichen) pigments are unique to the group and their attributes barely explored. Other than as textile and indicator dyes, few fungal pigments have been successfully exploited for their colorant properties for any length of time. Yet the several hundreds of fungal pigments so far described are among the most colourful of all biological systems.

Acknowledgements

Several colleagues have provided me with advice on their particular areas of expertise and I readily acknowledge, with thanks, the valuable discussions with Drs George Britton (animal carotenoids), Rod Cooke and Tony Lyon (fungal pigments), David Hill, Oliver Gilbert and Professor David Lewis (lichen pigments) and Professors Owen Jones and Stan Brown (tetrapyrroles).

I am also grateful to Dr Richard Winton for correcting my now rusty Latin and Greek.

References

1. Stearn, W.T. *Botanical Latin*. David & Charles. Newton Abbot (1973).
2. Flood, W.E. *The Origin of Chemical Names*. Chemical Society, London (1963).
3. Brown, S.B. *Introduction to Spectroscopy for Biochemists*. Academic Press, London (1980).
4. Harborne, J.B. *The Flavonoids, Advances in Research Since 1980*. Chapman & Hall, London (1988).
5. Britton, G. *The Biochemistry of Natural Pigments*. Cambridge University Press, Cambridge (1983).
6. Fox, D. *Biochromy*. University of California Press, Berkeley (1979).
7. Fox, H.M. and Vevers, G. *The Nature of Animal Colours*. Sidgwick & Jackson, London (1960).
8. Needham, A.E. *The Significance of Zoochromes*. Springer-Verlag, Berlin (1974).
9. Hendry, G.A.F. and Jones, O.T.G. *Journal of Medical Genetics* **17** (1980) 1–14.
10. Gill, M. and Steglich, W. *Progress in the Chemistry of Organic Natural Products* **51** (1987) 1–317.
11. Steglich, W. in *Pigments in Plants* (F.-C. Czygan Ed.), 2nd edn. Fischer, Stuttgart (1980) pp. 393–412.
12. Shibata, S., Naume, S. and Udagawa, S. *List of Fungal Products*. University of Tokyo Press, Tokyo (1964).
13. Ely, J.A., Whitton, A.A. and Sargent, M.V. *Progress in the Chemistry of Organic Natural Products* **45** (1984) 103–233.
14. Richardson, D.H.S. *The Vanishing Lichens*. David & Charles, Newton Abbot (1975).
15. Hale, M.E. *The Biology of Lichens*. Edward Arnold, London (1983).
16. Hawksworth, D.L. and Hill, D.J. *The Lichen-Forming Fungi*. Blackie & Son, Glasgow (1984).
17. Ingram, J.M. and Blackwood, A.C. *Advance in Applied Microbiology* **13** (1970) 267–282.
18. Jeffrey, S.W. Personal communication.

2 Natural food colours

B.S. HENRY

2.1 Summary

Natural colours have always formed part of man's normal diet and have, therefore, been safely consumed for countless generations. The desirability of retaining the natural colour of food is self-evident, but often the demands of industry are such that additional colour is required. Contrary to many reports, natural sources can provide a comprehensive range of attractive colours for use in the modern food industry. In particular, five natural colours—annatto, anthocyanins, beetroot, turmeric and carmine—are widely used in everyday foodstuffs. The factors affecting the stability of these and other permitted natural colours and their commercial applications are fully discussed.

2.2 The role of colour in food

The first characteristic of food that is noticed is its colour and this predetermines our expectation of both flavour and quality. Food quality is first judged on the basis of colour and we avoid wilting vegetables, bruised fruit, rotten meat and overcooked food.

Numerous tests have demonstrated how important colour is to our appreciation of food. When foods are coloured so that the colour and flavour are matched, for example yellow to lemon, green to lime, the flavour is correctly identified on most occasions. However, if the flavour does not correspond to the colour then it is unlikely to be identified correctly [1].

Colour level also affects the apparent level of sweetness; in one study, Johnson *et al.* [2] showed that sweetness appeared to increase between 2 and 12% with increasing colour of a strawberry flavour drink. The colour of a food will therefore influence not only the perception of flavour, but also that of sweetness and quality. It is also important not to underestimate aesthetic value. The best food with a perfect balance of nutrients is useless if it is not consumed. Consequently, food needs to be attractive. Domestic cooking has traditionally attempted to enhance or preserve food colour. Pies are glazed with beaten eggs, and lemon juice is used to prevent the browning of fruit.

Recently, new varieties of peppers of different colours, yellow and purple, have become available.

Likewise, there is a need for processed foods to be visually appealing. Colour may be introduced in several ways. The raw materials—the fruit, vegetables, meat, eggs—have their own intrinsic colour, the processing conditions may generate colour or a colour may be added. Some food products have little or no inherent colour and rely on added colour for their visual appeal.

However, although the colour of fruits and vegetables can vary during the season and processing can cause colour loss, food manufacturers need to ensure uniformity of product appearance from week to week—a factor that does not concern the domestic cook where a slight variation in recipe or cooking time is usual. For the manufacturer, colour consistency is seen as visual proof of absolute consistency of the manufacturing process.

Colours may be added to foods for several reasons, which may be summarized as follows [3]:

1. To reinforce colours already present in food but less intense than the consumer would expect;
2. To ensure uniformity of colour in food from batch to batch;
3. To restore the original appearance of food whose colour has been affected by processing;
4. To give colour to certain foods such as sugar confectionery, ice lollies and soft drinks, which would otherwise be virtually colourless.

The continued use of colour in food is acknowledged by the Food Additives and Contaminants Committee (FACC) who concluded that 'if consumers are to continue to have an adequate and varied diet, attractively presented, the responsible use of colouring matter, the safety-in-use of which has been fully evaluated, still has a valid part to play in the food industry' [3].

2.3 Classification of food colours

It is only in the last 100 years or so, following the discovery of the first synthetic dye by Sir William Perkin in 1856 and the subsequent development of the dyestuffs industry, that synthetic colours have been added to food. For centuries prior to this, natural products in the form of spices, berries and herbs were used to enhance the colour and flavour of food. During this century, the use of synthetic colour has steadily increased at the expense of these products of natural origin, due principally to their ready availability and lower relative price. In the last 15 years following the delisting of several synthetic colours, notably that of amaranth in the USA in 1976 and that of all synthetic colours by Norway also in 1976, there has been an increase in the use of colours derived from natural sources. Generally three types of organic food colours are recognized in the literature: synthetic colours, nature-identical colours and natural colours.

2.3.1 *Synthetic colours*

These are colorants that do not occur in nature and are produced by chemical synthesis (*e.g.* sunset yellow, carmoisine and tartrazine).

2.3.2 *Nature-identical colours*

These are colorants that are manufactured by chemical synthesis so as to be identical chemically to colorants found in nature, for example β-carotene,[1] riboflavin and canthaxanthin.

2.3.3 *Natural colours*

These are organic colorants that are derived from natural edible sources using recognized food preparation methods, for example curcumin (from turmeric), bixin (from annatto seeds) and anthocyanins (from red fruits).

This description of a natural colour would exclude caramels manufactured using ammonia and its salts and also copper chlorophyllins, since both of these products involve chemical modification during processing using methods not normally associated with food preparation.

2.4 Legislation

Natural colours are widely permitted throughout the world. However, there is no universally accepted definition of this term and many countries exclude from their list of permitted colours those substances that have both a flavouring and a colouring effect. Thus spices are generally not regarded as colours. Sweden, for example, states that 'turmeric, paprika, saffron and sandalwood shall not be considered to be colours but primarily spices, providing that none of their flavouring components have been removed' [5]. Italian legislation states that 'natural substances having a secondary colouring effect, such as paprika, turmeric, saffron and sandalwood, are not classed as colours but must be declared as ingredients in the normal way' [6]. Similar comments are included in the food legislation of Holland, Switzerland and Norway.

The European community (EC) permits a wide range of colours that may be of natural origin and these are listed in Table 2.1. However, it is important

[1] β-carotene is fundamental as a precursor of vitamin A and also acts as an antioxidant in the diet as a result of its ability to scavenge singlet oygen [4].

Table 2.1 Natural colours (and colours of natural origin) listed by the EC

E100	Curcumin
E101	Riboflavin
E120	Cochineal/carminic acid/carmines
E140	Chlorophyll
E141	Copper complexes of chlorophyll and chlorophyllins
E150	Caramel
E153	Vegetable carbon
E160	(a) α-, β-, γ-carotene
	(b) Annatto extracts, bixin, norbixin
	(c) Paprika extract, capsanthin, capsorubin
	(d) Lycopene
	(e) β-apo-8′-carotenal (C30)
E161	(a) Flavoxanthin
	(b) Lutein
	(c) Cryptoxanthin
	(d) Rubixanthin
	(e) Violaxanthin
	(f) Rhodoxanthin
	(g) Canthaxanthin
E162	Beetroot red, betanin
E163	Anthocyanins

to note that the following colours are not available commercially:

- E160(d) Lycopene
- E161(a) Flavoxanthin
- E161(c) Cryptoxanthin
- E161(d) Rubixanthin
- E161(e) Violaxanthin
- E161(f) Rhodoxanthin

A proposal is before the Commission to replace E161(a) to (f) with a general E161 classification, which would encompass mixed xanthophylls [7].

In addition, three of the colours in Table 2.1 are only available commercially as nature-identical products. These are:

- E101 Riboflavin
- E160(e) β-apo-8′-carotenal
- E161(g) Canthaxanthin

Finally, although E160(a) β-carotene is available as a natural extract it is the nature-identical form that is most widely used.

The USA has a different set of 'natural' colours and those currently permitted by the Food and Drug Administration (FDA) are listed in Table 2.2;

Table 2.2 Natural colours (and colours of natural origin) listed by the FDA for food and beverage use

Annatto extract
β-apo-8′-carotenal[a]
β-Carotene[a]
Beet powder
Canthaxanthin[a]
Caramel
Carrot oil
Cochineal/carmine
Cottonseed flour, toasted
Fruit juice
Grape colour extract
Grape skin extract
Paprika and paprika oleoresin
Riboflavin (NI)
Saffron
Turmeric and turmeric oleoresin

[a] Nature-identical forms only.

these do not require certification (therefore do not have FD & C numbers) and are permanently listed.

One of the advantages of using natural colours is that they are generally more widely permitted in foodstuffs than synthetic colours. It should be remembered that colour usage may be controlled in three distinct ways as follows:

1. National legislation lists those colours that may be used in foods;
2. Colour use within a country is limited by the type of food that may be coloured; and
3. The maximum quantity of colour that can be added to a food may also be specified.

Thus for example, beetroot extract is permitted in Sweden. However, its use is limited to specified food products only, such as sugar confectionery, flour confectionery and edible ices. There is a maximum dose level limit of 20 mg/kg as betanin in the first two categories of food and 50 mg/kg in edible ices. 20 mg/kg equates to 0.4% of beetroot extract containing 0.5% betanin, which is the standard strength for such an extract. Its use, however, is not permitted in dry mix desserts, milk shakes or soups.

It is therefore essential to consult the legislation relating to the particular food product before using any colour. It is obviously insufficient just to check the simplified list of permitted colours since this relates solely to one aspect of colour regulation.

2.5 Factors affecting colour choice

When deciding which natural colour to use in a specific application it should be noted that there are several factors to influence that choice as follows:

1. Colour shade required. It is often necessary to blend two or more colours together in order to achieve the desired shade.
2. Legislation of the countries in which the food is to be sold.
3. Physical form required. Liquid natural colours are generally more cost effective than powder forms.
4. Composition of the foodstuff—in particular whether it is an aqueous system or whether there is a significant level of oil or fat present. The presence of tannins and proteins may limit the use of some colours such as anthocyanins. It should also be determined whether a cloudy or crystal clear colour is required.
5. Processing conditions—particularly the temperatures used and the times for which these temperatures are held.
6. pH. The pH of a food will often determine the suitability of a particular colour for a given application. The stability or colour shade of most natural colours are affected by pH.
7. Packaging. This will determine the amount of oxygen and light that reaches the product and hence the suitability of such colours as carotenoids and curcumin.
8. Required shelf-life and storage conditions.

2.6 Factors affecting colour application forms

Once these factors have been defined it will enable the most suitable natural colour to be selected by reference to the relevant sections of this chapter. It should also be remembered that natural colours are a diverse group of colorants with widely differing solubility and stability characteristics. Consequently each colorant is available in several different application forms, each formulated to ensure that the colour is compatible with a particular food system. A product application form is a formulation that enables a specific food additive to be easily and efficiently incorporated into a manufactured food product. It may, for example, be a very fine spray-dried powder of low pigment content for mass coloration. Or it may be a more complex emulsion incorporating an oil-soluble colour dissolved in citrus oils, which is subsequently emulsified into an aqueous phase containing emulsifiers and stabilizers. There are, therefore, several factors relating to these application forms that should be considered by the food technologist.

2.6.1 *Solubility*

Anthocyanins and beetroot are water soluble. Curcumin, chlorophylls and xanthophylls are oil soluble. Some blends of curcumin and annatto may be both oil and water miscible.

2.6.2 *Physical form*

Colours are available in the form of liquids, powders, pastes and suspensions. When using suspensions it is important to realize that colour shade will change if the pigment becomes dissolved during processing. An increase in temperature is often sufficient to dissolve a suspended pigment and so effect a colour change; this particularly applies to β-carotene and annatto suspensions. The viscosity of liquids and pastes is temperature dependent and their ease of dispersibility into a foodstuff will decrease with decreasing temperatures.

2.6.3 *pH*

Most liquid, water-soluble natural colours are manufactured with their pH close to that of their maximum stability. Thus extracts containing norbixin are alkaline, whereas anthocyanins are acid. Addition of such products to an unbuffered solution is likely to alter the pH of that solution.

2.6.4 *Microbiological quality*

Oil-soluble pigments tend to have low moisture contents and are therefore not generally susceptible to microbiological spoilage. Anthocyanin and beetroot extracts contain significant levels of water and sugars and thus steps must be taken to ensure good microbiological quality.

2.6.5 *Other ingredients*

Some oil-soluble colours such as curcumin and β-carotene need the addition of gums, stabilizers or emulsifiers in order to render them water miscible. It is important that these ingredients are compatible with the food system to which the colour is being added.

2.7 Performance and consumption of natural colours

As has been stated, natural colours are a very diverse group of compounds and it is therefore extremely difficult to make general comments about their

Table 2.3 Relative absorptivities of some natural and synthetic dyes

Natural dyes			Synthetic dyes		
Dye	$E^{1\%}_{1\,cm}$	Wavelength (nm)	Dye	$E^{1\%}_{1\,cm}$	Wavelength (nm)
Norbixin	2870	482	Sunset yellow	551	480
Curcumin	1607	420	Tartrazine	527	426
Betanin	1120	535	Amaranth	438	523

nature and performance. However, the literature contains many such statements that, over the years, have led to general misunderstandings.

One of those frequently repeated statements says that natural colours have a lower tinctorial strength than synthetic colours and thus require higher levels of addition. However, in reality the reverse is generally true. Beetroot, β-carotene, bixin and curcumin, for example, are all intense colours and their use in food generally results in a *decrease* in colour-dose level. It is interesting to compare the absorptivities of some natural colours with azo-dyes of a similar shade.

Table 2.3 clearly illustrates that these natural colours are considerably more intense than some commonly used synthetic colours—a fact that is barely mentioned in the literature.

When natural colours are added to food it is also quite common for the dose level to be adjusted so that a pastel, more 'natural' appearance is achieved. This is particularly noticeable when considering the colour strength of yoghurts and fruit squashes where colour inclusion rates have been reduced following the introduction of natural or nature-identical colour. Thus colour inclusion levels in food are tending to reduce on account of both the strength of many natural colours and the move to more pastel shades.

Another view is that natural colours are only available in small amounts and that excessive areas of land would be required to cultivate commercial quantities. In reality, the colour supply industry has responded quickly to any increase in demand, although the quantities of most natural colours required by industry are small in relation to those produced by nature. The consumption of natural colours as an integral part of the diet is far in excess of the quantities added as food colour. Kuhnau [8] stated that the normal daily intake of anthocyanins in food was around 215 mg in summer and 180 mg during winter. Assuming an annual consumption of 70 g per person this equated to approximately 4000 tonnes of anthocyanin being consumed each year. The quantity of anthocyanins added as colour to food was certainly less than 5 tonnes per year. A similar comparison can also be made for carotenoids, chlorophylls and beetroot.

2.8 Annatto

2.8.1 Source

Annatto (Figure 2.1) is the seed of the tropical bush *Bixa orellana*. The major colour present is *cis*-bixin, the monomethyl ester of the diapocarotenoic acid norbixin, which is found as a resinous coating surrounding the seed itself. Also present, as minor constituents, are *trans*-bixin and *cis*-norbixin. The annatto bush is native to Central and South America where its seeds are used as a spice in traditional cooking. In Brazil, substantial quantities of processed annatto seeds are sold in retail outlets often blended with other ingredients

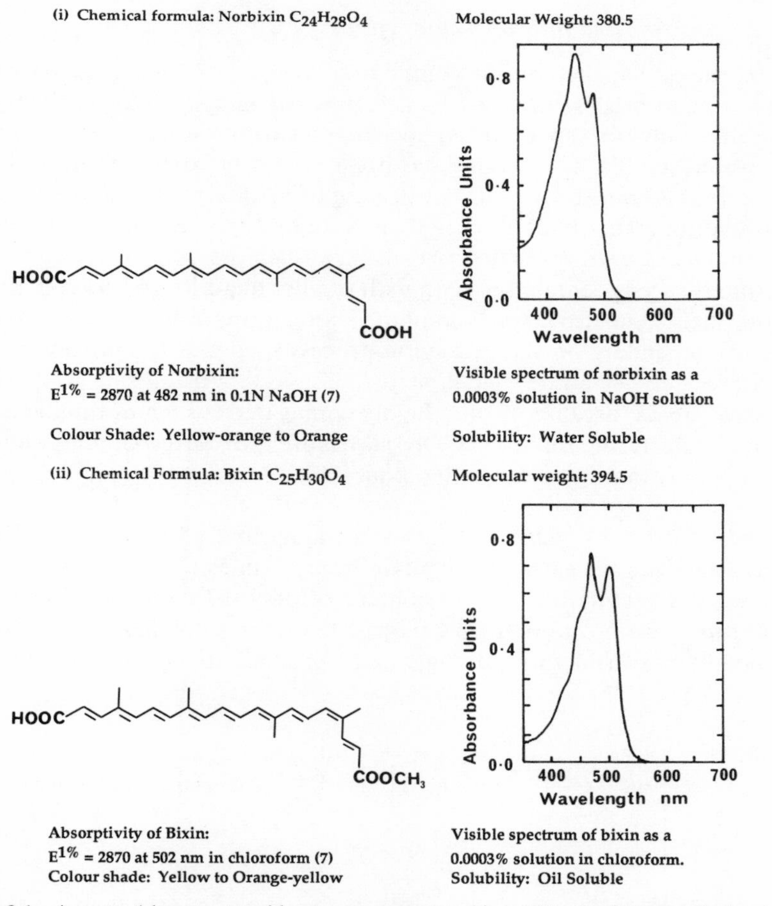

Figure 2.1 Annatto (class carotenoid; synonyms rocou, achiote, CI natural orange 4): structural and physical characteristics of the constituents (i) norbixin and (ii) bixin.

for addition to soups and meat dishes similar to the use of paprika seasonings in Europe.

Although annatto seed is harvested in many tropical countries including Bolivia, Ecuador, Jamaica, the Dominican Republic, East Africa, India and the Philippines, it is Peru and Brazil that are the dominant sources of supply. The principal export form of annatto is the seed, although to increase export values, several seed suppliers now also manufacture extracts. It is likely that around 7000 tonnes of annatto seed is used annually as a food colour worldwide and, assuming an average colour content of 2%, this equates to 140 tonnes of bixin. The main maket for annatto is the USA and Western Europe, although there is also considerable inter-trade between the Central and South American suppliers.

2.8.2 Extracts and their application forms

Bixin, the principal colouring compound, is oil soluble, although only sparingly so, and occurs in *cis* and *trans* forms. Alkali hydrolysis of bixin yields the salts of the corresponding *cis* and *trans* dicarboxylic acid norbixin, which are water soluble. On heating, *cis*-bixin is converted to the more stable, more soluble *trans*-bixin. Further heating of bixin leads to the formation of degradation compounds (Figure 2.2). Note that the heating of *cis*-norbixin does not yield *trans*-norbixin.

Annatto seeds contain *cis*-bixin with smaller quantities of *trans*-bixin and *cis*-norbixin. Annatto extracts contain varying proportions of colouring compounds depending on the extraction process used and the processing temperature.

Carotenoids, because of their highly conjugated system of double bonds, are particularly intense colours. Dose levels are thus correspondingly low and often in the range 5 p.p.m. to 10 p.p.m.

2.8.2.1 Oil-soluble extracts. Extracts of annatto seed produced using hot vegetable oil as the extractant contain *trans*-bixin as the major colour component and usually have a dye content of only 0.2 to 0.3%. Higher concentrations are not possible because of the poor solubility of bixin in oil. Although of low tinctorial strength, such products are ideally suited for use

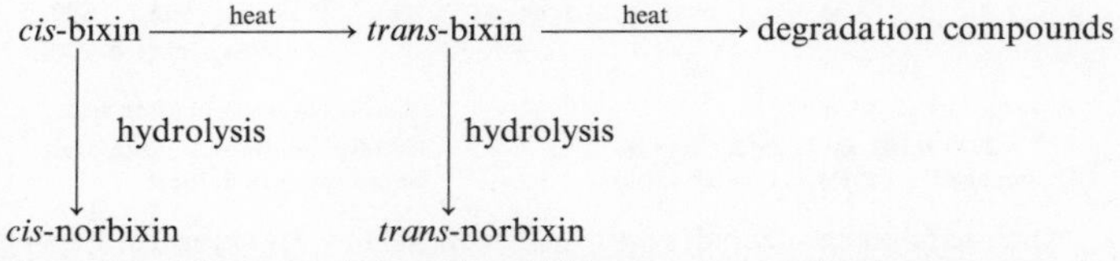

Figure 2.2 The interrelationship of different annato colouring compounds.

in foods where there is a substantial quantity of oil present, such as dairy spreads, salad dressings and extruded snack foods.

2.8.2.2 Suspensions in oil. A more cost effective way of incorporating bixin is to suspend the undissolved pigment in vegetable oil. By this method products with 4% bixin are manufactured, both *cis*- and *trans*-bixin being present. However, it is important to remember that the colour shade obtained depends upon the amount of bixin that dissolves in the oil phase which, in turn, depends upon dose level and the temperature attained during processing. A suspension of bixin is orange in colour whereas a solution of bixin in vegetable oil is yellow and so, as the end-product temperature is increased, the bixin dissolves and the colour becomes more yellow.

2.8.2.3 Water-soluble extracts. Norbixin, the *cis* form in particular, is very water soluble and solutions with more than 5% dye can be achieved. It is usual commercial practice to manufacture a range of alkaline solutions containing 0.5% to 4.0% norbixin by alkaline hydrolysis of annatto. Extraction of norbixin from the seeds is not commercially viable because of its low level of occurrence. In the UK, an aqueous solution containing 0.5% norbixin is often termed 'single strength' annatto and 1.0% norbixin 'double strength'. However, in the USA, 1.4% norbixin is termed 'single strength' and 2.8% 'double strength'. All such products are often termed 'cheese colour' and use dilute potassium hydroxide as the diluent.

Norbixin may also be spray dried using a carrier such as gum acacia, maltodextrin or modified starch, to give a water-soluble powder colour. Norbixin concentrations range from approximately 1% to 14% with 1%, 3.5%, 7% and 14% being the most commonly manufactured products. Due to their large surface area, such products are particularly prone to oxidation.

2.8.2.4 Process colours. It is also possible to blend together bixin and norbixin extracts by the use of suitable carriers such as propylene glycol and polysorbates to produce a colour that can be added to water- or oil-based foods. Bixin content is usually in the range 1 to 2%. Other extracts, notably those of turmeric and paprika, may also be blended with such products to allow a wider range of colour shades to be obtained. These blends find application in flour confectionery and dairy products where some oil or fat is present.

2.8.3 Legislation

Annatto extracts are permitted virtually worldwide including the EC, Scandinavia and North America. Some countries impose dose-level limitations on colour usage, including that of annatto; such countries include Belgium, Denmark, France, Sweden and Spain.

2.8.4 *Factors affecting stability*

2.8.4.1 pH. Norbixin will precipitate from an acid solution as the free acid. Thus a cheese colour should not be used in water ice, acid sugar confectionery or soft drinks. Bixin, however, is not affected by pH, and special application forms often based on bixin can be used in acidic products.

2.8.4.2 Cations. Divalent cations (particularly calcium) will form salts with norbixin. Since calcium norbixinate has very poor water solubility, norbixin extracts are incompatible with products containing high calcium levels. It is therefore not recommended to make dilutions of cheese colour using hard water. Particular application grades of norbixin have to be used where high levels of salts may be present such as in the processing of cured fish.

2.8.4.3 Heat and light. When bound to proteins or starch, norbixin is stable to heat and light. Nevertheless, stability is significantly reduced when norbixin is present in dilute aqueous systems.

Bixin and norbixin are reasonably stable to heat, although degradation of bixin can occur at temperatures over 100°C leading initially to a more lemon yellow shade and subsequently to loss of colour. Exposure to light will result in a gradual loss of colour.

Aqueous annatto extracts should not be allowed to freeze otherwise norbixin may be thrown out of solution.

2.8.4.4 Oxygen. Because of the conjugated double-bond structure, all carotenoids are susceptible to oxidation. The addition of ascorbic acid as an oxygen scavenger is beneficial.

2.8.4.5 Sulphur dioxide (SO_2). Colour intensity will reduce in the presence of sulphur dioxide thus alternative preservative systems are recommended when using annatto as a colour.

2.8.5 *Applications*

Since annatto extracts are available in so many different application forms—more so than any other natural colour—virtually any yellow to orange food product may be successfully coloured with annatto. The most difficult task is to establish which application form is best suited to a given food system. The simplest answer to the problem is to consult an annatto colour manufacturer.

2.8.5.1 Dairy products. Dairy products are the single most important application for annatto extracts. Cheese, particularly hard cheese such as Red Leicester and Cheshire, has traditionally been coloured with a solution of norbixin. Norbixin binds to the milk protein and forms a stable colour, which is not lost during the separation of the whey. For Cheddar cheese production, dose levels of norbixin are usually in the range 0.75 to 1.25 p.p.m. in the milk. Similar norbixin extracts are also used to colour other dairy products, notably vanilla-flavour ice cream where a blend of norbixin and curcumin (extracted from turmeric) is commonly used. A typical dose level would be 10 p.p.m. norbixin together with 15 p.p.m. of curcumin. Bixin is used to colour dairy spreads at a dose in the range 1 to 5 p.p.m.

2.8.5.2 Flour confectionery. Norbixin is ideally suited for the colouring of such products since it binds to the flour and forms a stable colour that will not leach. Dose levels are approximately 4 to 8 p.p.m. for sponge cake. Other related applications include breadcrumbs, ice-cream wafers, biscuits and snack foods. For flour confectionery containing a significant level of fat, then a solution of bixin in oil or a process annatto colour can be used.

2.8.5.3 Fish. Annatto extracts, in the form of special application grades of norbixin, are widely used in the coloration of smoked herrings and mackerel. Prior to smoking, the prepared fish are dipped in a brine to which has been added norbixin at 200 to 300 p.p.m. The norbixin binds well to the protein such that no significant amount of colour is lost during the subsequent processing and the final cooking by the consumer. The final norbixin content of the fish is of the order 20 to 40 p.p.m.

2.8.5.4 Sugar confectionery. Norbixin is used in a variety of sugar confectionery products, although special application forms need to be used in acidic products where a crystal-clear colour is required, such as high boilings.

2.8.5.5 Soft drinks. Norbixin extracts that are acid and light stable, therefore suitable for use in soft drinks have been developed. Dose levels for norbixin usually lie within the range 1 to 10 p.p.m. in the ready-to-drink product. This is one of the most demanding applications for any colour.

2.8.5.6 Meat products. Norbixin extracts are often used in combination with carmine in the manufacture of chicken dishes. A variety of colour shades can be obtained to enable different ethnic styles to be reproduced.

2.8.5.7 Snack foods. Annatto extracts are used in many snack food products. The colour is incorporated in several different ways depending on the nature of the food. Norbixin can be added to the cereal starch prior to extrusion or bixin may be added to the oil slurry that is used to flavour the snack after extrusion.

2.8.5.8 Dry mixes. Norbixin in powder form is ideally suited for the coloration of both savoury and sweet dry mixes. For powders that are acid on reconstitution, special application forms are required.

2.9 Anthocyanins

2.9.1 Source

Anthocyanins (Figure 2.3) are those water-soluble compounds responsible for the red to blue colour of a wide range of fruits and vegetables. The list of sources is very long and includes grapes, redcurrants and blackcurrants, raspberries, strawberries, apples, cherries, red cabbages and aubergines [9]. Chemically, they are glycosides of flavylium or 2-phenylbenzopyrylium salts and are most commonly based on 6 anthocyanidins: pelargonidin, cyanidin, peonidin, delphinidin, petunidin and malvidin. The sugar moiety present is most commonly one of the following: glucose, galactose, rhamnose, arabinose.

In addition, the sugar may be acylated with a phenolic or aliphatic acid. Some 300 anthocyanins are found in nature, some fruits (*e.g.* strawberry) contain just one or two while others such as 'Concord' grapes have at least 15 present.

Chemical Formula: Malvidin-3-glucoside $C_{23}H_{25}O_{12}$

Molecular Weight: 529

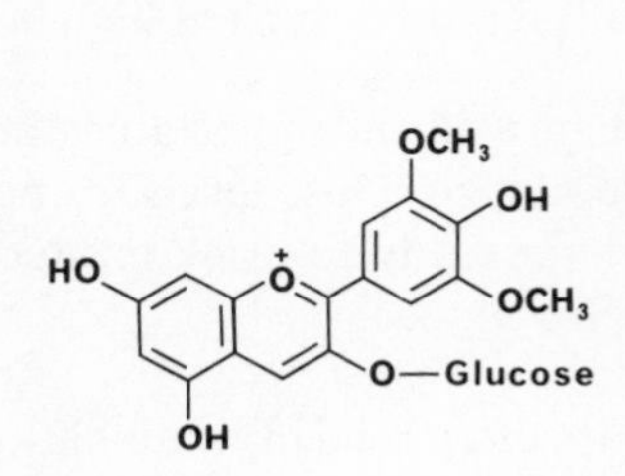

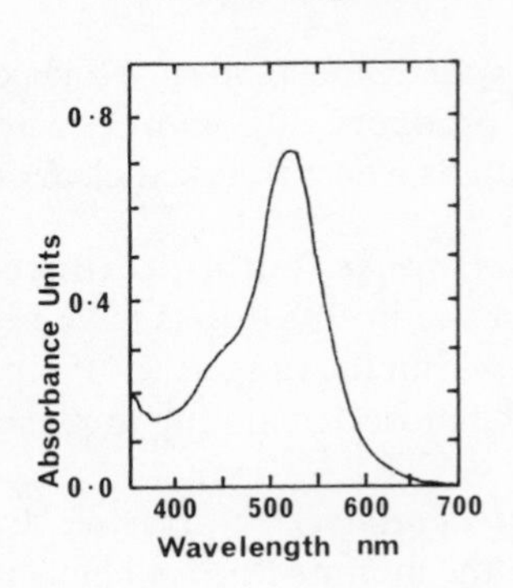

Visible spectrum of grape skin extract as a 0.002% aqueous solution at pH 3.0.

Colour Shade: Red at pH 2
Blue-red at pH 4

Solubility: Water Soluble

Absorptivity of Grape Skin Extract: $E^{1\%}$ = 500 at pH 1.5 at λ max. closest to 520 nm.

Figure 2.3 Anthocyanins (class anthocyanin; synonyms grape skin extract, grape colour extract, enocianina): structural and physical characteristics of malvidin-3-glucoside.

It is possible to extract colour from any of these raw materials but for economic reasons, grape skins, a by-product of the wine industry, are the usual source.

Grapes are the single most abundant fruit harvested in the world from which an annual consumption of about 10 000 tonnes of anthocyanins has been estimated [10]. Grapes with highly pigmented skins such as Ancellotta, Lambrusco, Alicante and Salamina have sufficient colour remaining after wine production to justify colour extraction [11]. It is estimated that some 10 000 tonnes of grape skins are extracted annually in Europe, yielding approximately 50 tonnes of anthocyanins. Extraction is carried out using a dilute aqueous solution of an acid, usually sulphurous acid, and yields a product containing sugars, acids, salts and pigments all derived from the skins. It is normal to concentrate this extract to 20 to 30° Brix, at which strength the anthocyanin content is usually in the range 0.5 to 1%. It is technically possible to concentrate further but the cost of doing so is not usually justified.

2.9.1.1 Grape extracts and their colour strength. It is important to remember that natural food colours, as used by the food industry, are commonly in the form of a standardized dilute solution of a particular colour or blend of colours. Thus when calculating pure dye inclusion levels, it is essential to know the dye content of the colour. It is usual for the specification for a natural colour to include a measurement of colour strength based upon the absorbance of a solution of the colour measured spectrophotometrically. This is often expressed as the calculated absorbance of a 1% w/v solution in a 1-cm cell. However, this is often not translated into a dye content.

With anthocyanin extracts, the most common method is to calculate either the absorbance per gram of colour or the absorbance of a 1% w/v solution of the colour in a pH 3.0 citric/citrate buffer.

Grape skin extracts. Most liquid anthocyanin extracts derived from grape skins have a colour strength (absorbance/g) of between 150 and 300, Table 2.4, which equates to 0.5% and 1.0% anthocyanin colour respectively.

Table 2.4 Colour strengths of grape skin extracts

Grape skin extract	$E^{1\%}_{1\,cm}$	Absorbance per gram	Approximate % anthocyanin
Standard	1.5	150	0.5
Double strength	3.0	300	1.0
Concentrate	6.0	600	2.0
Powder	12.0	1200	4.0

Since grapes contain from at least four to over sixteen separate anthocyanins, depending on the variety, it is not possible to calculate the exact colour content by a simple analytical technique. To do so you would need to know the exact percentage anthocyanin composition of the extract. However, by measuring colour strength at pH 1.0 when the anthocyanins are essentially monomeric and by taking an average value for the $E^{1\%}_{1\,cm}$ of anthocyanins of 500 at 520 nm then a calculation [12] can be made, giving the results shown in Table 2.4.

Such extracts can be oven- or spray-dried, using malto-dextrin as the carrier if necessary, to yield a water-soluble powder. Such a product usually contains 4% anthocyanin, although more dilute grades are produced for specific applications.

Thus there are essentially only two application forms of grape anthocyanin extract—a water-soluble liquid and a water-soluble powder. Nevertheless, different varieties of grapes are extracted and extraction techniques vary; this in turn produces extracts with different properties. These differences will affect colour shade and stability such that those extracts with a higher level of non-pigmented phenolic compounds are likely to be bluer in shade. However, too high a level may well induce a haze or precipitation if used in a soft drink.

Grape colour extracts. It is also possible to extract anthocyanins from the lees derived from grape juice production. Some grape juices, particularly those produced from 'Concord' grapes (*Vitis labrusca*), have high levels of anthocyanins and tartrates.

On storage, some of the tartrates precipitate bringing down with them a proportion of the colour. Aqueous extraction of this precipitate (the lees) will yield an anthocyanin, which, after ion-exchange to convert the insoluble tartrates into soluble tartaric acid, is suitable for use as a food colour. Since this type of extract consists of monomeric pigments and a low level of phenols, it tends to be redder in shade than *Vitis vinifera* extracts. In the USA, they are termed 'grape colour extracts' and are classified separately from 'grape skin extracts'.

2.9.1.2 Other sources. There are also several food products that have significant levels of anthocyanin present. These include the concentrated juice of blackcurrant, elderberry, cranberry and cherry, the aqueous extract of roselle (the calyces of hibiscus) and of red cabbage. These products contain all the colouring and flavouring principles of the raw material, no attempt having been made to concentrate the colour at the expense of the flavour. Such products are therefore food ingredients and not colour extracts.

Colour, however, is commercially extracted from red cabbage and this flavour-free extract, although considerably more expensive than grape colour, is used in the food industry because of its impressive stability to heat and light [13].

2.9.2 *Legislation*

Anthocyanins, particularly those extracted from grapes, are widely permitted. In the USA, grape colour extracts (from *V. labrusca*) are permitted in non-beverage foods, whereas grape skin extracts (from *V. vinifera*) are permitted in beverages only. In the EC the proposed specification for anthocyanins states that they should be obtained by extraction with water, methanol or ethanol from vegetables or edible fruits. Maximum limits on levels of colour addition are imposed by many countries; however, the use of anthocyanins often has no numerical limit and maximum dose is simply that sufficient to give the desired colour strength.

2.9.3 *Factors affecting stability*

2.9.3.1 pH. Anthocyanins act as pH indicators. They gradually change from red through blue-red, purple, blue and green to yellow as the pH increases from pH1 through 4, 6, 8, 12 to 13 respectively. From a practical point of view, anthocyanins are only used in acidic products where the pH is 4 or below. Not only does colour shade change with pH but colour intensity is also pH dependent being greatest at pH 1.0 and decreasing rapidly as pH rises. As seen from the plot of absorbance against pH (Figure 2.4), the colour intensity at pH 3.5 is half that at pH 2.0. Also, the gradient is steepest in the range pH 2.0 to 3.5. Since this is the pH range in which colour intensity is usually measured, it is necessary to take great care when preparing the relevant buffer.

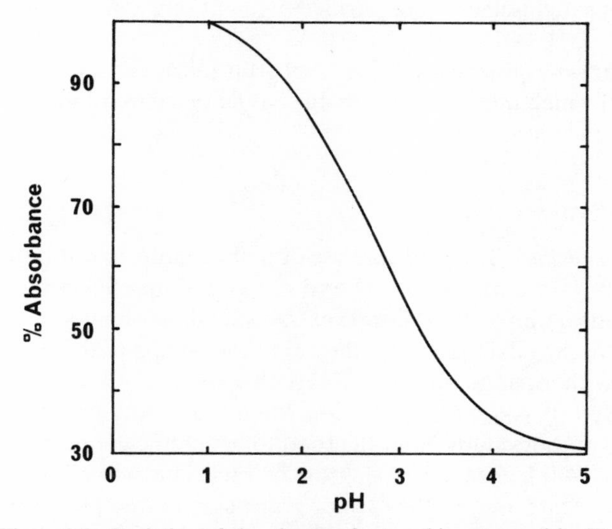

Figure 2.4 Variation of absorbance of grape skin extract with pH.

2.9.3.2 Cations. Some cations, particularly di- and trivalent metal ions, will cause a bathochromic shift of wavelength of maximum absorption. This is seen as a distinct 'blueing' of the colour and may result eventually in precipitation of the pigment. Iron, mild steel and copper surfaces should be avoided and tinplate containers should be lacquered.

2.9.3.3 Heat and light. Stability to heat is good and is adequate for processes such as jam and sugar boiling and fruit canning. Acylation of the sugar moiety will increase stability to both heat and light. Colours from red cabbage contain a significant quantity of mono- and di-acylated anthocyanins and are particularly stable.

2.9.3.4 Oxygen. Anthocyanins will slowly oxidize when in aqueous solution. Ascorbic acid does not improve stability under such conditions.

2.9.3.5 Sulphur dioxide (SO_2). Sulphur dioxide reacts with anthocyanins to form a colourless addition product. This reaction is reversible and on heating SO_2 is released to yield the anthocyanin and so restore the colour—a process that is demonstrated whenever sulphited fruit is boiled to make a preserve.

Sulphur dioxide should not be used as a preservative in a product containing anthocyanin, where a mixed benzoate/sorbate preservative is recommended.

2.9.3.6 Proteins. Some grape extracts will react with proteins (such as gelatin) to form a haze or even a precipitate. This reaction appears to be caused by non-pigmented phenolic compounds present in the extract rather than the anthocyanins themselves, since purified pigments are compatible with gelatin.

2.9.3.7 Enzymes. Enzyme treatment of fruit juices can cause loss of anthocyanins [14] which may, in part, be due to the presence of glucosidase in the enzyme preparation.

2.9.4 Applications

2.9.4.1 Soft drinks. The principal use of anthocyanin colour is in soft drinks. Clear drinks with a pH below 3.4 and not containing SO_2 as a preservative present an ideal application. Grape extracts, high in oligomeric or polymeric colours, have the advantage that they are more stable in the presence of SO_2 than those with monomeric colours since the position of attack of the sulphite ion is blocked. It is a wise precaution when evaluating natural colours and particularly anthocyanins to let the food to which the colour has been added to stand for 24 h before appraising the colour. This will allow the pigments time to equilibrate and, in the case of grape skin extract, it is possible to see an increase in colour during this period as anthocyanin is released from its sulphite derivative. Dose levels of around 30 to 40 p.p.m. anthocyanin in a ready-to-drink beverage are usually sufficient to give a deep-red colour. This is a relatively low level bearing in mind that blackcurrants contain 2000 to 4000 p.p.m. of anthocyanins.

Storage at elevated temperature (25°C or above) or exposure to sunlight will cause a significant loss of colour.

Anthocyanins are not usually suitable for use in a cloudy beverage. The presence of the cloud causes a very noticeable 'blueing' of the colour because of adsorption onto the cloudifier and the 'thin layer' effect.

2.9.4.2 Fruit preserves. Anthocyanins are also used in fruit preparations, jams and preserves. The nature and quality of the fruit is important—fresh or frozen being preferable to sulphited or canned fruit. Canned fruit, in particular, can be rather brown in shade and this is hard to mask using anthocyanins since they themselves absorb in the brown region (420 to 440 nm). Dose levels in such applications vary widely depending on the natural anthocyanin content of the fruit and the degree of brownness but are typically in the range 20 to 60 p.p.m.

2.9.4.3 Sugar confectionery. Acid sugar confectionery, particularly high boilings, and pectin jellies are an ideal application for anthocyanins where a variety of red shades can be obtained.

As mentioned previously, some anthocyanin extracts, particularly those derived from grape, are incompatible with gelatin and thus care must be taken to select the correct application form to ensure good clarity of the end product. When a concentrated grape anthocyanin colour is added to a solution of gelatin, a haze or precipitate is likely to occur. The higher the concentration of the extract, the more severe the problem. Thus it is always a wise precaution to dilute the colour prior to use and to establish its gelatin compatibility before carrying out production trials.

2.9.4.4 Dairy products. It is uncommon to colour dairy products with anthocyanins since their pH is such that a violet to grey colour would be achieved. In addition, the presence of the suspended fat particles increases the visual blueness of the colour. Acid dairy products such as yoghurt can, however, be successfully coloured although the shade achieved is distinctly purple. Thus black-cherry flavoured yoghurt is an intense colour because of the presence of anthocyanins derived from either the grape skin extract added as colour or the cherry juice present in the formulation.

2.9.4.5 Frozen products. Ice cream is not usually coloured with anthocyanins because its pH is too high, beetroot red being the preferred red colour. However, although water ice with a pH of around 3.0 would seem an ideal application, when frozen it is distinctly bluer than the red solution before freezing. This is probably due to internal reflections creating a 'thin-layer' effect similar to that seen at the meniscus of a glass of young red wine, which appears bluer than the bulk of the wine.

2.9.4.6 Dry mixes. A variety of acid dessert mixers and drink powders can be successfully coloured with spray-dried anthocyanin extracts.

2.9.4.7 Other applications. It is also technically possible to colour alcoholic drinks and products containing vinegar with anthocyanins, although commercial applications are limited. Red wine would usually be used in the former example and very few vinegar-based products require the addition of a red colour. The prime requirement for the successful use of anthocyanin is a low pH and absence of a cloud; under such conditions attractive red shades are obtained.

2.10 Beetroot

2.10.1 Source

The red beetroot has been cultivated for many hundreds of years in all temperate climates. The pigments present are collectively known as betalains and can be divided into two classes, the red betacyanins and yellow betaxanthins; both are very water soluble. The betalains have a limited distribution in the plant world and it would appear that betalains and anthocyanins are mutually exclusive. Plants producing betalains do not contain anthocyanins.

Most varieties of beetroot contain the red betacyanin, betanin (Figure 2.5) as the predominant colouring compound and this represents 75 to 90% of the total colour present. Vulgaxanthin I and II are the principal yellow betaxanthins.

Beetroot is an excellent source of colour and some varieties contain up to 200 mg/100 g fresh weight of betacyanins representing up to 2% of the soluble solids. Very large quantities, over 200 000 tonnes, of beetroot are grown annually in Western Europe, most of which is either consumed as the boiled vegetable or as a conserve [15]. Of this quantity some 20 000 tonnes are converted into juice and colour. As a proportion of this is exported, the quantity of beetroot colour consumed as a food additive is small compared with that consumed as a constituent of the vegetable. Another way of looking at this is that when eating 100 g of beetroot the quantity of betanin consumed is 200 mg, whereas when eating 100 g of strawberry yoghurt containing beetroot as the added colour the quantity of betanin consumed is only 0.5 mg.

Chemical Formula: Betanin $C_{24}H_{26}N_2O_{13}$ Molecular Weight: 550.5

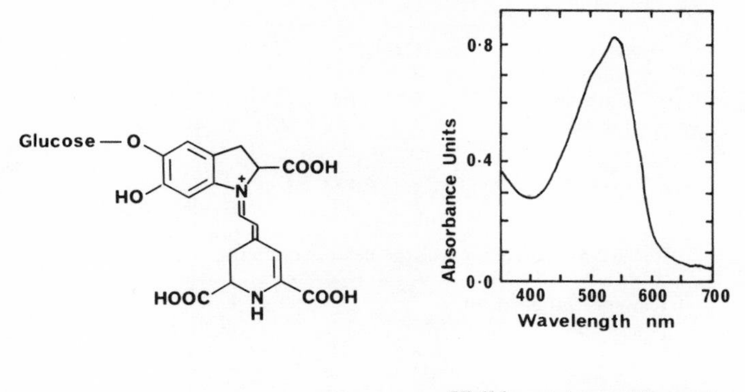

Visible spectrum of beet red as a 0.00075% aqueous solution at pH 5.0.

Colour shade: Red to Blue red

Solubility: Water Soluble

Absorptivity of Betanin: $E^{1\%}$ = 1120 in pH 5 buffer at λ max. closest to 530 nm

Figure 2.5 Beetroot (class betalain; synonyms beetroot red, beet red, betanin): structural and physical characteristics of betanin.

2.10.2 Extracts and their application forms

Beetroots are processed into juice in a manner very similar to that used for fruit-juice production, using either pressing or diffusion techniques. The juice is then centrifuged, pasteurized and concentrated to yield a viscous liquid concentrate with approximately 70% sugar and 0.5% betanin. This product, the concentrated beetroot juice, is widely used as a food ingredient. It is possible to produce a colour extract (*i.e.* to increase the colour content and so reduce the flavour) by fermenting some of the sugar to alcohol and removing the alcohol during concentration. The benefits gained are limited since, for many applications, the juice is a satisfactory ingredient.

The juice can be spray-dried to a powder although maltodextrin has to be added as a carrier since the high sucrose content precludes direct drying of the juice.

Betanin is a particularly intense colour and is stronger than many synthetic colours (Table 2.5). Thus dose levels for betanin in, for example, yoghurt are very low at around 5 p.p.m. and for strawberry ice cream around 20 p.p.m.

It should be mentioned that concentrated beetroot juice is usually offered

Table 2.5 Comparative colour strengths of betanin and some equivalent synthetic colours

Colour	$E^{1\%}_{1\,cm}$	λ max (nm)
Betanin	1120	537
Amaranth	438	523
Carmoisine	545	515
Ponceau 4R	431	505

Table 2.6 Calculation of the betanin content

Commercial beetroot concentrate	$E^{1\%}_{1\,cm}$	True % betanin
'1% betanin' liquid	5.60	0.5
'0.66% betanin' powder	3.60	0.33

for sale with a specification that includes a betanin content. The figure quoted is usually 1% betanin.

When spray dried to give a powder, the betanin content is lower since the amount of maltodextrin required is greater than the quantity of water removed during drying. Betanin figures quoted for beetroot juice powder are usually in the range 0.4 to 0.7%.

The betanin content is calculated by measuring the spectrophotometric absorption at 535 nm in a 1 cm cuvette and using 550 as the absorptivity of betanin. However, as mentioned previously, the absorptivity is really 1120 [16] and thus the true betanin content of a standard juice concentrate is 0.5% (Table 2.6). This is an important point to remember when calculating dye content.

2.10.3 Legislation

Beetroot-juice concentrate is universally permitted as a food ingredient. Beetroot red is widely permitted in Europe and North America, although it is not listed by a number of countries such as Egypt, Iceland, India and Thailand. The EC has proposed [7] a specification for beetroot red of 1.2% betanin minimum (both liquid and powder forms), which would distinguish the purified colour from the vegetable juice. The specification prepared by the Joint Food and Agriculture Organization/World Health Organization Expert Committee on Food Additives (JECFA) [16] is very similar, with a minimum limit of 1.0% for liquids and 4% for powders and both calculations use 1120 as the $E^{1\%}_{1\,cm}$ for betanin.

2.10.4 *Factors affecting stability*

When considering the stability of beetroot colour, it is betanin stability that is being measured. Thus although other pigments are present the measurements of colour strength made are at the wavelength of maximal absorbance of betanin, about 535 nm. Vulgaxanthin is less stable than betanin and is also present at relatively low levels and absorbs maximally at 477 nm.

2.10.4.1 pH. Beetroot colour shows greatest stability at pH 4.5. At pH 7.0 and above the betanin degrades more rapidly and thus is not recommended for alkaline applications. Colour shade is not significantly affected by pH change in the range 3.0 to 7.0. In very acidic conditions, the shade becomes more blue-violet as the red anionic form is converted to the violet cation. In alkaline conditions, the colour rapidly becomes yellow-brown due to the loss of betanin.

2.10.4.2 Heat. Beetroot pigments are susceptible to heat degradation and this is the most important factor that determines their use as food colorants. The rate of loss of colour at different temperatures is also dependent on several factors including pH and water activity.

At high sugar levels, beetroot pigments will withstand pasteurization but not retorting. When pasteurizing at 60% solids, a colour loss of around 5% would be expected assuming rapid cooling after the heat process. Generally, a high temperature for a short period of time is acceptable but it is most important to cool the product immediately after processing. Often it is not the heat process itself that causes colour degradation but the heat input during a slow (unforced) cooling cycle. It is interesting to note that betanin loss during heating is partially reversible since some betanin is regenerated from its two degradation products betalmic acid and cyclodopa-5-O-glycoside. Thus when evaluating beetroot in a food system it is important to allow the end product to fully equilibrate before evaluating the final colour shade and strength. This important point relates to many other natural colours—particularly anthocyanins.

2.10.4.3 Oxygen. Betanin is susceptible to oxidation and loss of colour may be noticeable in some long-life dairy products. Oxidation is most rapid in products with high water activity. Some antioxidants such as ascorbic acid have been shown to be beneficial though β-hydroxyanisole (BHA) and β-hydroxytoluene (BHT) are ineffective.

2.10.4.4 Light. Light does cause degradation of beetroot pigments; however other factors have a more limiting effect on product applications.

2.10.4.5 Water activity. It is important to remember that water activity (in this case the amount of free water available for the dissolution of a colour) significantly affects stability. Beetroot-juice powder stored in dry conditions is very stable even in the presence of oxygen. Carotenoid powders under similar conditions will oxidize more rapidly. In aqueous solution, the higher the solids content, the more stable the colour.

2.10.4.6 Cations. Some di- and trivalent metal ions, particularly iron and copper, will accelerate betanin oxidation. Sequestration of metal ions will improve colour stability.

2.10.4.7 Sulphur dioxide (SO_2). SO_2 will completely decolourize beetroot pigments. Thus if a preservative is required in a foodstuff containing beetroot it should be either benzoate or sorbate.

2.10.5 Applications

The susceptibility of betanin to heat, oxygen and high water activity restricts its use as a food colorant. Its applications are therefore confined to products that receive limited heat processing, have either a low water activity or a short shelf-life and contain no SO_2. Beetroot pigments are thus particularly suited to use in powder mixes and most dairy and frozen products.

2.10.5.1 Ice cream. This is the most important application of beetroot colour. Most pink ice cream in Germany and the UK contains betanin either from the use of beetroot juice or beetroot colour. Betanin levels are usually in the range 15 to 25 p.p.m. (0.3 to 0.5% beetroot juice).

The exact colour shade obtained using a beetroot juice will depend upon the variety of beetroot used and the extraction and processing conditions employed. The differences will be evident in the degree of blueness of the colour. When using a 'blue' beetroot juice it may be necessary to add a yellow or orange colour to achieve an acceptable strawberry colour. A water-soluble annatto extract, at a dose level equivalent to 10 p.p.m. norbixin, would be sufficient.

2.10.5.2 Yoghurt. Again, beetroot is the preferred colouring, and annatto may also be added to achieve the correct shade. Since the colour associated with flavoured yoghurt is paler than that of ice cream, the betanin dose level is lower—usually in the range 4 to 8 p.p.m. The microbiological quality of a beetroot colour used in a set yoghurt is particularly important. Care must be taken to ensure that yeast counts are extremely low—preferably less than 10/g. Beetroot juice is particularly susceptible to yeast and mould growth because of its high sugar content and lack of added preservative. Good quality is dependent upon a production process designed to ensure any yeasts and moulds are destroyed, carried out under hygienic conditions, a high (above

66%) sugar content, storage at below 5°C and careful handling at all times to avoid contamination.

2.10.5.3 Dry mixes. Beetroot-juice powders are ideal for such applications because of their excellent solubility characteristics and good stability. They are commercially used in both instant desserts and soup mixes. Again, colour shade may be too blue for strawberry and tomato varieties, and thus a yellow or orange colour may need to be included in the formulation.

2.10.5.4 Sugar confectionery. The fourth of the commercially important uses of beetroot extracts is in the coloration of some sugar confectionery products, principally those products based on sugar pastes having no added acid. Fondants, sugar strands, sugar coatings and 'cream' fillings are all suitable applications.

2.10.5.5 Other applications. Beetroot is used to colour some snack foods. The colour is applied in stripes onto an extruded maize snack immediately after extrusion, giving the snack the appearance of a strip of bacon.

Beetroot may also be used in those comminuted meat products with a low moisture content and which do not contain SO_2 such as salami-style sausages. However, the coloration of meat products is not permitted in several countries. Anthocyanins are also unsuitable because of the pH, thus carmine is commonly used as a colorant for meat products. The above comments about end-product applications are all applicable to both beetroot juices and the further processed beetroot colours. The distinctive flavour and low pigment strength of the juice may, on occasion, limit its application. However, since betanin is a very intense pigment its presence at only 0.5% in a juice concentrate is still sufficient to provide an attractive colour when the juice is used at relatively low levels.

2.11 Cochineal and carmine

2.11.1 Source

Carmine is the word used to describe the aluminium chelate of carminic acid. Carminic acid (Figure 2.6) is the colour extracted from the dried female coccid insect *Dactylopius coccus costa* (*Coccus cacti L.*). The word cochineal is used to describe both the dried insects themselves and also the colour derived from them.

Coccid insects of many species have been used for thousands of years as a source of red colour. Each insect is associated with a specific host plant and each is the source of a particular colour such as Armenian red, kermes, Polish cochineal, lac dye and American cochineal. The Spanish conquest of Central and South America brought the latter product to Europe and now American

Chemical Formula: Carminic Acid $C_{22}H_{20}O_{13}$

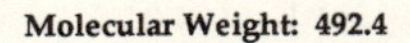

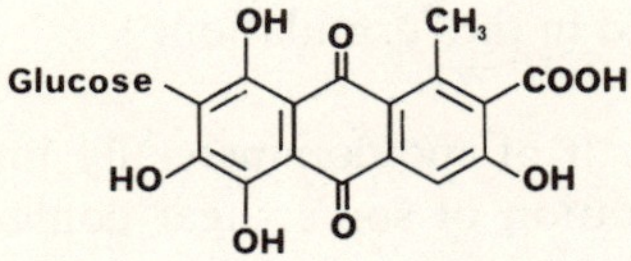

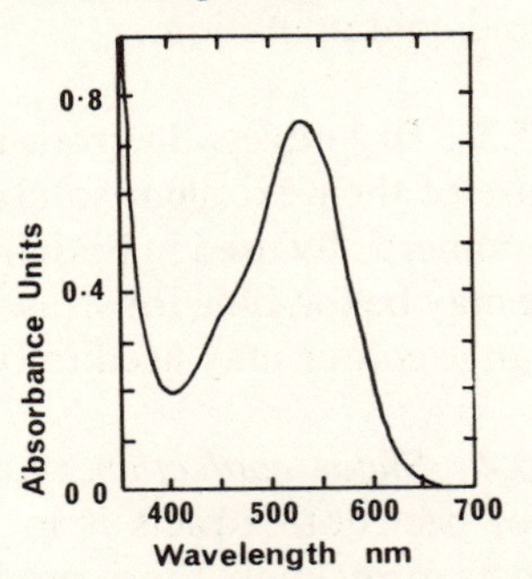

Molecular Weight: 492.4

Visible spectrum of carminic acid as a 0.004% solution in pH 7 buffer.

Colour Shade: **Orange at pH 3**
Red at pH 5.5
Purple at pH 7

Solubility: Water Soluble

Absorptivity of Carminic acid: **$E^{1\%}$ = 174.7 in 2N HCl at λ max. closest to 494 nm**

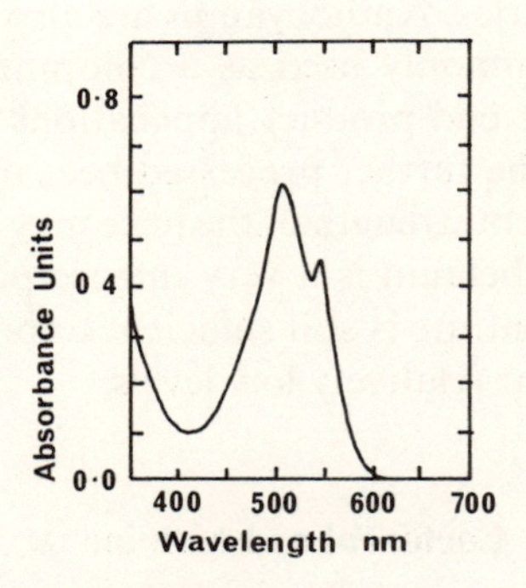

Visible spectrum of carmine as a 0.0025% solution in pH 7 buffer

Figure 2.6 Cochineal and carmine (class anthroquinone): structural and physical characteristics of carminic acid.

cochineal is the only cochineal of commercial importance, although Lac dye (from *Laccifera lacca*) is available in the Far East.

The current principal source of cochineal insects is Peru, although these insects are also available in the Canary Islands where they live on various species of cacti (principally *Nopalea cochenillifera*). About 300 tonnes of dried cochineal is produced annually, all of which is used to produce colour, principally carmine. A significant proportion of the carmine produced is actually used in the cosmetics industry.

The extraction of colour from the cochineal takes place principally in Peru, France, UK, USA and Japan.

2.11.2 Extracts and their application forms

2.11.2.1 Carminic acid. Carminic acid is very water soluble and its colour shade in solution is pH dependent. It is orange in acid solution and violet when alkaline, with a rapid colour change through red as the pH increases from 5.0 to 7.0. Its colour intensity is relatively low, having an $E^{1\%}_{1\,cm}$ of only 175; thus the number of commercial applications is limited.

2.11.2.2 Carmine. The ability of carminic acid to complex with metals, such as aluminium, is used in the manufacture of the more intense pigment known as carmine. Carmine is a chelate of carminic acid with aluminium and calcium and is precipitated out of solution by the addition of acid at the final stage of production. It is soluble in alkaline solutions but insoluble in acids. Its colour is essentially independent of pH, being red at pH 4.0 and changing to blue-red at pH 10.0. The intensity of carmine is almost twice that of carminic acid and it is thus a more cost-effective colour.

2.11.2.3 Commercially available forms. Carminic acid is usually available as an aqueous solution with a dye content of below 5% and from this spray-dried powders can be prepared.

Carmine is manufactured as a sparingly water-soluble powder with a carminic acid content in the range 40 to 60%. This product is used by both the food and cosmetic industries. Two widely accepted specifications exist for carmine. One is detailed in the British Pharmaceutical Codex and equates to approximately 60% carminic acid. This measures carmine content by its absorbance in dilute ammonia solution. The other is detailed in the Food Chemicals Codex, 2nd Edition, which specifies a minimum of 50% carminic acid; and also in the more recent Food Chemicals Codex III, where a revised assay procedure gives a more accurate figure of minimum 42% carminic acid for the same absorbance value. These both measure carminic acid content following treatment with boiling 2N hydrochloric acid. It is still the 50% figure that is widely quoted in the trade.

Due to its insolubility in acid solutions, carmine is usually produced as an alkaline solution with a carminic acid content in the range 2 to 7%. Traditionally, ammonia solutions have been used as the solvent since carmine is particularly soluble in ammonia solutions. However, because of the unpleasant nature of ammonia, other more acceptable diluents such as potassium hydroxide are now in use. It is possible to spray-dry these solutions using maltodextrin as the carrier to give very water-soluble powder colours with carminic acid contents in the range 3.5 to 7%.

2.11.3 Legislation

Carmine is widely permitted in Europe and North America, although in Denmark its use is limited to alcoholic drinks and it is not permitted in Finland. Sweden has recently extended its use to sugar confectionery with a limit of 200 p.p.m. [17].

2.11.4 Factors affecting stability

The usual product used by the food industry is an alkaline solution of carmine and the following description relates to this solution.

2.11.4.1 pH. Colour shade is fairly constant with changing pH. However carmine will precipitate out of solution when the product pH is below 3.5. The exact point of precipitation depends on a number of factors including viscosity, water content and pH.

2.11.4.2 Heat, light and oxygen. Carmine is very stable to heat and light and is resistant to oxidation.

2.11.4.3 Sulphur dioxide (SO_2). SO_2 does not bleach carmine at levels usually found in foodstuffs.

2.11.4.4 Cations. Cations will affect colour shade, generally increasing the blueness of the colour.

Carmine is thus a very stable colour and ideally suited to a variety of applications.

2.11.5 Applications

The only significant technical limitation on the use of carmine is low pH. However, carmine is less cost-effective in use than both beetroot and anthocyanins, principally because it is less intense and hence more needs to be added in order to obtain the same visual effect. Based on 1990 prices, beetroot is approximately half the cost in use of anthocyanin extracts and one third that of carmine when considering the dose levels required to obtain the same visual colour strength in food. In the past, supplies and prices of carmine have been adversely affected by a shortage of raw material. However, there has recently been a steady increase in cochineal availability and prices have stabilized; this has led to a gradual increase in demand.

Historically, carmine was widely used as a textile dye but it has been totally replaced by synthetic dyes during the past 100 years, principally because of their lower cost and ready availability. As previously mentioned, carmine is used in cosmetics where the insoluble pigment finds application because of its consistency of colour and its stability.

Other important applications include the coloration of certain alcoholic aperitifs and use within the food industry where the solubilized colour is widely used.

2.11.5.1 Meat products. Being somewhat blue-red in shade and stable in the presence of SO_2 it is ideally suited for use in sausages and other comminuted meat products, where inclusion levels are usually in the range 10 to 25 p.p.m. as carminic acid. With the increasing popularity of further processed poultry products, carmine finds application in ethnic chicken dishes where a blend of yoghurt and spices coloured with carmine is used as a marinade for chicken portions. In order to obtain a variety of different shades other colours, often annatto, are used in combination. Being heat stable, the colour is retained during the cooking process.

2.11.5.2 Jams and preserves. For this application, beetroot is usually insufficiently heat stable and anthocyanins may not be totally effective, either because of the length of the heat process or the underlying brownness of the preserve. Under these conditions, carmine is used to give a bright-red colour with good stability.

2.11.5.3 Gelatin desserts. It is possible to use beetroot for gelatin products that are designed to have a short shelf-life. Nevertheless, carmine is the preferred colour for products stored at ambient temperatures for an extended period, since under these conditions beetroot may oxidize. Anthocyanin extracts, particularly those derived from grapes, are often incompatible with gelatin.

2.11.5.4 Flour confectionery. Carmine, being heat stable, is ideally suited for use in baked goods. At a level equivalent to 40 p.p.m. carminic acid, carmine will produce a pink-coloured sponge. At a similar dose level, carmine is also used to produce a pink icing for the decoration of cakes and biscuits.

2.11.5.5 Dairy products. Beetroot is suitable for most dairy applications; however, carmine finds application in long-life flavoured milk because of its resistance to oxidation during storage. For strawberry flavour varieties a yellow colour would also be added in order to achieve the correct shade.

2.12 Curcumin

2.12.1 Source

Curcumin (Figure 2.7) is the principal colour present in the rhizome of the turmeric plant (*Curcuma longa*). Turmeric has been used as a spice for many thousands of years and is today still one of the principal ingredients of curry powder.

Turmeric is cultivated in many tropical countries including India, China, Pakistan, Haiti and Peru and is usually marketed as the dried rhizome, which is subsequently milled to a fine powder. This imparts both flavour and colour to a food product. Several types of turmeric are recognized but the most important are Madras, Alleppy and West Indian.

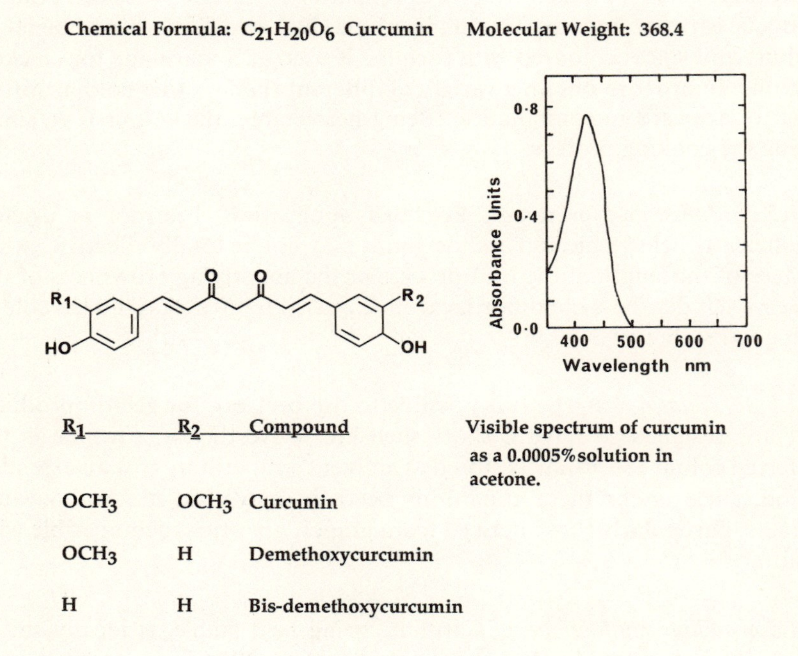

R_1	R_2	Compound
OCH_3	OCH_3	Curcumin
OCH_3	H	Demethoxycurcumin
H	H	Bis-demethoxycurcumin

Colour Shade: Lemon Yellow at pH 3
Orange at pH 10

Solubility: Oil Soluble

Absorptivity of Curcumin: $E^{1\%}$ = 1607 at 425 nm in ethanol

Figure 2.7 Curcumin (class phenalone; synonyms turmeric yellow, diferoylmethane, natural yellow 3): structural and physical characteristics of curcumin.

Ground turmeric powder is insoluble in water and imparts colour either by dispersion throughout the food or by dissolution of the curcumin into vegetable oil. Curcumin, the major pigment present, is accompanied by smaller quantities of related compounds, and all are insoluble in water.

India is the largest producer with annual quantities usually in the range of 250 000 to 300 000 tonnes. Most of this is consumed in the ground form as a spice and only a small amount, 1000 to 1500 tonnes, is converted into extracts.

The quantity of curcumin colour produced annually is very small—probably of the order of 30 tonnes—when compared with the quantity of curcumin consumed as a spice. Taking only the Indian production of 250 000 tonnes of turmeric and assuming an average curcumin content of 3% gives an annual consumption of 7500 tonnes of curcumin.

2.12.2 *Extracts and their application forms*

There are three principal types of turmeric extract, namely essential oil of turmeric, turmeric oleoresin and curcumin.

2.12.2.1 Essential oil of turmeric. This is the oil obtained by the steam distillation of turmeric powder and contains all the volatile flavour components of the spice and none of the colour. It is present in the turmeric at levels of 3 to 5%. There is only a small commercial demand for this product.

2.12.2.2 Turmeric oleoresin. This is the most commonly produced extract and contains the flavour compounds and colour in the same relative proportion as that present in the spice. It is obtained by solvent extraction of the ground turmeric, a process identical to that used in the production of other spice oleoresins.

Spice oleoresins have several advantages over ground spices, principally their excellent microbiological quality, standardized organoleptic properties and freedom from contaminants; for these reasons the use of oleoresins has increased steadily during the past 30 years. Turmeric oleoresins usually contain 37 to 55% curcumin [18].

2.12.2.3 Curcumin. This is the pure colouring principle and contains very little of the flavour components of turmeric. It is produced by crystallization from the oleoresin and has a purity level of around 95%, which is the standard commercially available quality. The proposed EC specification for curcumin [7] states that the dye content must not be below 90% when measured spectrophotometrically at 425 nm.

It is important to note that the distinction between these three products lies in the ratio of colour to flavour. A spice oleoresin contains the total sapid, odorous and related characterizing principles normally associated with the spice. Thus the ratio of flavour components to curcumin in ground spice and oleoresin is the same. A colour, however, results when an attempt is made to reduce the flavour of the product and increase the relative concentration of colour such that the ratio of flavour to colour is altered in favour of the colour. Thus, in the case of turmeric, the ratio of curcumin to volatile oil is of the order 50 : 50, usually lying within the range 40 : 60 to 60 : 40. However, curcumin made to the proposed EC specification has a ratio of 99 : 1 whereas the spice oleoresin ratio lies within the same range as that of the ground spice. Perhaps this technique could be applied to other extracts to distinguish between the total extract and the colour additive.

2.12.3 Application forms

Pure 95% curcumin is not an ideal product for direct use by the food industry since it is insoluble in water and has poor solubility in other solvents. Thus it is usual for curcumin to be converted into a convenient application form. In many countries, this is achieved by dissolving the curcumin in a mixture of food-grade solvent and permitted emulsifier. In this form, the product contains 4 to 10% curcumin and is easily miscible in water. Polysorbate 80 is the favoured emulsifier/diluent for such products since it is an ideal carrier for curcumin.

Other forms are also commercially available, including suspensions of curcumin in vegetable oil and dispersions onto starch but these are not so commonly used.

2.12.4 Legislation

Turmeric oleoresin is permitted universally as a spice oleoresin. The Swedish legislation, for example, specifically states that 'turmeric, paprika, saffron and sandalwood shall not be considered to be colours but primarily spices, providing that none of their flavouring components have been removed' [5]. It is not a permitted colour in the EC.

Curcumin is specifically permitted as a colour in the EC; however many countries simply list turmeric without a specification for its colour strength.

2.12.5 Factors affecting stability

All the following comments relate to solubilized curcumin dissolved in an aqueous medium.

2.12.5.1 pH. Curcumin gives a lemon yellow colour in acidic media with a distinct green shade. As the pH increases, so the green shade becomes less distinct. In alkaline conditions (above pH 9.0) the colour becomes distinctly orange.

2.12.5.2 Heat. Curcumin is essentially stable to heat and will withstand baking.

2.12.5.3 Light. Curcumin is sensitive to light and this factor is the one that usually limits its application in foods. Curcumin as the suspended pigment is more stable than the solubilized colour.

2.12.5.4 Cations. In general, cations will tend to induce a more orange-brown shade.

2.12.5.5 Sulphur dioxide (SO_2). SO_2 reduces the colour intensity of solubilized curcumin, particularly when present at over 100 p.p.m.

2.12.6 Applications

Curcumin is an intense colour with a very bright yellow appearance even at low doses. It is noticeable that the colour easily becomes saturated and at levels over 20 p.p.m. it is hard to perceive a small increase in colour-dose level. Thus when using curcumin it is important to determine the minimum dose level required to give the desired colour. Dose levels are often very low usually in the range 5 to 20 p.p.m. Its colour shade is very similar to that of tartrazine and its use has increased in the USA, following the requirement to specifically identify the presence of tartrazine.

For many products where an egg yellow colour is required, its shade is too green and another more orange colour has to be used in combination. It is usual for annatto to perform this function and numerous blends are available commercially, particularly for the dairy and flour confectionery industries.

2.12.6.1 Dairy industry. This is one of the principal applications of curcumin. Vanilla ice cream is often coloured with a combination of curcumin and norbixin and usually contains about 20 p.p.m. curcumin together with 12 p.p.m. norbixin. In yoghurt, 5 p.p.m. curcumin will give an acceptable lemon yellow colour. Curcumin colours are generally fairly viscous products and thus care must be taken to ensure that they are fully mixed into the milk. A premix made using a high-speed mixer is often a preferred method to ensure complete homogeneity.

Dairy spreads can successfully use either vegetable oil extracts or the standard water-miscible form, the curcumin, however, will always migrate to the oil phase. The only dairy-products application where curcumin may not be suitable is in the coloration of long-life flavoured milks that are packed in transparent containers. Here, light stability is the limiting factor.

2.12.6.2 Flour confectionery. It is traditional to use a yellow colour in cakes and biscuits. The required colour is achieved using a blend of curcumin and annatto similar to that used in ice cream but at a lower dose level—usually 10 to 15 p.p.m. curcumin together with 5 to 10 p.p.m. norbixin. Any cut surfaces should be protected from prolonged exposure to light by suitable packaging [19].

2.12.6.3 Sugar confectionery. Curcumin at 20 p.p.m. will impart a deep, bright-yellow colour to high boilings. It is recommended to use more dilute propylene glycol based curcumin colours for wrapped confectionery where polysorbate may not be compatible. Generally, curcumin is acceptable in all sugar confectionery products not excessively exposed to light.

It is quite common, especially in the sugar confectionery industry, to prepare stock solutions of colour such that these can be easily measured out and mixed into the sugar mass. However, curcumin is not particularly soluble in a concentrated water-based stock solution and it is likely to crystallize out if left overnight. If stock solutions have to be made, this should be discussed with the colour supplier so that the solubility limits can be ascertained.

2.12.6.4 Frozen products. Sorbets and water ice can be successfully coloured with 5 to 15 p.p.m. curcumin and their low pH does not pose any problems. Nevertheless, it should be remembered that curcumin solutions should not, themselves, be frozen since the curcumin may precipitate out of solution.

2.12.6.5 Dry mixes. Spray-dried curcumin colour using gum acacia as the carrier is ideal for use in dessert mixes and instant puddings. Such products usually contain up to 8% curcumin. When using powders based on gum acacia it is necessary to use warm water when reconstituting the product in order to fully solubilize the gum and so release the colour. Alternatively, dispersed starch based colours with 1% curcumin or less may be used. The lower the curcumin content the less susceptible the colour to specking. When a concentrated colour is present at a low level in a dry mix and water is added, each individual particle of colour slowly dissolves creating a temporary speck of deep colour that is very noticeable. However, if a powder colour of low tinctorial strength is used at a relatively high level in a dry mix, and if that colour is bound to the starch, then coloration is by mass effect and there are consequently no specks of bright colour when the water is added. It should also be pointed out that powders using a spray-dried 5% curcumin would be white in appearance with a few yellow specks visible whereas the second product, using a weaker dispersed colour would appear an even yellow colour.

2.12.6.6 Savoury products. If turmeric flavour is an acceptable part of a seasoning then turmeric oleoresin can be used as a spice to give both colour and flavour to savoury products. This would be true of some spiced chicken dishes or soups. There are many applications of turmeric oleoresin but the use of such products is outside the scope of this chapter. However, it should be remembered that turmeric oleoresin is widely used in the food industry and thus curcumin is present in many savoury foods.

2.13 Other colours

The five natural colours previously discussed are those most widely used by the food industry in Europe and North America. They are available in substantial quantities and at an affordable price. While carmine is more expensive in use than beetroot, this disadvantage is offset by its excellent stability. Several other natural colours are permitted but for reasons of price, availability or colour performance these are not so widely used.

2.13.1 Chlorophyll

Chlorophyll is a vitally important pigment in nature and is present in all plants capable of photosynthesis. In the food industry, much effort is put into retaining the chlorophyll naturally present in the green vegetables (see chapter 3). However, the addition of chlorophyll as a colour to foodstuffs is very limited, principally because of its poor stability.

Chlorophyll is an oil-soluble colour that can be extracted from a range of green leaves, but usually grass, nettles or alfalfa is used. Chlorophyll degrades easily, particularly in acidic conditions, losing its magnesium ion to yield phaeophytin, which is yellow-brown in colour. Chlorophyll colours tend to be rather dull in appearance and of an olive green-brown colour. This obviously limits their food applications. Chlorophyll extracts can be standardized using vegetable oil for oil-soluble products or blended with a food solvent or permitted emulsifier to give a water-miscible form. An extract of chlorophyll will contain approximately 10% chlorophyll together with other colouring compounds, principally lutein and carotenes as well as fats, waxes and phospholipids. Raw-material quality and the extraction techniques used will affect the ratio of chlorophyll to carotenoids and hence the colour shade of the extract.

Water-miscible forms can be used in sugar confectionery, flavoured yoghurts and ice cream if the colour shade obtained is acceptable. However, most chlorophyll extracted is used in cosmetics and toiletries, with only very small quantities in foods. In the USA chlorophyll is not a permitted food ingredient, whilst in the UK consumption of chlorophyll as an added colour in food is probably less than 400 kg per year.

2.13.2 Copper complexes of chlorophylls and chlorophyllins

The replacement of the central magnesium ion with copper produces a more stable complex with greater tinctorial strength. The subsequent removal of the phytol chain by hydrolysis with dilute alkali renders a water-soluble product termed copper chlorophyllin. This particular complex, in the form of its sodium or potassium salt, is the most widely used green colour of natural origin and it is usually manufactured from either grass or alfalfa. One of the purification steps involves the precipitation of the chlorophyllin, which ensures the elimination of the yellow carotenoids.

Copper chlorophyllin is permitted widely as a food colour throughout Europe, although its use in the USA is limited to dentifrice. Copper chlorophylls and copper chlorophyllins are chemically modified natural extracts and cannot truly be termed natural.

2.13.2.1 Application forms. Copper chlorophyll extracts are oil-soluble viscous pastes often standardized with vegetable oil. Dye contents are of the order of 5 to 10%. There is no internationally recognized method of chlorophyll content estimation though White *et al.* [20] have reported a method for the determination of chlorophylls and copper chlorophylls, which, although difficult mathematically, seems to give reproducible results.

In contrast, salts of copper chlorophyllin are available in both liquid and powder forms. There are many grades of powder product available with dye contents varying from 10 to 100%. Colour content is determined by a widely used method involving measuring the absorption at 405 nm in a pH 7.5 buffer and calculating the dye content using 565 as the absorptivity of sodium copper chlorophyllin. However, it is now realized that the true absorbancy is actually higher than 565 but no new figure has been agreed. The powders will dissolve in water to give an alkaline solution but special application forms are required if the powder needs to dissolve in acid solution.

The other widely used forms are alkaline solutions containing up to 10% chlorophyllin, the strength of solution being related to the intended use of the colour. In addition, some special application forms that are both soluble and stable in acid conditions are available.

2.13.2.2 Principal factors affecting stability. So far as pH is concerned, all chlorophylls are most stable under alkaline conditions. In dilute acids, chlorophylls will hydrolyze and lose colour rapidly and copper chlorophyllins will precipitate. In addition, copper chlorophyllins are relatively heat stable but will lose colour on exposure to light.

2.13.2.3 Applications. Green is not a widely used colour in the food industry and hence applications are limited principally to lime flavour sugar confectionery and pistachio flavour ice cream. Dose levels for copper chlorophyllin lie between 30 and 50 p.p.m. for clear sugar confectionery products and 50 to 100 p.p.m. for ice cream. In most foodstuffs, copper chlorophyllin gives a mint green (blue-green) shade, therefore it is necessary to add a yellow colour to achieve the required shade of yellow-green. The addition of an orange colour would tend to give a brown cast to the colour therefore a yellow, preferably a green-yellow, is necessary. Curcumin meets this requirement perfectly and is often used in combination with copper chlorophyllin at a dose level approximately half that of the copper chlorophyllin.

Other applications include cucumber relishes, dessert mixes and some cheeses either to give a green vein or at very low levels (less than 1 p.p.m), to whiten a soft cheese.

2.13.3 Carotenoids

Carotenoids are very widely distributed in nature and it has been estimated that nature produces some 3.5 tonnes of carotenoids evey second [21]. Over 400 different carotenoids have been identified and many of these are present in our diet (see chapter 5). Lutein is found in all green leaves and β-carotene is an essential source of vitamin A. Several of these carotenoids can be extracted and used as colorants. Bixin has already been described but lutein, β-carotene, paprika extracts and crocin are also available commercially.

2.13.3.1 Lutein. Lutein is an oil-soluble colour either produced as a by-product of chlorophyll extraction or extracted from marigolds. Its principal use is in the fortification of the xanthophyll content of poultry feed and its food applications are very limited. Annatto and turmeric are the most cost-effective source of natural yellow-orange colours and lutein only finds application where the light sensitivity of turmeric limits its use. Lutein is used commercially in some lemon-flavoured cloudy soft drinks, selected sugar confectionery products and emulsified salad dressings. Consumption in Europe as an added colorant to foodstuffs, is estimated at less than 1000 kg per year.

A concentrated extract contains 5 to 12% lutein and this extract can be solubilized in citrus or vegetable oils for use in the above mentioned products.

2.13.3.2 β-carotene. β-carotene is an oil-soluble colour that can be extracted from either carrots or algae. However, virtually all the β-carotene used as a colorant in foodstuffs is the nature-identical form produced by synthesis. Naturally extracted β-carotene is significantly more expensive in use and is mainly used in the pharmaceutical/health food industry as a dietary supplement. Special application forms of nature-identical β-carotene have found extensive use in the soft drink and dairy industries.

Natural β-carotene is available as 20 to 30% suspensions, solutions in vegetable oil and as emulsified, water-miscible forms. Spray-dried powders can also be manufactured. Consumption in Europe of the natural food colour is very low and is estimated at less than 500 kg per year.

Projects to produce more cost-effective β-carotene extracts from *Dunaliella* algae are under way; only if these are successful will the use of natural β-carotene as a food colour become more widespread.

2.13.3.3 Paprika. The total extract of paprika (*Capsicum annuum*), containing all the flavour and colour components present in the spice, is widely used as a spice oleoresin. Several carotenoids, principally capsanthin, capsorubin and β-carotene, are responsible for its orange-red colour. It also has a distinctive flavour, which is responsible for the characteristic taste of such products as goulash and chorizo. Its use as a colour is very much limited by its flavour and there is no significant production of the purified flavour-free colour.

The extract is available as oil-soluble solutions standardized with vegetable oil to a colour strength that is quoted in multiples of 10 000 with 40 000, 60 000, 80 000 and 100 000 being the most commonly available. A 100 000 colour-strength extract equates to approximately 10% carotenoids.

Water-miscible forms are usually prepared by incorporating polysorbates during production, although emulsions can also be formed by using gum acacia. Of the 400 tonnes of spice extract manufactured annually in Europe, most is used as a spice in meat products, soups and sauces. Very little is used as colour in sweet products because of the flavour carry-over, although occasional use in sugar confectionery and gelatin desserts is encountered.

2.13.3.4 Crocin. Crocin is the water-soluble yellow carotenoid found in saffron and in gardenia fruits. It is responsible for the yellow colour of paella and saffron rice. It is not currently permitted by the EC as a food colour but it has extensive use in the Far East. The use of saffron as a source of crocin is limited by its distinctive flavour and high cost.

2.13.4 Caramel

Caramel colour is, in volume terms, the most widely used colour in the food industry. It is totally derived from sugar and most is modified chemically by the use of ammonia, ammonium salts and sulphites. The sugary, aromatic product obtained from heating sugar solutions without the addition of any other substance is caramel syrup, sometimes called burnt-sugar syrup. This material falls outside the scope of the colour directives. Crème caramel is a good example of a natural caramel product. Caramels are very stable colours but usually carry an electrical charge and thus may cause precipitation in the presence of oppositely charged molecules.

2.13.5 Carbon black

This is derived from vegetable material, usually peat, by complete combustion to the insoluble carbon. It is widely used in Europe in sugar confectionery but is not a permitted colour in the USA. The powder colour has a very fine particle size, usually less than 5 μ, and is consequently very difficult to handle. It is therefore usual for carbon black to be sold as a viscous paste where the carbon is suspended in glucose syrup. Carbon black is a very stable pigment.

2.14 Conclusions

Colour is essential to the full enjoyment of our food. Our diet contains a wide range of colours that are naturally present in the food we eat. Chlorophylls, carotenoids and anthocyanins are consumed virtually every day.

In order to ensure that our processed food is attractive, it is often necessary, for reasons stated previously, to enhance their appearance by the addition of colour. It is common practice, especially in Europe, to use natural colours for this purpose and five product groups in particular, annatto, anthocyanins, beetroot, cochineal and curcumin, dominate the natural colour industry. These five groups represent over 90% of the market for natural colour extracts. It may seem that the choice of natural colour is limited to just a handful of products. However, it should be remembered that each colour is available in numerous forms and it is the choice of the correct application form that requires careful consideration.

In the future, it is unlikely that new colours will be added to the permitted lists since the cost of the necessary toxicological testing is extremely high. Future development will be aimed towards the production of more stable forms of the currently permitted colours. If you consider that natural colours are quite stable to direct sunlight for many days when present in the plant itself, the researcher has the comfort of knowing that colour stability is possible.

As the number of permitted artificial colours gradually reduces and as consumers express a preference for products of a natural origin, so the range of natural colours available to the food industry has increased. This, in turn, has led to an increased awareness of their many attributes and how best to exploit these qualities to the advantage of both the food manufacturer and the consumer.

References

1. DuBose, C.N., Cardello, A.V. and Maller, O.J., *Food Science* **45** (1980) 1393–1399.
2. Johnson, J., Dzendolet, E. and Clydesdale, F.M., *J. Food Protect.* **46** (1983) 21–25.
3. Food Advisory Committee, *FdAC/REP/4*, HMSO, London (1987).
4. Ames, B.N. *Science* **221** (1983) 1256.
5. Swedish Food Regulations, *Food Additives* (1985) SLV FS 1985:1.
6. Italian Official Gazette **28** (1968) 159.
7. Commission of the European Communities, Report COM (85) 474 final (1985) and working document 111/9266/90 (1990).

8. Kuhnau, J., *World Rev. Nutr. Diet* **24** (1976) 117.
9. Timberlake, C.F. and Bridle, P., 'Anthocyanins' in *Developments in Food Colours*, Vol. 1, *J. Walford* (ed.), Applied Science Publishers, London (1980) 115–149.
10. Timberlake, C.F. *Food Chem.* **5** (1980) 69–80.
11. Timberlake, C.F. and Henry, B.S. 'Anthocyanins as Natural Food Colorants' in *Plant Flavonoids in Biology and Medicine III*, A.R. Liss Publishers, New York (1988) 107–121.
12. Knuthsen, P. *Lebensm-Unters Forsch* **184** (1987) 204–209.
13. La Bell, F. *Food Processing* (*USA*), April (1990) 69–72.
14. Jiang, J., Paterson, A. and Piggot, J.R. *Int. J. of Food Science Technol.* **25** (1990) 596–600.
15. Verniers, C., NATCOL *Quart. Inform. Bull.* No. 3 (1987) 4–10.
16. Food and Agriculture Organization, *Food and Nutrition Paper* 31/1 (1984).
17. Swedish Food Regulations, *Food Additives* (1989), SLV FS 1989. 31.
18. Food and Agriculture Organization, *Food and Nutrition Paper* **49** (1990).
19. Knewstubb, C.J. and Henry, B.S., *Food Technology International* (*Europe*) (1988) 179–186.
20. White, R.C., Jones, I.D., Gibbs, E. and Butler, L.S., *J. Agricul. Food Chem.* **25** (1977) 143.
21. Weedon, B.C.L., in *Carotenoids*, O. Isler (ed.), Birkaüser Verlag, Basel (1971).

3 Chlorophylls and chlorophyll derivatives

G.A.F. HENDRY

3.1 Summary

Chlorophyll derivatives form a significant and growing element in the range of natural pigments used as food colorants. An account is provided of the methods of production and of the products formed under industrial conditions. Consideration is given to the formation of chlorophyll under biological conditions, together with a comprehensive summary of the structure and occurrence of all known natural chlorophylls. Brief details are provided of the function and natural environment of chlorophylls in biological systems, with information on their relative stability *in vivo* and *in vitro*, together with a description of the known major products of natural and artificial degradation. The economic value of chlorophyll derivatives is described, together with future prospects for exploitation.

3.2 Introduction

The chlorophylls are a family of naturally-occurring pigments present in the photosynthetic tissues of all living plants, including algae, and in some photosynthetic bacteria. As an integral part of vegetable foodstuffs, chlorophylls have formed a constant component of the natural diet of animals, including humans, from earliest times. Throughout recorded history, the presence, or absence, of chlorophylls has been used as a sensitive indicator of the health and ripeness of vegetables and fruit and the freshness of harvested produce. As a guide to ripening it may be valid but to freshness it can be misleading. Once plants are harvested the chlorophylls are invariably degraded, in some species within a few hours, in others over several weeks. Prolonged storage, particularly if associated with several forms of deliberate preservation, usually completes the destruction of chlorophyll to leave, often at best, a grey-brown colour indicative or suggestive of loss of freshness. Further processing, particularly heating, involves the breakdown of any remaining chlorophylls. The wish to modify food to restore the colour of the freshly-harvested crop is understandable and has, over the years, given rise to the practice of deliberately adding back natural, inorganic and later organic synthesized pigments.

Chlorophylls are almost the only natural green plant pigment and certainly the only ones in super-abundance. Unfortunately, the inherent instability of chlorophylls in isolation has been a drawback to their more widespread application as food colorants. The instability and degradation of chlorophylls is a natural process in senescing leaves and ripening fruits [1]. Whatever the evolutionary significance of this natural degeneration on death, were chlorophylls not to be rapidly broken down in plants, the billions of tonnes of chlorophylls produced each year on land and in the oceans would accumulate with unthinkable consequences.

Given the long history of animal (and human) consumption of chlorophylls and of the many natural breakdown products of chlorophyll, there is little evidence to suggest that this family of pigments is in any way harmful following digestion, at least to the *healthy* herbivore. The few recorded adverse effects of ingested chlorophylls arise in diseased animals [2] or those with genetic defects such as albinoism [3]. The practice of improving on nature's bounty, particularly the natural pigmentation of vegetables and fruits, has evolved with the sophistication of the market place in the area of preserved and processed foods. In part, this has been achieved by adding back naturally-occurring plant pigments, notably the yellow-red carotenoids, the subject of several recent reviews [4, 5, 6, 7]. Similarly the use of the water-soluble red-purple anthocyanins and betalaines is a growing practice [8, 9]. Chlorophylls as food colorants have had less-detailed attention, although short reviews have been provided in recent years [4, 10].

The aim of this chapter is to provide a concise but up-to-date synopsis of the principal chemical, biochemical and biological characteristics of the chlorophylls, with some consideration given to the stability and breakdown of chlorophyll derivatives, both in natural systems and as a food colorant. An outline description is given of the preparation of the known chlorophyll derivatives for industrial purposes, together with a discussion of the likely future prospects for the commercial exploitation of these pigments.

Finally, the opportunity is taken to introduce the revised nomenclature for porphyrins (and chlorophylls) approved by the IUPAC-IUB Joint Commission on Biochemical Nomenclature [11]. There may be merit in reconsidering the current trade terminology surrounding chlorophyll-derived food colorants, if only to bring these into line with current scientific trends towards simplicity with precision in chemical nomenclature. Within the narrow context of chlorophylls as food colorants, the important name changes are few. To those involved in research into the chemistry and biochemistry of the porphyrins, the changes in nomenclature have involved the demise of many cherished, if rather quirky friends, but are being increasingly adopted in the current literature.

3.2.1 Nomenclature

Before considering the natural and unnatural processes to which chlorophylls are exposed, it is necessary to introduce the changes in porphyrin nomenclature made by the IUPAC-IUB Joint Commission [11].

These changes make clear the relationship between the chlorophyll derivatives found in nature and those formed industrially. In the narrow context of this chapter, the most notable change is the abandoning of the porphyrin ring-numbering system (1 to 8 and α to δ) originated by Fischer and colleagues [12]. In its place, a complete carbon numbering system (from 1 to 24) has been recommended. The two systems are shown in Figure 3.1.

Instead of a mixed system of Roman numerals for the outer carbons of the pyrroles and Greek letters for the methine bridge carbons, the new system numbers each and every carbon in sequence. The pyrrole rings previously numbered I to IV are now lettered A to D, the fifth or cyclopentanone ring being lettered E. Side-groups take their number in sequence from the originating alkyl group. Thus the two carbons of ring E, previously called positions 9 and 10, are now 13^1 and 13^2.

Chlorophyll *a* and many of its derivatives have long been known as chlorins, which, by definition, are dihydroporphyrins. In biology, the reduction occurs on ring D, thus chlorophyll *a* is a 17,18 dihydroporphyrin. All of the naturally-occurring chlorophylls have a fifth or cyclopentanone ring E (Fischer's ring V) and are based on the structure known as phaeophytin. Thus, again, chlorophyll *a* would be based on 17,18-dihydrophaeophytin. All natural chlorophylls are magnesium (Mg) co-ordination complexes and the structures are defined, by earlier IUPAC Rules, by adding the ending 'ato', followed by the name of the metal and its oxidation number. Thus chlorophyll *a* is now, 17,18-dihydrophaeophytinato Mg(II) and if the Mg

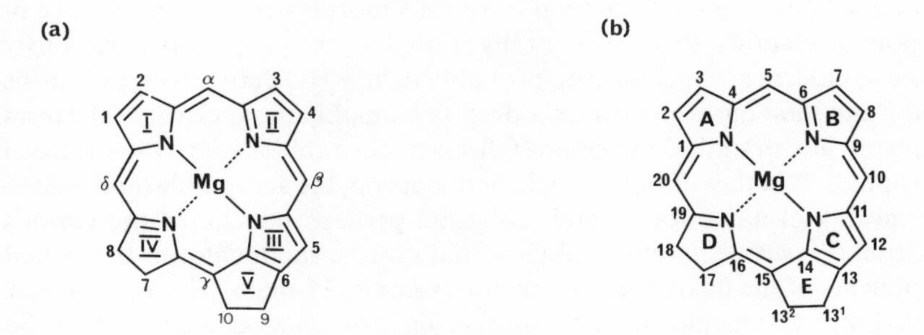

Figure 3.1 The numbering system of chlorophyll *a* showing: (a) the recently abandoned Fischer system; and (b) the recommended IUPAC-IUB system.

Table 3.1 The nomenclature of chlorophylls and derivatives as recommended by the IUPAC-IUB Joint Commission on nomenclature [11]

Retained trivial names	Based on
Chlorophyll	17,18-dihydrophaeophytinato Mg(II)
Chlorophyllide	17,18-dihydrophaeophorbideato Mg(II)
Phaeophytin	17,18-dihydrophaeophytin
Phaeophorbide	17,18-dihydrophaeophorbide

Abandoned names	Recommended names
Protochlorophyllide	Didehydrophaeophorbideato Mg(II)
Phyllin e_6 Chlorin e_6 Chlorin e_4 Purpurin 7 Purpurin 5 The Rhodin equivalents of the above Chlorophyllin	Didehydrorhodochlorin with a specific description of the substitution at C13 and 15

were to be replaced by copper (Cu), the pure compound would be 17,18-dihydrophaeophytinato Cu(II). These names would come into play principally in naming unusual porphyrins and, in this context, the derivatives of chlorophyll. Fortunately, a substantial body of long-established trivial names (as they are called) have been retained, while some of the less-common derived porphyrins, often sequentially and chronologically named at the bench in Fischer's laboratory 50 years ago, have gone. Among these abandoned names are the various chlorins, rhodins and erythrins. Examples of retained and discarded trivial names, in the special context of this review, are shown in Table 3.1.

The term chlorophyllin, widely used in the food colorant industry, is not an accepted name. Indeed the food colorant 'chlorophyllin' covers a range of compounds identical to, or structurally related to, the porphyrins, previously known as chlorin e_6, isochlorin e_4, probably their 3-HO derivatives, purpurins 5 and 7 and their corresponding rhodins. Presumably the survival of the term 'chlorophyllin' in trade descriptions reflects its desirable similarity to the word chlorophyll. Whether onomatopoeia or malaprop, the acceptable (and indeed more accurate) name for several, although perhaps not all, of the various industrial derivatives of chlorophyll would now be rhodochlorin. Individual components of the mixture might be, for example, 15-methylrhodochlorin or 13-ethylester rhodochlorin. If the unspecific term rhodochlorin were to be adopted as the name for water-soluble 'chlorophyll(in)s', it would at least have the merit of greater chemical accuracy and meaningfulness, particularly to new generations of porphyrin chemists.

3.3 Chlorophylls under natural conditions

3.3.1 Function

To make clear the difference between chlorophylls in their natural and their artificial (man-made) environments, it is useful to consider the function of these pigments in nature.

In the intact plant, more particularly in subcellular organelles known as chloroplasts, different groups of chlorophylls are associated structurally with particular proteins or polypeptides, for the most part buried in the hydrophobic environment of the chloroplast lipid membranes or thylakoids. The greater part (mass) of the individual chlorophyll molecules are so arranged, physically, to trap or harvest incoming particles of light in the form of photons or visible-light energy. At full sunlight, groups of chlorophyll molecules will capture about 40 to 45 photons per second. This trapped energy causes each of the electron-rich chlorophyll molecules to wobble or resonate, passing the resonance onto neighbouring chlorophyll molecules, ultimately concentrating the energy into a smaller number of specialized reaction centres that contain an epimer of chlorophyll *a* and phaeophytin.

In these activated reaction centres, the now intensely-concentrated energy flips an electron to a higher energetic state before ejecting it. Through an elaborate circuit of quinones and other electron carriers, the electron, with associated protons, yields chemical reducing power. Almost simultaneously, the electron imbalance in the reaction centres are restored by further electrons formed by the splitting of water

$$2H_2O \rightarrow 4H^+ + O_2 + 4e^- \qquad (3.1)$$

which in turn also yields protons used to drive the formation of storable chemical reducing power, together with oxygen, which is released into the atmosphere. The chemical reducing power of nicotinamide adenine dinucleotide phosphate (NADPH) and adenosine triphosphate (ATP) are then consumed in the reduction of CO_2 (NO_3 or SO_4) to give rise to simple carbohydrates, phosphorylated fructose and glucose, and sucrose together with amino acids used largely in protein synthesis. The whole process, photosynthesis, sustains life on earth by capturing energy from an external source, the sun.

The role of the chlorophylls is central to the process, of photosynthesis, although other pigments, notably the xanthophylls, certain carotenes and bile pigments, play an auxiliary but essential, role. Photosynthesis itself is at least 2600 million years old and, in all but its fine details, has probably not altered significantly in that time. Considering the powerful selective forces of evolution this is a remarkable longevity and says much about its efficiency. Perhaps solely for this reason, the structure and biosynthesis of the chlorophyll molecule has been highly conserved since earliest times. The relatively modest differences between one type of chlorophyll and another do not alter the basic function of the pigment in photosynthesis.

3.3.2 *Structure of chlorophylls and chlorophyll-protein complexes*

The chlorophyll molecules of higher plants consist of a reduced porphyrin chromophore with a colourless 20C terpenoid or phytyl side-group (Figure 3.2). The chlorophyll molecule itself is relatively large; the porphyrin head being about 1.5 × 1.5 nm in size and the phytol chain being overall about 2 nm in length. The 10 double bonds and the extensive ring structure allow electrons to become delocalized (see chapter 1) over three of the rings, giving numerous oxidation-reduction states that are important for the transfer of energy from one excited chlorophyll molecule to its neighbour. The bonding of Mg (rather than any other metal) is critical to photosynthesis, the electronic structure of Mg enabling it to change the electron distribution of the porphyrin ring to produce powerful excited states. The side-groups, particularly the vinyl and methyl groups at positions 3 and 7, dictate the predominant wavelengths of light that will be absorbed (and so alter the perceived colour of the pigment). Substitutions such as a formyl group at these two positions, as in chlorophylls *b* and *d* are sufficient to increase or decrease the blue and

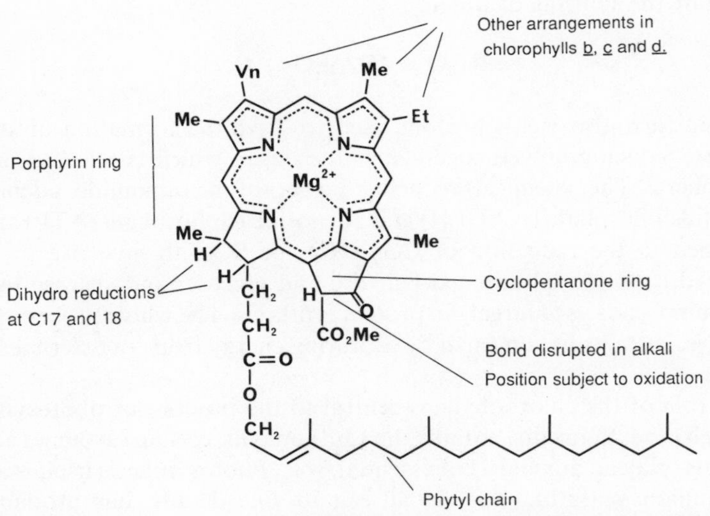

Figure 3.2 The structure of chlorophyll *a*. The two positive charges on the central magnesium (Mg^{2+}) ion are balanced by two negative charges shared randomly among the four pyrrole-nitrogens. The arrangements of the ten double bonds within the ring system may also vary. Vn = $CHCH_2$; Me = CH_3; Et = CH_2CH_3. The delocalized system of electrons is shown as - - - -.

red light-absorption patterns to give rise to a more yellow-green or more blue-green coloured pigment, respectively.

Similar colour changes can be obtained by the oxidation of carbons 17 and 18 (so forming didehydroporphyrin) and disruption and oxidation of the cyclopentanone ring, although these are biologically destructive changes. Reduction of ring B to yield the tetrahydrochlorins (or bacteriochlorophylls) drastically alters the area of delocalized electrons by removing a carbon double bond. The effect is to shift the red absorption peak into the infra-red spectrum. Such pigments are no longer green but, to our eyes, grey-pink.

In the living cell, within the discrete chloroplasts, the chlorophylls are complexed with, but not covalently bonded to, one of a series of polypeptides. Here, the whole chlorophyll-polypeptide complex is ordered in particular locations in the chloroplast membranes or thylakoids, from which water is excluded.

The chlorophyll-polypeptide (or protein) complexes are closely associated with several types of carotenes or xanthophylls (depending on the species of organism) together with a family of hydrophobic molecules the tocopherols (in the human diet better known as vitamin E). These are present at a concentration of about one molecule for every ten or so chlorophylls. The carotenoids act to quench excess chlorophyll excitation energy and also act as scavengers (or antioxidants) of highly reactive oxygen radicals and singlet oxygen. The tocopherols (or tocophenylquinones) also scavenge singlet oxygen and act as a terminal trap for various damaging radicals and lipid peroxides. The tocopherols, in essence, prevent the unsaturated-lipid environment of the chlorophylls from becoming peroxidized or rancid. The oxygen radicals or high-energy forms of oxygen, are an inevitable by-product of photosynthesis arising, in part, from the photo-splitting of water. They have the potential to irreversibly oxidize and to destroy (decolorize) the chlorophylls in a process known as photo-bleaching or photo-oxidation. In the intact healthy cell, photo-oxidation is strictly controlled and damage limited to diseased, environmentally stressed or ageing plants [13].

In the chloroplast thylakoid membranes, the greater part of chlorophyll (at least in higher plants) is associated with a group of proteins that make up two types of light-harvesting complexes (LHCs), designated LHCI and LHCII [14]. The latter, LHCII, on isolation, resolves into about seven polypeptides, three of which are known as LHCII. This complex alone holds more than half of all the chlorophylls and about one-third of the total protein in higher plant thylakoids. The remainder of the chlorophylls are associated with some seven other complexes whose characteristics are less well-known. Within the major complex, the LHCII, each of the three polypeptides is associated with about 10 chlorophyll molecules, approximately 6 chlorophyll *a* and 4 chlorophyll *b*, together with 1 to 2 xanthophylls (the precise number varies in different species). The chlorophyll *a* molecules appear to surround the densely-packed chlorophyll *b*. The three polypeptides, each with their complement of chlorophylls and xanthophylls, form a single unit folded in a structurally constant way from one side of the lipid membrane across to the other. Those parts of the polypeptide buried within the hydrophobic environment of the lipids appear to have a helical structure and to these the chlorophyll molecules are linked. The precise form of the ligand bonding between the chlorophylls and the amino acids of the helices is not known with certainty. The hydrophobic phytyl tail probably dictates the precise orientation of each chlorophyll molecule in the lipid membrane.

This brief outline should be sufficient to indicate that the environment in which chlorophyll functions, during photosynthesis, is elaborate, highly protected and comparatively stable. In this environment, the chlorophyll molecule is relatively immune from photo-bleaching and from oxidation. It remains green. It should be no surprise then that when such a photo-reactive molecule as chlorophyll is extracted from the plant and isolated from its associated radical and singlet oxygen scavengers (antioxidants), the pigment becomes unstable and is rapidly destroyed in air.

3.3.3 Biosynthesis of chlorophyll a *and other natural porphyrin pigments*

An outline of the biosynthetic pathway of the porphyrins and chlorophyll is provided in Figure 3.3. Fuller accounts are available elsewhere [15, 16].

Among the various steps in the pathway, it is worth noting that the first six, from 5-aminolaevulinic acid to protoporphyrin, are common to all living organisms, bacteria, fungi, animals and plants, if only because all such organisms contain haemproteins. Plants and some photosynthetic bacteria contain chlorophyll in addition. Protoporphyrin is the branch point of the haem and chlorophyll pathways, and may be chelated to iron to form protohaem—the prosthetic group of many haemproteins. Most of these are a deep-red colour and include the familiar red chromophore of haemoglobin and myoglobin of blood and muscle. In plants, protoporphyrin is also chelated to magnesium before undergoing a series of elaborations including cyclization of the side-groups at positions 13 and 15 to form the cyclopentanone ring E. In the presence of light, the porphyrin is reduced (at positions 17 and 18) to form the class of compounds known as chlorins. Final elaboration takes place in the form of esterification with the acyclic diterpenoid phytol (geranylgeraniol or farnesol in some chlorophylls). At this, or possibly at an earlier step, the chlorophyll molecule undergoes its final insertion into the polypeptide-lipid environment of the thylakoid membranes.

Regulation of the biosynthetic pathways is complex and involves internal control by simple inorganic molecules, by porphyrins themselves as well as the external stimulus, light. In addition, many steps in the pathway operate in close association with the synthesis of otherwise unrelated molecules. The formation of chlorophyll *a* is coordinated to the synthesis of several thylakoid polypeptides, to the tocopherols, phytol and carotenes.

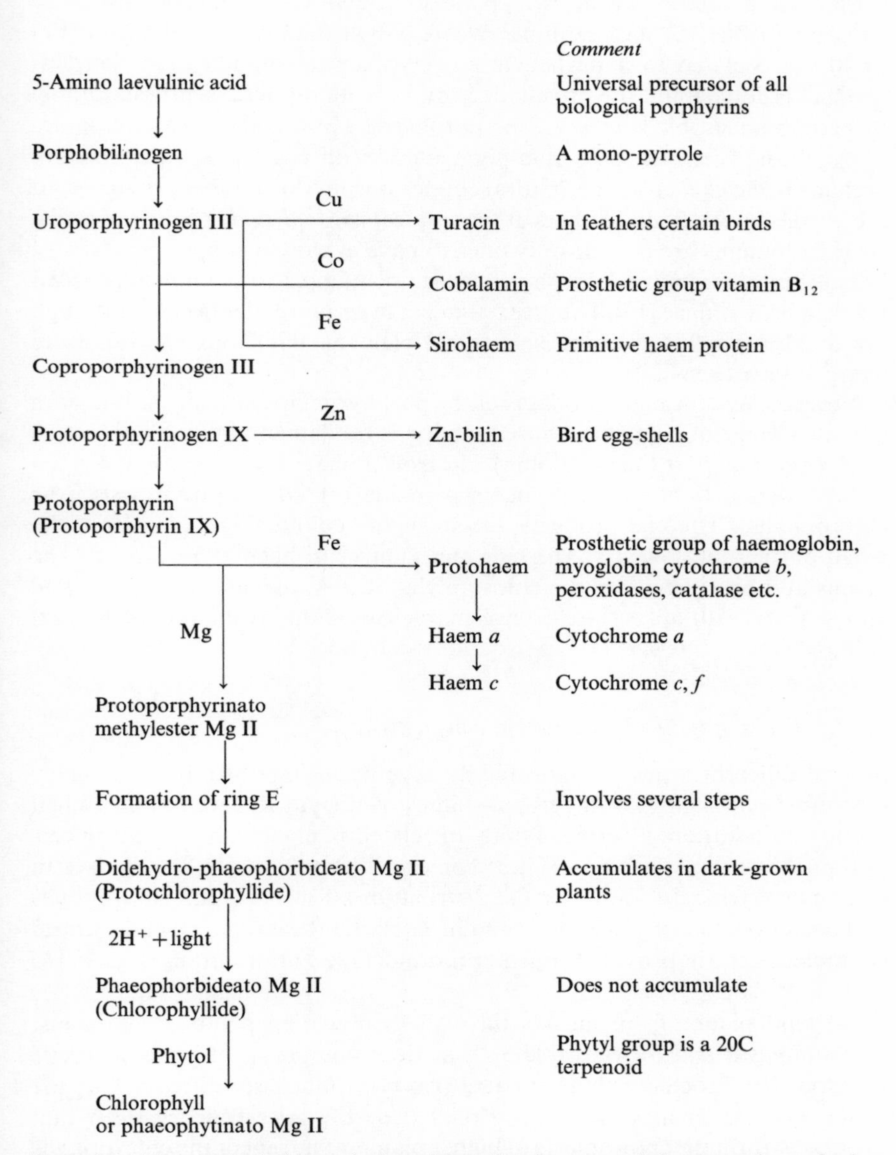

Figure 3.3 The porphyrin biosynthetic pathway showing some of the alternative and minor variations.

With such a long-conserved, ancient, biosynthetic pathway, it is not surprising that evolution has thrown up the occasional variation on an otherwise constant theme [15]. For example, where cobalt (Co) rather than iron (Fe) or Mg is chelated to a particular porphyrin, after modification the end-product is cobalamin, the prosthetic group of vitamin B12, synthesized only by certain anaerobic bacteria. Zinc porphyrins are found in several families of birds and form the pale-blue pigmentation on egg shells. More bizarre perhaps is the existence of a natural copper porphyrin confined, however, to the purple-brown flight feathers of the tropical *Musophagidae* family of birds. At least humans are not the only ones to have exploited copper porphyrins! Numerous other unusual porphyrins exist, including a vanadium-substituted haem in brown algae [17]. The green pigment in certain water beetles, although not a chlorophyll derivative, is a related pigment, biliverdin 1X [18], more likely to be a haem derivative (see chapter 4).

Nevertheless, the main products of the porphyrin biosynthetic pathway are a limited range of pigments whose function is fundamental to such processes as oxygen transport (haemoglobin), electron transfer (cytochromes), oxygen radical destruction (through haem-peroxidases) to photosynthesis (the chlorophylls). The end-products are all highly coloured, ranging from the avian purples and blues, to the reds and mauves of haem pigments and the greens and blue-greens of the chlorophylls. It is no wonder that poets [19] and scholars [20] alike should have marvelled at this paint box of natural colours.

3.3.4 Types of naturally-occurring chlorophylls

Several different types of chlorophylls have been described and it is likely that more will follow with further studies of certain less well-known algal groups. In addition, a further family of related pigments are present in certain photosynthetic bacteria. These bacteriochlorophylls are 7,8,17,18-tetra hydroporphyrins. In Table 3.2 the distribution of all recorded chlorophylls and bacteriochlorophylls is given and in Table 3.3 the structures and spectral characteristics are provided. Further more-detailed information is available [21–23, 25–28].

All land plants, from mosses through to flowering plants, contain just chlorophylls *a* and *b* (see Table 3.2), as does one group of photosynthetic bacteria, the Prochlorophyta. (There has been much speculation that the genus *Prochloron* may be directly related to the ancestral organism that evolved to form the chloroplasts of higher plants.) The remaining chlorophylls *c*, *d* and *e* have been found only in algae, including the brown and red seaweeds, as well as the single-celled marine algae making up the phytoplankton of the oceans. The chlorophylls are also found in certain photosynthetic bacteria, including the widespread blue-green Cyanophyta; other bacteria possess tetrahydroporphyrins or bacteriochlorophylls.

Table 3.2 Distribution of naturally-occurring chlorophylls and bacteriochlorophylls[a]

Chlorophyll/ Bacteriochlorophyll	Organism
Chlorophyll:	
a	All oxygen-evolving photosynthetic organisms including all higher plants, all algae and the photosynthetic bacteria Cyanophyta and Prochlorophyta
b	Higher plants, the algal Chlorophyta and Euglenophyta and the bacterial Prochlorophyta
c	Algal Phaeophyta (brown algae), Pyrrophyta (dinoflagellates), Bacillariophyta (diatoms), Chrysophyta, Prasinophyta and Cryptophyta
d	In some Rhodophyta (red algae) and Chrysophyta
e	Reported from algal Xanthophyta
Bacteriochlorophyll:	
a and *b*	Purple bacteria including Chromatiaceae and Rhodospirillaceae
c, *d* and *e*	Green and brown sulphur bacteria including Chlorobiaceae and Chloroflexaceae

[a] Nomenclature follows Reference 24.

Structurally (see Table 3.3), the variations in the ten known chlorophylls and bacteriochlorophylls are, relatively, modest. The most common variations are at positions 3, 7 and 8 and when compared to chlorophyll *a*, many of these variations involve an oxidation. Alternatives to the phytol group at

Table 3.3 Structures of all known chlorophylls and bacteriochlorophylls[a–c]

	Position (see Figure 3.1)					
Pigment	3	7	8	17	18	20
Dihydroporphyrins:						
Chlorophyll *a*	Vn	Me	Et	Phy H	Me H	—
Chlorophyll *b*	Vn	CHO	Et	Phy H	Me H	—
Chlorophyll *d*	CHO	Me	Et	Phy H	Me H	—
Tetrahydroporphyrins:						
Bacteriochlorophyll *a*	$COCH_3$	Me H	Et	Phy H	Me H	—
Bacteriochlorophyll *b*	$COCH_3$	Me H	$CHCH_3$ H	Phy H	Me H	—
Bacteriochlorophyll *c*	HO-Et	Me H	CH_2CH_3 H	Farn H	Me H	Me or Et
Bacteriochlorophyll *d*	HO-Et	Me H	CH_2CH_3 H	Farn H	Me H	—
Bacteriochlorophyll *e*	HO-Et	CHO H	CH_2CH_3 H	Farn H	Me H	Me or Et
Porphyrins:						
Chlorophyll c_1	Vn	Me	Et	Acr	Me	—
Chlorophyll c_2	Vn	Me	Vn	Acr	Me	—

[a] Positions: 2, 12, 18—Me throughout, except B'chl *e* 12 position Et; 13, 15—CP throughout; other unspecified positions unsubstituted.
[b] Abbreviations: Me, $-CH_3$; Et, $-CH_2CH_3$; Vn, $-CHCH_2$; Acr, -CHCHCOOH; Cp, cyclopentanone; Phy, phytol; Farn, farnesol.
[c] Original data derived and modified from Reference 1.

position 17 are found in some photosynthetic bacteria and algal groups. The overall impression, however, is that much the greater part of the original Mg protoporphyrin structure has been conserved throughout all known photosynthetic organisms.

3.3.5 Turnover and degradation of chlorophylls in biology

For all its relative stability in undiseased tissue, chlorophyll is subject to natural turnover, involving simultaneous destruction and synthesis, probably throughout the life of the organism. The rates of degradation are considered to be fastest at the initial greening-up phase of the seedling [29] and at the end of the life-span during leaf senescence [1] and fruit ripening [30]. The latter processes, involving a net loss of chlorophyll, are accompanied by profound physical changes to the immediate environment of the pigments. These changes will include alteration to the fluidity of the thylakoid membrane and peroxidation of membrane-bound thylakoid proteins [31], almost certainly as a result of increased attacks from various forms of activated oxygen. Just as the assembly of chlorophyll into the photosynthetic pigment is elaborate, so is its disassembly. Unfortunately, much less is known about the order in which disassembly takes place and even less about the precise mechanisms by which chlorophyll is turned from a stable green pigment to a colourless compound, often over a few hours or days. The phase of destruction is usually much shorter than the period of net synthesis. For example, in the north temperate areas, deciduous trees develop green leaves over a period of several days, often weeks, and may bear fully green leaves for up to 5 months, during which time chlorophyll turnover is probably extremely slow. But in early October most, if not all, of that chlorophyll is rapidly destroyed with a half-life of as little as 3 to 4 days [1].

The nature of the colourless end-products are unknown. In aquatic systems, and particularly under anoxic conditions, the destruction of chlorophyll is considerably slower, sometimes with identifiable products, often phaeophytin, being formed in substantial amounts.

3.3.6 Natural, and some unnatural, chlorophyll derivatives

Despite the rapid destruction of the chlorophylls in senescence and on fruit ripening, naturally-occurring breakdown products (as opposed to artifacts) can sometimes be detected under certain conditions. For example, grasses when cut for hay, rapidly form the blue-grey colour of phaeophytin. Grazing of chlorophyll-containing algae, in the oceans, gives rise to phaeophorbides, detected mainly in the faecal pellets of the grazers. Some of these phaeophorbides survive to sink to the ocean floor, where they may in time contribute to the most long-lived of chlorophyll derivatives, the geo- or petroporphyrins of oils and coal deposits [32]. In addition, numerous *in vitro* experiments have documented the conditions required to bring about particular modifications to the original chlorophyll molecule. Piecing all this evidence together, it is possible to draw up a somewhat simplified flowchart (see Figure 3.4) of the routes by which chlorophylls in natural and man-made conditions become altered and ultimately destroyed.

The action of weak acids on chlorophylls *in vitro* will readily bring about the loss of Mg to form phaeophytins. *In vivo* there may be an enzyme responsible for this demetallation [33]. Under natural and industrial conditions, in some plant species but not others, the action of the lipid-protein chlorophyllase readily cleaves the phytyl ester to yield chlorophyllide (dihydrophaeophorbidato Mg(II)).

Acidification would give rise to the demetallated phaeophorbide, the starting material for most of the chlorophyll food colorants. In biology, however, probably the greater part of chlorophyll destruction is either initiated or ultimately promoted through a series of disruptions of the cyclopentanone ring to yield a plethora of ill-described isomeric and epimeric forms of chlorophyll. Traces of these and more degraded products have often been recorded in chromatographic analyses [34–37] where they may individually represent minor intermediates in a pathway leading to the formation of low molecular-weight colourless but otherwise uncharacterized compounds. An account of the likely form of these degradation products has been given [1].

Much of the global production of chlorophylls is ultimately grazed and pass through one, or more, animals' guts. In the oceans, grazing leads to the formation of phaeophorbide, although again the major product appears to be a colourless unknown (Figure 3.4). On land, among those strict herbivores whose diet is composed of the green parts of plants, much of the ingested chlorophyll is excreted as the decarboxylated phaeophorbide derivative phytoporphyrin. Having survived grazing and photodestruction, some remaining chlorophylls will be deposited in oxygen-depleted conditions, such as the ocean floor or on a water-saturated peat bed. Undisturbed over several tens of millions of years, given the right physical and chemical environment, including prolonged heating, eventually those chlorophylls may emerge with considerable modifications, to contribute to the make-up of lignite, coal, shale, bitumen and oils. Characteristically, the porphyrin skeleton, having long lost its Mg, is chelated with nickel (Ni) or, under certain conditions, with $V = 0$.

Much is known of such compounds [32] because these contaminants can add to the costs of cracking during petroleum production, as well as having a diagnostic significance for the exploration for new reserves of fossil fuels. To all these many products of degradation, the food, cosmetic and pharmaceutical industries have added a few more, particularly during the early stages of isolation and concentration of chlorophyll derivatives. Figure 3.4 also provides a highly condensed account of some of these events (end-products names in square brackets). Phaeophorbide itself, in the appropriate

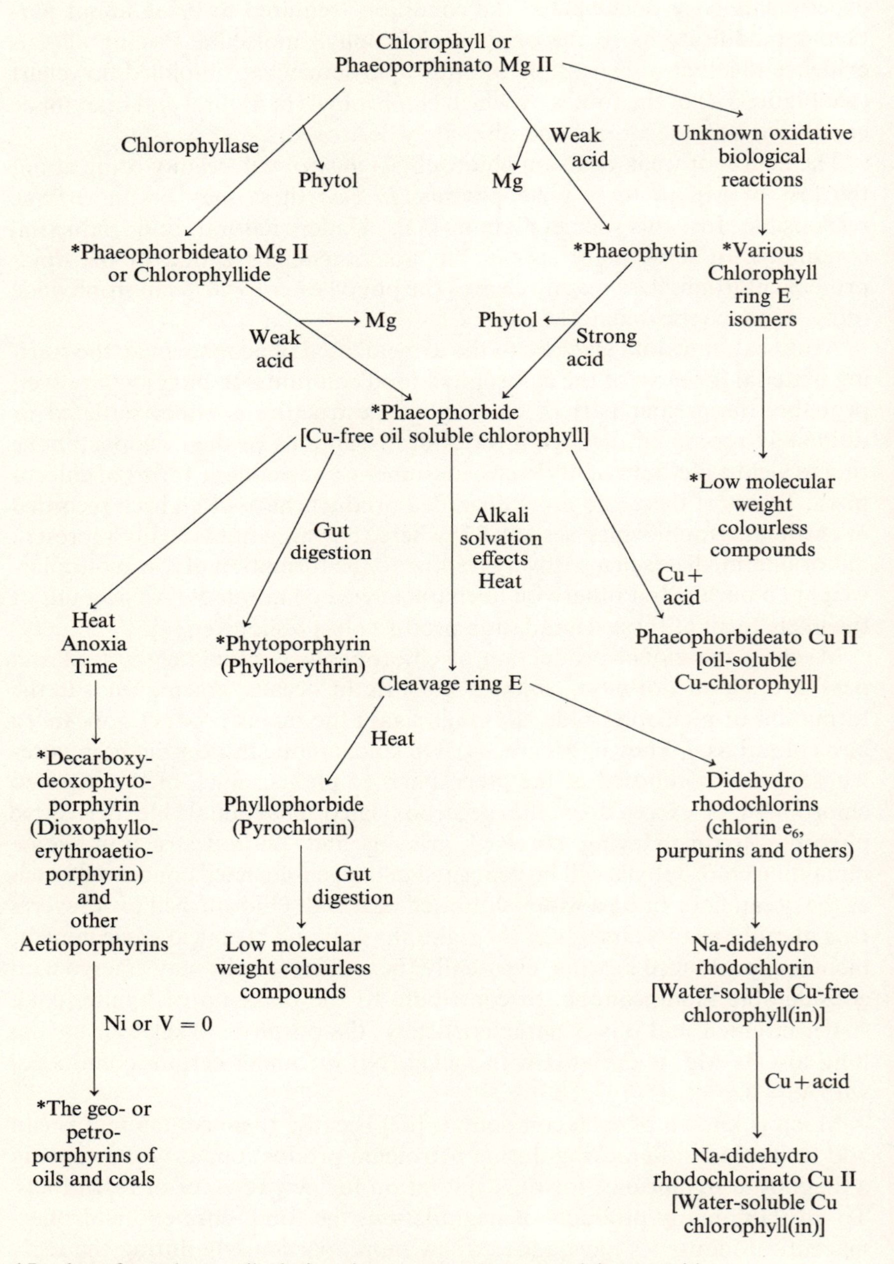

* Products formed naturally during plant senescence or natural decomposition.

Figure 3.4 Natural and unnatural pathways of the degradation of chlorophyll. Fischer nomenclature in parentheses, industrial chlorophyll-derivatives in brackets.

solvent, may be marketed as an oil-soluble Mg-free pigment. Acidification in the presence of copper (Cu) acetate or other salts gives the oil-soluble Cu phaeophorbide (dihydrophaeophorbideato Cu(II)). The preparation of the water-soluble derivatives involves several steps, which aim to achieve a controlled cleavage of the cyclopentanone ring. Among the many products are a class of porphyrins that may best be described, generally, as di(de)hydrorhodochlorins. This mixture on acidification ultimately yields the soluble sodium (or potassium) salt or copper rhodochlorins (the so-called copper chlorophyllins).

In porphyrin chemistry, few degradative steps, however well controlled, yield only one product. Part of the reason for the multiplicity of products may be due to the ease with which chlorophyll molecules in solution form aggregates that can be both complex and highly variable. Such randomized aggregates will have different side-groups exposed to (or hidden from) the external effects of solvents and will respond, chemically, in different and seemingly random ways. An analogous situation arises in the random formation of lignins from lignin radicals where any two molecules are highly likely to be structurally different from each other. Perhaps, as a consequence, it is commonly observed in the laboratory that the recovery of derivatives from an original concentration of chlorophyll is rarely complete. Much, often a substantial part, of the original chlorophyll disappears (spectrally) to form unknown colourless products. This alone seems to be one of the few consistent patterns in studies of the degradative pathway of the chlorophylls.

3.4 Chlorophyll derivatives as food colorants

3.4.1 Sources of chlorophylls for food colorants

To date, most, if not all, chlorophylls entering the food-colorant market are derived from land plants which, in recent decades, have included lucerne, or alfalfa (*Medicago sativa*), nettles (*Urtica dioica*) or several high-yield pasture grasses. One notable exception has been the use in Japan of the green faeces of the silk-worm where the larvae feed on mulberry leaves and excrete 13^2HO-chlorophyll, among other pigments [38]. Convenience of harvesting, ease of pigment extraction, yield of useful derivatives of chlorophyll and the value of waste by-products as cattle feed, dictate the choice of raw material. However, there is no evidence that the current selection of species necessarily gives the highest yields. An additional problem, in the temperate regions, is that the supply of raw material is restricted, often being confined to a few weeks in the year. Outside the summer period, the availability of suitable raw material is limited. In contrast, by far the greater synthesis of the chlorophylls takes place in the oceans, mainly in single-celled phytoplankton and, in the north Atlantic, over much of the year.

Estimates of the global production of chlorophyll are shown in Table 3.4.

Table 3.4 Total net annual global production of chlorophylls[a]

Environment	Tonnes $\times 10^8$
Terrestrial	2.92
Aquatic	8.63
Total	11.55

[a] From Reference [1].

Some 75% of the total annual production of chlorophyll, globally, takes place in the aquatic (and mainly marine) environment. Interestingly, the estimated figure of approximately 11×10^8 tonnes of chlorophyll production compares quite well with an independent estimate of 1×10^8 tonnes given for carotenoids [39]. In plants, the ratio of chlorophylls to carotenoids by weight (not molarity) is about 5:1, giving a value for the global natural production of carotenoids as about 2×10^8 tonnes annually. Much the greater part, probably well in excess of 99%, of both the chlorophylls and carotenoids is degraded within days of the death of the organism. The traces that remain to be recovered in geological deposits are an exceedingly small part of the whole [32]. Nevertheless, in terms of pigment production in living plants, there is a super-abundance of both chlorophylls and carotenoids. The supplies, whether terrestrial or aquatic in origin, are self-renewing and, in terms of their economic value, have been barely exploited.

3.4.2 *Alternatives to chlorophyll* a

Much the greater part of the industrially produced chlorophyll colorants are derived from chlorophyll *a*, the principal (occasionally the only) form of chlorophyll in all oxygen-evolving photosynthetic organisms. Some 20 to 30% of chlorophyll colorants may also be derived from chlorophyll *b*. In all, some ten chlorophylls or bacteriochlorophylls have been described (see Table 3.3) from biological systems. Each pigment has its own distinct colour and these are listed, together with the spectral characteristics, in Table 3.5.

Given adequate access to a supply of algae and with suitable isolation techniques, it should be possible to prepare chlorophylls with spectral characteristics ranging from yellow-green to blue-green. Derivatives of these chlorophylls would be likely to produce orange or, under drastic chemical conditions, even red colours.

Two other photosynthetic pigments exist, orange and blue coloured, which are structurally related to the chlorophylls. Among the algae, the Rhodophyta and Cryptophyta, together with the photosynthetic bacteria, the Cyanophyta, produce large quantities of highly coloured phycobilins (see chapter 4). These linear tetrapyrroles are derived not from chlorophylls but from protohaem

Table 3.5 Spectral characteristics of the naturally-occurring chlorophylls and bacteriochlorophylls

				Spectra			
				Soret		Red	
References	Pigment	Solvent	Colour	max	mM	max	mM
25	Chlorophyll *a*	80% acetone	Blue-green	431.2	95.8	663.2	86.3
25	Chlorophyll *b*	80% acetone	Green	459.0	135.4	646.8	49.2
28	Chlorophyll c_1	90% acetone	Yellow-green	443.2	318.0	630.6	44.8
28	Chlorophyll c_2	90% acetone	Yellow-green	443.8	374.0	630.9	40.4
21	Chlorophyll *d*	Ether	Blue-green	445.0	97.8	686.0	117.8
26	Bacteriochlorophyll *a*	Methanol	Grey-pink	365.0	53.9	772.0	42.0
21	Bacteriochlorophyll *b*	Acetone	Brown-pink	368.0	—	795.0	—
27	Bacteriochlorophyll *c*	Acetone	Green	428.0	—	660.0	—
27	Bacteriochlorophyll *d*	Acetone	Green	424.0	—	654.0	—
27	Bacteriochlorophyll *e*	Acetone	Green	456.0	—	646.0	—

[40], and are related to the mammalian bile pigment bilirubin. In the living cell, these phycobilins are highly stable and covalently bonded to certain proteins. Cleavage from these proteins yields two classes of pigments, the orange-red erythrobilins and the blue cyanobilins, of which there are several variants [41]. They have immediate interest as they are water soluble and relatively stable. Very large concentrations of these pigments are formed in algal cells; 20% or more of the dry weight may consist of the phycobilin plus its protein.

Finally, for the sake of completion, there exists a third class of linear tetrapyrroles, which are present in all plants including algae. This blue pigment, called phytochrome, has a highly specific and dominant function as a light receptor, controlling many critical aspects of plant metabolism and development. Its concentration in plants is exceedingly low and for this reason it is unlikely to be of commercial interest.

3.4.3 Extraction, isolation and derivatization

Chlorophylls are usually extracted, not from freshly-harvested plants, but from bulk-harvested material previously sun-dried or subjected to artificial, rapid, heat—usually at closely controlled temperatures. The subsequent yield of pigment from the dried plants, pellets or powder varies with the stage of development of the plant species itself, the duration and temperature of the drying process and the length of storage of the dehydrated raw material. The drying process itself leads to the formation of several degraded (euphemistically called 'altered') chlorophylls. Heating for 5 min at 70°C was found to produce at least six derivatives with substantial amounts of colourless degradation products [36]. Different products were formed at different pH and were dependent on the availability of O_2. Most of the identified products show alterations to the 13^1- and 13^2-position side-groups on the cyclopentanone ring, indicative of oxidative attack.

Primary extraction, under strict control, using an aqueous solvent such as acetone or a chlorinated hydrocarbon, is followed by washing, concentration and solvent recovery. The degradative effects of organic solvents have been described by many authors [34, 36, 42]. For example, plant species rich in the enzyme chlorophyllase readily form chlorophyllides, particularly during extraction with aqueous acetone. Other products, similar to the 'altered' chlorophylls of heat treatment, are also formed during contact with certain water-organic solvent mixtures in the presence of air. For this reason, the ratio of solvent to plant material is critical during industrial extraction. Overall, the yield of chlorophyll (and green derivatives) may be no better than 20%, although much higher and lower values are quoted [43]. During these early steps, other plant products, including resins, waxes and fats released by the solvent extraction process, are removed by differential fractionation.

The partially purified extracts containing phaeophytin, and other degraded chlorophylls are subjected to further processing depending on the desired end-product. Oil-soluble colorants can be secured after further washing against water-immiscible solvents and are marketed as the metal-free phaeophytin, or further acidified in the presence of copper salts to form oil-soluble copper phaeophytin. The possibility of producing authentic Mg phaeophytin (that is, chlorophyll) exists, although the extraordinary care needed to produce this makes the product comparatively costly and creates further problems in maintaining its stability.

Water-soluble derivatives are prepared by saponification of the crude mixtures of degraded chlorophylls and phaeophytins allowing the hydrophobic phytyl group to be displaced by sodium (Na) (or potassium (K)). The acidic groups at 13^1 and 13^2 of the cyclopentanone ring will also be converted to the Na (or K) salt. After further fractionation and washing to remove much of the lipophilic contamination, the chlorin (or porphyrin) salt may be marketed as the metal-free grey-green water-soluble pigment. Conversion to a stable green colorant is achieved by acidification in the presence of copper salts. Outlines of the process have been provided [10, 43], although most producers of chlorophyll-based colorants will have procedures that vary in detail and in sequence from those described. While the logic of the design of the procedures is clear, the actual process of extraction of chlorophyll derivatives has been described as more of an art [10]. In this, it is in good company with many other complex and ill-defined chemical processes.

3.4.4 Structure of the derivatives

Chlorophyll derivatives, as food colorants, are usually sold as 'chlorophylls' or 'chlorophyllins' with, or without copper, soluble in water or insoluble, in various strengths (concentrations) and diluents. Several products are marketed with a subsidiary description based on the word 'phaeophytin', others use the base-word 'chlorophyllin'. The term chlorophyllin is broadly understood to refer to an unspecified product of alkaline hydrolysis of chlorophylls [44] and may be further equated with chlorophyllide, also known as dihydrophaeophorbideato Mg(II) [10], or the term 'phytochlorin' [43]. The implication is certainly that chlorophyllin is one substance. However, the number of products formed both during the drying of the raw material, in the initial extractions and on saponification is potentially large [45] and inevitably the term 'chlorophyllin' has only a generic meaning. This is even more true of the word 'chlorophyll' in the sense used in the food trade.

To summarize a complex series of reactions, the products marketed as chlorophyllins are water-soluble salts of derivatives of phaeophorbide *a* and *b* or allomerized forms of Mg-free chlorophylls *a* and *b*, characterized by disruption of the cyclopentanone ring with the presence of nil, one or two carboxyl groups at positions 13^1 and 13^2 with the oxidation of ring D and

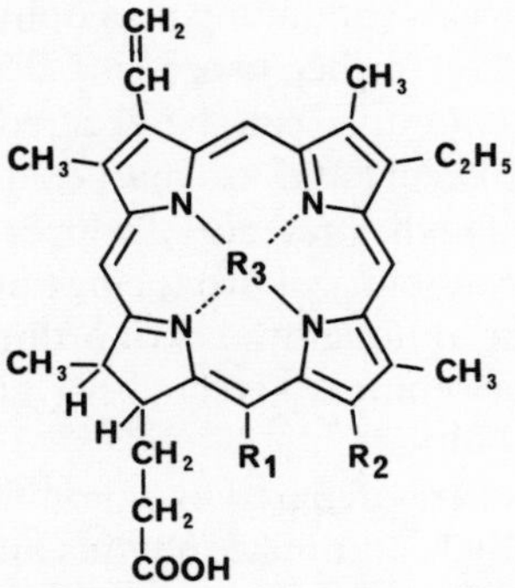

Figure 3.5 The structure of a rhodochlorin where R_1 and R_2 are simple derivatives or substitutions to the positions 13^1 and 13^2 of the disrupted cyclopentanone ring. R_3 would be either $2H^+$ or a metal ion such as Cu^{2+}.

possible reduction of the vinyl group at position 3. The products marketed as oil-soluble chlorophylls are perhaps less-varied and, under carefully controlled conditions, may be based substantially on phaeophytin *a* and *b* and their epimers and isomers with more severe re-arrangements around positions 13^1 and 13^2. In so far as any name could accurately describe this varied family of pigments, the most appropriate, and perhaps most accurate term would be one based either on phaeophytin (the oil-soluble pigments) or rhodochlorin (the water-soluble form) (see Figure 3.5). Further precision, where necessary, could be obtained by additional qualifying terms such as disodium didehydrorhodochlorin 13,15 diacetic acid.

3.4.5 *Stability and instability*

The origin of most of the industrial chlorophyll derivatives lies in the demand for a green pigment, of natural origin, closely related to authentic chlorophyll, but by necessity restructured to form a pigment that is stable in isolation. This stability has been achieved by restoring not magnesium but copper to the metal-free pigment. The Cu-substituted derivatives are relatively unresponsive to light, due to the electronic structure of the Cu ion. Neither are they readily decomposed by mineral acids [43]. In contrast, the isolated and purified Mg-chelated derivatives are unstable, particularly the presence of even dilute acids, increasingly so in the light. Even the metal-free derivatives such as phaeophytin and phaeophorbide with an intact cyclopentanone ring have a propensity to become oxidized, particularly on exposure to light. However desirable (from the point of view of food colorant regulations) it may be to provide authentic Mg-chelated 'natural' chlorophylls, these pigments are inherently unstable and acquire short-term stability only at pHs of more than 7.0, under anoxic conditions in the absence of light. Degradation to the level of phaeophytins is to be expected, with the presence of oxygen actually hastening the decomposition.

The biological explanation for the apparent instability of the Mg-chelated porphyrins is that, in the excited state, these metalloporphyrins are strong reducing agents and thereafter readily oxidized. Mg (and Zn) porphyrins are, essentially, porphyrin di-anions and so lose an electron following excitation by light as in photosynthesis (the biological reason) or in photo-bleaching (the artifactual phenomenon). Following excitation, the resulting oxidized chlorophyll (Cl^{+}) undergoes a series of electronic rearrangements that further add to excitation, or instability. If the electron deficiency is not rapidly balanced, as it is in photosynthesis, the chlorophyll anion species will combine with oxygen. The precise details of the order (if there is order) in the subsequent oxidative reactions are unknown; the products degenerate to colourless compounds, usually rapidly. However, a variety of intermediates of chlorophyll oxidative breakdown have been detected from time to time [34–37], several of which suggest that the first site of oxidative attack is the cyclopentanone ring. Certain transition metals, such as Cu or Fe, are readily chelated to porphyrins but the resulting complexes (copper porphyrins and protohaem) are largely inert photo-chemically, that is they do not readily loose electrons on exposure to light energy. The apparent instability of Mg porphyrins *in vitro* is a direct consequence of their functional, and usually stable, photoreactivity *in vivo*.

The stability of *some* porphyrins is legendary, however. Proof of this lies in the occurrence of nickel and vanadyl porphyrins, derived from chlorophylls *a* and *b*, in ancient oil deposits [32]. Mg coordinate petro-porphyrins have, however, never been reported. Nevertheless, from time to time claims are made for the discovery of 'fresh' green plants exposed during archaeological investigations. Some of these plant remains have been preserved under the most exotic conditions, in one case buried in waterlogged conditions (near anoxia) under thousands of tonnes of chalk (thus no light and $pH > 7.0$) at the heart of the neolithic structure Silbury Hill, Wiltshire, UK [1]. None of these archaeological finds, to the author's knowledge, has ever been shown to contain the original unaltered chlorophyll in more than what *may* be barely detectable traces. Much the greater part of such long-preserved pigments consist of the metal-free grey-coloured phaeophytin *a*. Indeed, the fresh greenness, jubilantly reported on exposure of archaeological plant remains, frequently turns to a dull-brown colour within minutes, strongly indicative of the presence of the oxidation of green-coloured phenols.

3.4.6 Economics of chlorophyll derivatives

Estimates of the worth of the world production of food colorants vary. A value of $150 to $200 m in 1985 has been given [46], increasing at a rate of about 10% annually. Other estimates [47] suggest that 10 to 20% of the market might be accounted for by natural colorants giving a value for these of say \$40 to \$50 m (at 1991 prices). This value covers carotenoids (much the most important of the 'natural' colorants), anthocyanins, betalaines, as well as chlorophylls. Informed estimates (see acknowledgements) place the UK production of chlorophylls as worth \$3.5 to \$5.0 m annually with the suggestion that this might represent up to one-third of the world production. Given these estimates, this places a world value on the sales of chlorophyll-derivatives at a little under \$15 m in 1991, a figure that includes both food colorants as well as for pharmaceutical and cosmetic usage.

From trade estimates, the major demand, in volume, is for the water-soluble chlorophyll derivatives with specialized interest in the more costly oil-soluble product. Much the greater part, perhaps 75% or more, of industrially prepared chlorophyll-derivatives are used in the growing demand for natural colorants in food and drinks, including dairy products, edible oils, soups, chewing gum and sugar confectionery. Some 20% of the industrial production, at least in Europe, is for the more volatile fashion-led cosmetic and toiletry market, including herbal bath foams and gels, shampoos, hair conditioners and soaps. A small but traditionally-based demand exists in the pharmaceutical trade for chlorophyll-derivatives as deodorants, mouthwashes and surgical dressings.

To put all this into perspective, if the estimate of \$15 m is accepted for the worth of the chlorophyll colorants, how does this compare with one other aspect of the economics of chlorophyll? Once a year, in early October, the destruction of chlorophyll in the leaves of the trees of the north-east United States of America creates a major tourist industry. Estimates for just one State, New Hampshire, put the value of fall tourism in 1990 as \$350 m! The chlorophyll colorant industry has some catching up to do.

3.4.7 *Future prospects*

There are three areas in which recent scientific and technological advances could have a significant effect both on production costs and on stimulating demand for chlorophylls as food colorants.

In recent years, the advances in molecular biology have resulted in the identification, isolation and engineering of particular genes that have properties open for exploitation. Considerable progress has been made in very recent years in the area of plant genetic transformations. In the context of chlorophyll degradation, a gene that controls one of the early steps in disassembly and degradation of chlorophyll, in the intact plant, has been identified [48]. This *sid*-gene (senescence inducing determinant) has been partially characterized in the grass *Festuca pratensis* of which a mutant (BF993) lacks the ability to express the gene. The consequence is that the mutant remains green on death, that is, it retains its chlorophyll following the natural senescence of the plant. While the wild-type turns yellow during senescence (following destruction of the chlorophyll), the mutant retains its deep-green colour. Although this may have interesting commercial possibilities for perhaps sports grounds in winter conditions, it may also have an application in the extraction of chlorophyll as a colorant. One immediate attraction would be that a supply of green plant material could be stored following the summer harvest but then used as the raw material for extraction throughout the rest of the year. A more long-term benefit might arise if the *sid*-gene from *F. pratensis* were to be isolated, copied *in vitro*, and inactivated by perhaps inverting part of the genetic sequence before transferring the now unexpressed or nonsense gene into new plant species such as nettles or high-yielding fodder grasses. Genetic lines from the seeds of these transformed plants would be selected to provide 'evergreen' plant material. The benefit would come in providing a year-round supply of desirable species of green plants and eliminating the need to rely on dehydrated plant tissues with their attendant slow oxidative losses of pigment during prolonged storage.

The second prospect lies in stabilizing the original Mg-chlorophyll (the authentic natural product). The prospect of achieving this might involve biochemically encapsulating the chlorophyll molecule in an environment designed to quench or scavenge for the destructive effects of activated forms of oxygen (radicals and singlet oxygen). The environment might, for example, retain one of the natural chloroplast LHC polypeptides, a complement of xanthophylls and tocopherols (vitamin E). Such a partial reconstitution of the environment of the photosynthetic membranes has already been achieved using photosynthetic bacteria [49]. The aim would be to provide a product that was stable in air, acid and light and which was *in its entirety* biologically 'natural'. Some encouragement for this prospect will come from the common laboratory observation that isolated thylakoid membranes remain remarkably green, in the light and air, for many days after the complete destruction of chlorophylls in organic solvents.

The third prospect would be in the exploitation of different chlorophylls, other than chlorophyll *a*, particularly where there is a possibility of improved stability. There are, for example, several reports [50, 51] of the unusual relative stability of chlorophyll *c* in biological systems and even in zooplankton faeces. Indeed the indications from the literature [1, 52] are that the order of stability of the chlorophylls in living and senescing systems is:

$$\text{chlorophyll } c > b > a \qquad (3.2)$$

It is perhaps ironic that chlorophyll *a*, the least stable of the three chlorophylls, has been the only one selected for exploitation. Chlorophyll *c* is a discarded (indeed combusted) product of the alginate industry, based on brown seaweeds. An alternative source of chlorophyll *c* is single-celled phytoplankton (see Table 3.2), which may be adaptable to large-scale continuous culture [53]. Chlorophyll *c*, unlike other natural chlorophylls, is not esterified at position 17. Without further modification, this pigment is readily soluble in methanol and other relatively polar solvents [26]. As to its suitability as a natural food colorant, chlorophyll *c*-containing brown seaweeds have long formed part of the traditional diet of Asiatic and some maritime European cultures.

The production and use of chlorophylls as food colorants has become increasingly important in recent years. The challenging opportunities for new production methods and new products may help to promote the wider use of this colourful group of pigments. The challenges will bear on chemists and biochemists and may provoke them to answer John Donne's perspective lines:

> 'Why grass is green or why our blood is
> red
> Are mysteries which none have reached
> unto.'
> John Donne, 1612 [19]

Acknowledgements

The author wishes to thank Dick Patrick, Fred Fost and Dr K. Helliwell for their helpful discussions.

References

1. Hendry, G.A.F., Houghton, J.D. and Brown, S.B., *New Phytol.* **107** (1987) 255.
2. Rimmington, C. and Quinn, J.I. *Ondersteepoort J. Vet. Sci. Anim. Ind.* **10** (1934) 421.
3. Lohrey, E., Tapper, B. and Hove, E.L. *Brit. J. Nutr.* **31** (1974) 159.
4. Eskin, N.A.M. *Plant Pigments, Flavors and Textures*, Academic Press, New York (1979).
5. Counsell, J.N. 'Some synthetic carotenoids as food colours' in *Developments in Food Colours—1* (J. Walford, Ed.), Applied Science Publishers, London (1980) pp.151–188.
6. Coulson, J. 'Miscellaneous naturally occurring colouring materials for foodstuff's: in *Developments in Food Colours—1* (J. Walford, Ed.), Applied Science Publishers, London (1980) pp.189–218.
7. Counsell, J.N. *Natural Colours for Food and Other Uses*, Applied Science Publishers, London (1981).
8. Taylor, A.J., 'Natural colours in food' in *Developments in Food Colours—2* (J. Walford, Ed.), Elsevier Applied Science Publishers, London (1984) pp.159–206.
9. Jackman, R.L., Yada, R.Y., Tung, M.A. and Speers, R.A. *J. Food Biochem.* **11** (1987) 201.
10. Humphrey, A.M. *Food Chem.* **5** (1980) 57.
11. Merritt, E. and Loening, K.L. *Eur. J. Biochem.* **108** (1980) 1.
12. Fischer, H. and Orth, H. *Die Chemie des Pyrrols—2*, Akademische Verlag, Leipzig (1937).
13. Price, A. and Hendry, G.A.F. *Plant Cell Environ.* (1991) (In press).
14. Thornber, J.P., Peter, G.F. and Nechustai, R. *Physiol. Plant* **71** (1987) 236.
15. Hendry, G.A.F. and Jones, O.T.G. *J. Med. Genet.* **17** (1980) 1.
16. Jones, O.T.G. 'Chlorophyll biosynthesis' in *The Porphyrins VI* (D. Dolphin, Ed.), Academic Press, New York (1979) pp. 179–232.
17. Vilter, H. *Phytochem.* **23** (1984) 1387.
18. Kayser, H. and Dettner, K. *Comp. Biochem. Physiol.* **77B** (1984) 639.
19. Donne, J. *On the Second Anniversary* (1612).
20. Brown, S.B. *Univ. Leeds Review* **30** (1987) 1.
21. Svec, W. 'The isolation, preparation, characterization and estimation of the chlorophylls

and bacteriochlorophylls' in *The Porphyrins VC* (D. Dolphin, Ed.), Academic Press, New Amsterdam (1979) pp.341–397.
23. Falk, J.E. *Porphyrins and Metalloporphyrins*, Elsevier, Amsterdam (1964).
24. Holmes, S. *Outline of Plant Classification*, Longmans, London (1983).
25. Lichtenthaler, H.K. *Methods in Enzymology* **148** (1987) 340.
26. Jones, O.T.G. 'Chlorophylls and related pigments' in *Data for Biochemical Research* 2nd edn. (R.M.C. Dawson, D.C. Elliott, W.H. Elliott and K.M. Jones, Eds.), Clarendon Press, Oxford (1969) pp.318–320.
27. Gloe, A., Pfennig, N., Brockman, H. and Trowitzsch, W. *Arch. Microbiol.* **102** (1975) 103.
28. Jeffrey, S.W. *Biochim. Biophys. Acta* **279** (1972) 15.
29. Hendry, G.A.F. and Stobart, A.K. *Phytochem.* **25** (1986) 2735.
30. Goldschmidt, E.E. 'Pigment changes associated with fruit maturation and their control' in *Senescence in Plants* (K.V. Thimann, Ed.), CRC Press, Boca Raton, Florida (1980) pp.207–217.
31. Thompson, J.C., Legge, R.L. and Barber, R.F. *New Phytol.* **105** (1987) 317.
32. Baker, E.W. and Palmer, S.E. 'Geochemistry of porphyrins', in *The Porphyrins IA* (D. Dolphin, Ed.), Academic Press, New York (1978) pp.485–551.
33. Ziegler, R., Dresken, M., Herl, B., Menth, M., Schneider, H. and Zange, P. *Deutsch. Bot. Gess. Meeting* (Freiburg) (1982) Abstract 472.
34. Maunders, M.J. and Brown, S.B. *Planta* **158** (1983) 309.
35. Merzlyak, M.N. and Kovrizhnikh, V.A. *J. Pl. Physiol.* **123** (1986) 503.
36. Bacon, M.F. and Holden, M. *Phytochem.* **6** (1967) 193.
37. Matile, Ph., Duggelin, T., Schellenberg, M., Rentsch, D., Bortlik, K., Peisker, C. and Thomas, H. *Plant Physiol Biochem.* **27** (1989) 595.
38. Nakatani, Y., Orisson, G. and Beck, J-P. *Chem. Pharmacol. Bull.* **29** (1981) 2261.
39. Kearsley, N.W. and Katsabaxakis, K.Z. *J. Food Technol.* **15** (1980) 501.
40. Brown, S.B., Holroyd, J.A., Troxler, R.F. and Offner, G.D. *Biochem. J.* **194** (1981) 137.
41. Rudiger, W. 'Plant biliproteins' in *Pigments in Plants* (F.-C. Czygan, Ed.), Gustav Fischer Verlag, Stuttgart (1980) pp.314–351.
42. Schoch, S. and Brown, J. *J. Plant Physiol.* **126** (1987) 483.
43. Kephart, J.C. *Econ. Bot.* **9** (1955) 3.
44. Strain, H.H. and Svec, W.A. 'Extraction, separation, estimation and isolation of the chlorophylls' in *The Chlorophylls* (L.P. Vernon and A.R. Seeley, Eds), Academic Press, New York (1966) pp.21–66.
45. Strell, M., Kalojanoff, A. and Zuther, F. *Arzneimittel-Forsch.* **6** (1956) 8.
46. Ilker, R. *Food Technol.* **41** (1987) 70.
47. Blenford, D. *Food* **7** (1985) 19.
48. Thomas, H. *Theor. Appl. Genet.* **73** (1987) 551.
49. Hunter, C.N., Van Grondelle, Holmes, N.G. and Jones, O.T.G. *Biochim. Biophys. Acta* **548** (1979) 458.
50. Jeffrey, S.W. and Allen, M.B. *J. Gen. Microbiol.* **36** (1964) 277.
51. Jeffrey, S.W. *Marine Biol.* **26** (1974) 101.
52. Brown, S.B., Houghton, J.D. and Hendry, G.A.F. 'Chlorophyll degradation' in *The Chlorophylls* (H. Scheer, Ed.), CRC Press, Boca Raton, Florida (1991) p. 465.
53. Dixon, G.K. and Syrett, P.J. *New Phytol.* **109** (1988) 289.

4 Haems and bilins

J.D. HOUGHTON

4.1 Summary

Haems and bilins provide some of the most visually apparent pigments in nature, from blood-red to the azure blue of marine algae. Their colours are associated with the natural world, with health and well-being, and have been used for centuries to reinforce these associations in traditional food and cosmetic preparations.

Man's interest in these pigments has also led to an accumulation of knowledge on their chemical and biochemical properties, and a growing research interest in the emerging field of molecular biology. Today, these have an important role to play in the modern use of natural colours, yet little of the scientific knowledge available concerning these pigments has been made accessible to the commercial or industrial user. Nomenclature of the natural or nature-alike pigments that are available is still inaccurate and confusing, and sometimes bears little relationship to scientifically accepted terms. In addition, much ignorance exists as to the true form of these pigments in nature, and the justification for their modification in commercial use.

It is hoped here to bring together the available information concerning haems and bilins, in a form that is accessible to the industrialist and research scientist alike, and to create a basis for interaction, which must surely offer future benefits to all.

4.2 Haems

4.2.1 Introduction

The term 'haem' is generally used to describe an iron chelate of the cyclic tetrapyrrole protoporphyrin IX (Figure 4.1). Haems are ubiquitous in nature and perform a central role in cellular energy transduction and metabolism in all known species. They are represented in this role by a great variety of apoprotein conjugates. All of these haemoprotein complexes are pink or red in colour by virtue of their common prosthetic group, but the structure most intensively studied is undoubtedly haemoglobin. It is the aim of this section to characterize the known physical and biochemical properties of the haemoproteins with particular emphasis on haemoglobin, and assess their present use and future potential as colorants.

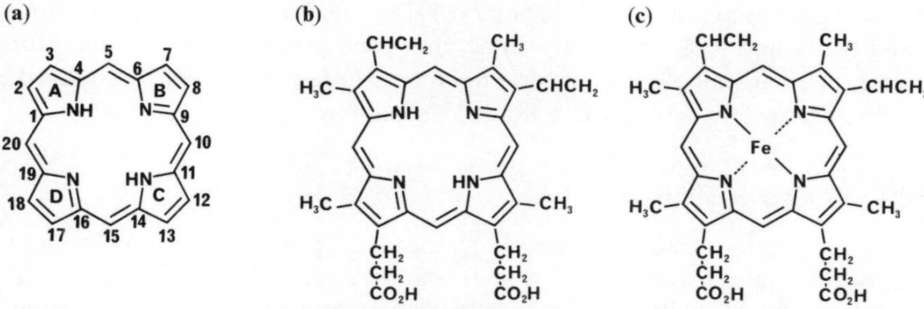

Figure 4.1 The structures of protoporphyrin IX (b) and haem (c) with that of their common porphyrin nucleus (a). The numbering of the porphyrin ring is according to the IUB-IUPAC recommendations [1].

4.2.1.1 Structure and nomenclature. The structure of haem is shown in Figure 4.1, together with the correct IUB-IUPAC system of numbering of the constituent carbon atoms [1]. The central iron atom of haem is co-ordinated to the four pyrrole nitrogens of protoporphyrin IX to form an almost planar metalloporphyrin ring.

The fifth and sixth co-ordination positions of the iron atom are perpendicular to the ring and may be occupied by several ligands, including a donor group from the surrounding apoprotein. Several trivial names persist, which describe the valency of the central iron atom and/or the ligands to which it is bound. Thus a haemin is a ferric (FeIII) haem in which a halogen or other anion is co-ordinated. The term haematin is also used to denote the ferric state of the central iron atom, and may be qualified to indicate the nature of the ligand bound, for example, alkaline haematin. This nomenclature is commonly used to describe the valency and ligand states of haems *in vitro*, and is summarized in Table 4.1. In nature, most haemoproteins perform their

Table 4.1 Nomenclature of some haems and haem derivatives [23]

Name	Synonyms	Valency	5th ligand
Haem	Ferroprotohaem IX Protohaem Ferroprotoporphyrin IX Iron protoporphyrin IX Reduced haematin	Fe II	H_2O
Haematin	Ferriprotohaem IX Hydroxyferriprotoporphyrin Alkaline haematin	Fe III	OH
Haemin	Protohaemin IX Chlorohaemin Chloroferriprotoporphyrin	Fe III	Cl

physiological functions in the ferrous (FeII) state, but may also exist in the FeIII form, and their names are sometimes prefixed to denote this fact, for example ferrihaemoglobin/ferrohaemoglobin. Other terms will be defined, where appropriate, within the body of the chapter.

4.2.2 Haems in nature

4.2.2.1 Function and occurrence. Haem forms the prosthetic group of a large family of proteins involved in several diverse functions throughout the animal and plant kingdoms. It is most visually prominent as the red blood pigment haemoglobin, which functions as an oxygen carrier, and also in muscle as myoglobin where it performs the same role. Haem is also central to the energy transducing mechanisms of the cytochromes, rsponsible for the use of energy from photosynthesis and respiration, and to several enzymes such as catalase and peroxidase, which use haem in their structures. The importance of haem in these vital processes is manifest and their role is so central to life processes that, along with the chlorophylls, they have been referred to as 'the pigments of life' [2].

The variety of functions that the haemoproteins as a whole perform are so wide as to be outside the scope of this chapter. It seems appropriate here to describe the function of haemoglobin as an example of one particular haemoprotein, and as an introduction to the structural and biochemical studies on haemoglobin described later.

The evolution of oxygen-carrying systems was a major step in the development of aerobic life on earth in overcoming the limitations to life of the low solubility of oxygen in water. The presence of haemoglobin in blood increases its oxygen carrying capacity by a factor of fifty. Haemoglobin also facilitates the removal of waste carbon dioxide from the blood. The binding of both of these adducts is a function of the haem moiety of the protein, and is achieved by a ligand bond to the sixth ligand position of the central iron atom. The binding of these ligands is stabilized by the three-dimensional structure of the surrounding apoprotein, in the presence of which the degree of binding is controlled over the physiological range of oxygen concentrations. This control is modified by the fact that haemoglobin exists as a non-covalently attached tetramer of four identical protein subunits. This tetrameric configuration allows co-operation of oxygen binding between subunits, and produces a sigmoidal curve of binding at increasing oxygen concentrations. By these means, haemoglobin is able to respond extremely sensitively to small changes in the concentration of available oxygen, and perform its role both as an acceptor and as a donor of oxygen in the different physiological compartments of the body [3].

The functions of the other haemoproteins are as various as the apoproteins to which they are bound. All, however, are involved in the binding either of oxygen or of reducing power in the form of electrons. Haemoproteins such as haemoglobin and the respiratory cytochromes act mainly as carriers of oxidation or reduction potential, while others such as the P-450 cytochromes, themselves use this potential to transform metabolic substrates. The subject of diversity and similarity of structure and function within the family of haemoproteins is one of great current interest, and it seems likely that as our knowledge of these proteins increases so will our understanding of the common themes in their make-up.

4.2.2.2 Biosynthesis. Haems are produced from the biosynthesis of protoporphyrin IX, in common with chlorophylls and bilins (see chapter 3 and section 4.3.2.2). In animals and most bacteria, the first committed step in porphyrin synthesis is the condensation of the amino-acid glycine with a metabolic precursor succinyl co-enzyme A, to form 5-aminolaevulinic acid (ALA). This reaction, carried out by the enzyme ALA synthase was characterized as long ago as 1958, and has become known as the classic Shemin pathway [4]. The next step in the biosynthesis of porphyrins is the formation of the monopyrrole porphobilinogen from two molecules of ALA by the enzyme porphobilinogen synthase. This step is followed by the cyclization of four molecules of porphobilinogen to produce the cyclic tetrapyrrole Uroporphyrinogen III. This intermediate contains six more hydrogen atoms than are found in porphyrins, the presence of which prevents conjugation of the electrons of the macrocyclic ring, rendering the porphyrinogens colourless.

Consecutive alterations to the constituents at the periphery of the porphyrinogen structure lead to the formation of protoporphyrinogen IX, which is then oxidized to produce the coloured protoporphyrin. This is a common precursor of both the haems and chlorophylls, and the pathway of conversion of ALA to form protoporphyrin IX is remarkably constant in the organisms in which it has been studied so far. A single enzyme, responsible for the chelation of iron into the centre of the porphyrin nucleus, results in the formation of free haem [5]. (This pathway is summarized in Figure 4.2.)

The free haem chromophore may then be specifically attached to its apoprotein in one of two possible ways [6]. Firstly several haemoproteins exist in which haem is covalently bound to the surrounding apoprotein to produce a highly stable pigment-protein complex. Among these are the *c*-type cytochromes, relatively ubiquitous in muscle tissue. In these structures, the two propionic acid substituents at positions 13 and 17 of the haem chromophore are esterified to amino-acid residues of the surrounding apoprotein, to form a stable covalent attachment. The remaining apoprotein structure is arranged around the chromophore to maximize hydrophobic and hydrophilic interactions between the two. This arrangement also presents a nitrogen atom from a histidine amino-acid of the apoprotein in the correct orientation to form a fifth ligand to the central iron atom of the haem, further stabilizing the complex.

In a second type of interaction found in *b*-type cytochromes and haemo-

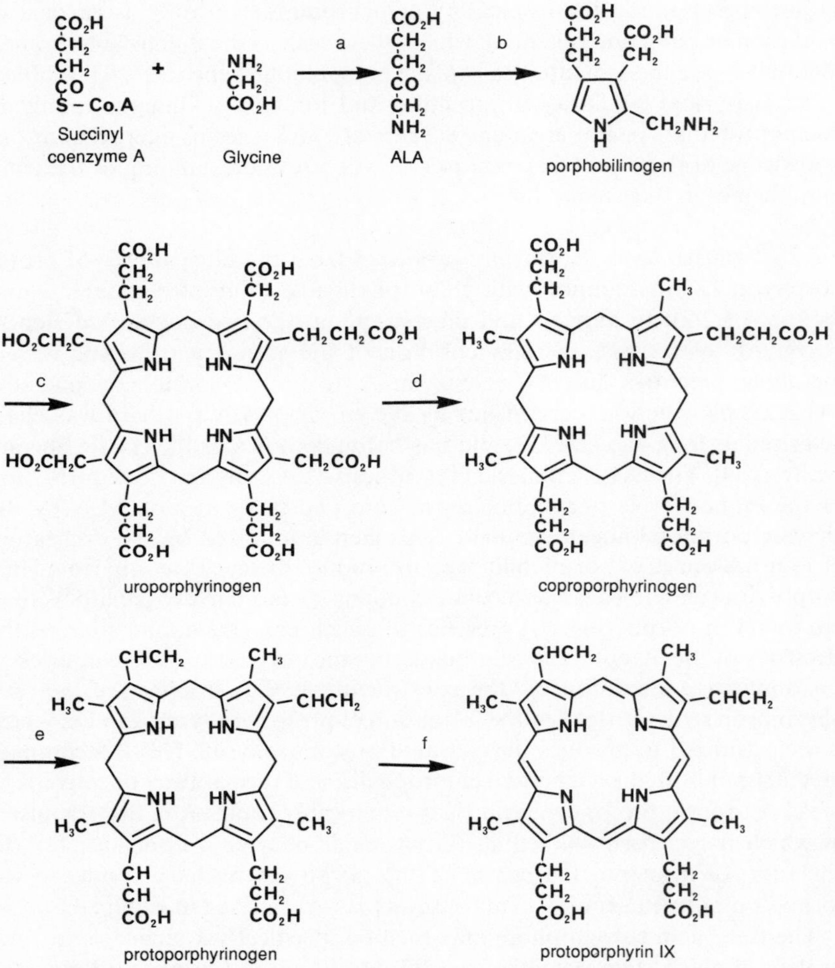

Figure 4.2 The biosynthesis of protoporphyrin IX from glycine and succinyl coenzyme A. The enzymatic steps shown are carried out by the following enzymes: (a) ALA synthetase, (b) ALA dehydratase, (c) uroporphyrinogen synthetase and cosynthetase, (d) uroporphyrinogen decarboxylase, (e) coproporphyrinogen oxidative decarboxylase, (f) protoporphyrinogen oxidase.

proteins such as haemoglobin and myoglobin, the chromophore protein complex is held together entirely by the non-covalent interactions described above, that is by hydrophobic and hydrophilic forces, and the presence of a ligand bond from a histidine nitrogen to the fifth ligand position of the iron atom in haem. This structure, although stable in nature, is considerably less resistant to chemical change *in vitro* than that of the *c*-type cytochromes, and this aspect of their chemistry will be discussed in section 4.2.3.1.

4.2.2.3 Metabolism. The metabolism of haem in mammalian tissues is a well-documented and active research area [7, 8]. In human beings and other vertebrates, endogenous haem is degraded by oxidative cleavage of the porphyrin ring to yield the linear tetrapyrroles biliverdin IX α and bilirubin IX α, in a process that releases carbon monoxide and iron (Figure 4.3). Ring cleavage of haem is a common step in the synthesis of both animal and plant bile pigments (see section 4.3.2.2), and is catalyzed by the enzyme haem oxygenase. The product of this reaction is the green pigment biliverdin IX α, a transient intermediate in the pathway of breakdown in mammals, and relatively unstable as a free compound. The transformation of this pigment by reduction to form the yellow bilirubin IX α is achieved by the enzyme biliverdin reductase, and is followed by conjugation to sugars to yield stable excretion products. The potential of animal bile pigments themselves as colorants and for medicinal purposes is worth mentioning here, and will be discussed further in section 4.2.3.3.

It should also be stated that the metabolism of ingested haem, as opposed to endogenous pigments, is most unlikely to occur via this pathway. It is probable that most ingested pigment chromophores—whether from haem, chlorophyll or bilins—pass through the human gut with few chemical changes and little or no absorption of nutrient. This has been proved extensively in the case of chlorophyll [9] except where degradation by light is possible, for example in the digestive tracts of transparent marine organisms close to the water surface. The metal-ion component of porphyrins may, however, be used—chlorophyll is reported as being a major source of magnesium in cows and haem is well-known as a valuable source of 'available' iron and, in this respect, must be entirely beneficial.

The protein component of the pigment-protein complexes will be digested in the normal way following ingestion and will itself have a nutritive value.

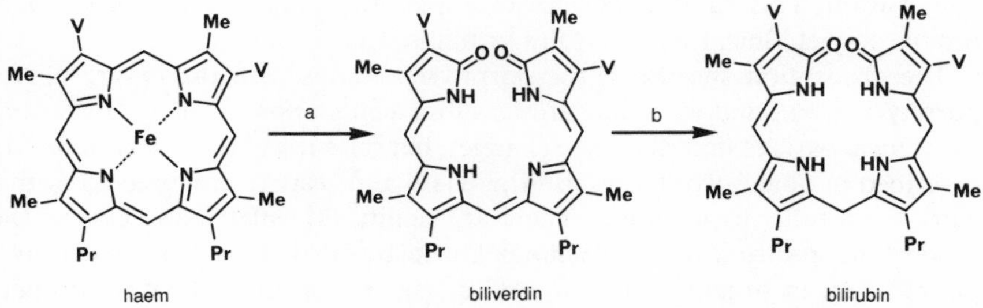

Figure 4.3 The degradation of haem to the bile pigments biliverdin and bilirubin. The reactions shown are carried out by the following enzymes: (a) haem oxygenase, (b) biliverdin reductase. V = $CHCH_2$, Me = CH_3, Pr = $CH_2CH_2CH_2OH$.

4.2.2.4 Physical properties. These can be summarized by their localization, absorbance and stability.

Localization. As has already been alluded to the distribution of haemoproteins both among species, and within a given organism is virtually comprehensive. Almost no single tissue within an animal is without its complement of haemoproteins, some soluble, some membrane-bound but all essential to the function of the organism. Some tissues, however, will obviously offer richer sources of haem pigments than others, and by far the greatest source of haem in any mammalian tissue must be that of haemoglobin. Comprising about 14 g in every 100 ml of blood, and readily available from abattoirs, it is tempting to overlook the possibility of extraction from other sources. Firstly, certain tissues such as spleen and liver have a very high haemoglobin content themselves, and methods for extraction from these tissues are sometimes more convenient than those for whole blood (see section 4.2.3.1). Secondly, the possibility exists of isolating different haem compounds from tissues, which may have advantages in terms of colour or stability, over haemoglobin itself. The *b*-type cytochromes, which are membrane bound and present generally in small concentrations in the tissues in which they are found, are unlikely to provide practical sources of haem in quantity. Also, in terms of stability and absorbance, they are rather similar in their properties to haemoglobin. The soluble *c*-type cytochromes may, however, offer certain advantages, and can be found in relatively high concentrations in specialized tissues. Thus, for example, heart muscle which is particularly rich in cytochrome *c*, could provide a source of highly stable, soluble haem of specific absorbance characteristics rather different to those of haemoglobin itself.

Absorbance. The absorbance properties of all porphyrins are generated by the conjugated system of double and single bounds that are characteristic of the molecules. Thus the aromatic porphyrin nucleus of haems and chlorophylls produces strongly coloured pigments with high extinction coefficients for their main absorption bands. In addition to the absorption bands in the visible region of the spectrum, there is also an intense band around 400 nm called the 'Soret' band, which is a characteristic of the macrocyclic ring conjugation. This band is lost, for example, following cleavage of the porphyrin, to yield linear tetrapyrroles or bilins [10].

The absorption spectra of the porphyrins varies with the nature of the porphyrin itself, and with the various substituents upon it. There is a strong theoretical basis to describe these changes, but the subject has been admirably described in qualitative terms elsewhere [10] and readers are directed to this work for a fuller explanation. There are significant differences between the absorption spectrum of say haemoglobin in blood with that of its chromophore haem in organic solution, and consideration of absorption properties *in vitro* will be given in section 4.2.3.1.

The absorbance characteristics of haems *in vivo* depend on the nature of the apoproteins to which they are attached, the means by which this attachment is achieved, and the presence of additional ligands binding to the central iron atom. Thus haemoglobin has a characteristic absorption spectrum similar to that of myoglobin, but relatively dissimilar to that of the *c*-type cytochromes [3]. In addition, the spectrum of haemoglobin even *in vivo* is dependent on the presence of oxygen (or carbon dioxide) occupying the sixth ligand position of the central iron atom. Thus partial pressures of oxygen sufficient to oxygenate all of the haemoglobin present will result in a shift of the absorbance to form two discrete maxima, one at higher and one at lower wavelength to that of the single peak of deoxyhaemoglobin (Figure 4.4). Under normal circumstances *in vivo*, a proportion of the haemoglobin would be present in both forms, and the observed spectrum would be a rather broad peak representing an average of the two states. It is normal practice for the purpose of quantification to measure the absorbance spectra of haems in the reduced form, or as a difference spectrum of the reduced minus the oxidized form. The absorbance characteristics of the *b*- and *c*-type cytochromes obtained by these methods are given in Table 4.2 for comparison with those of haemoglobin.

Stability. The stability of haems *in vivo* is of some interest, as in mammals there appear to be two separate pools of haem, subject to turnover at different rates. The first pool has a life-span of between 1 and 3 h and is thought to consist of haems present in cytochromes and the haem-containing enzymes. The second pool represents haemoglobin haem, and is remarkable in its long life-span of 60 days for rats and 120 days for humans [8], which illustrates the great natural stability of biomolecules *in vivo*. The task for the biochemist must be to equal this achievement in the laboratory.

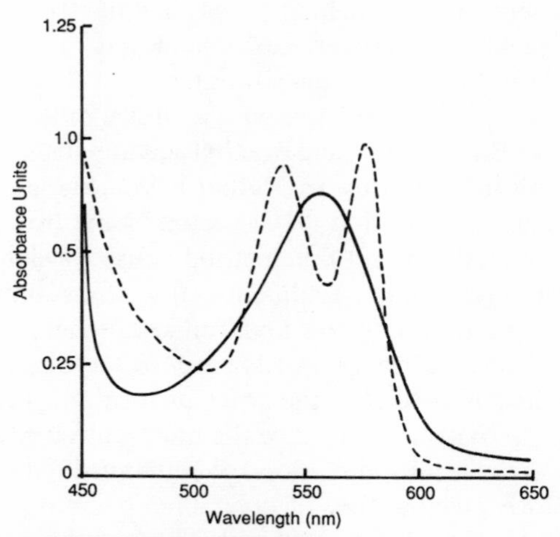

Figure 4.4 The visible absorption spectra of haemoglobin as oxyhaemoglobin (----) and deoxyhaemoglobin (——).

Table 4.2 Absorbance characteristics of globins and cytochromes [3, 111]

Name	Spectrum	λmax, λmin (nm)	EmM
Haemoglobin	reduced	555	12.5
	oxidized	535	14
		575	15
Myoglobin	as haemoglobin		
Cytochrome *b*	reduced-oxidized	559, 570	20
Cytochrome *c*	reduced-oxidized	550, 540	20
Cytochrome *a*	reduced-oxidized	605, 630	16

4.2.3 *Free haems*

4.2.3.1 Physical properties. These are divided into (i) extraction and purification, (ii) absorbance, (iii) stability, and (iv) quantitative analysis.

Extraction and purification. It should be mentioned here that the absorbance properties of haems are largely maintained during processes such as freeze-drying of tissues and of blood itself, without the necessity for extraction or purification procedures. As many of these sources are themselves foodstuffs, it may sometimes be appropriate to use the colorants directly from the crude product. The disadvantages of this approach are usually associated with problems of stability, and the associated changes in spectral properties. The advantages of increased stability and the reproduction of particular spectral characteristics often justify the use of extraction and purification, and these processes will now be described.

Extraction of haems from proteins such as haemoglobin, to which they are not covalently attached is relatively easily achieved by the use of acidified organic solvents [10]. Several biological materials such as blood and tissue samples may be treated this way, and with a choice of solvents appropriate to the proposed use. Ether-acetic acid or ethyl acetate-acetic acid are the most commonly used solvent mixtures, with ether having the advantage of being easier to remove by evaporation. HCl-acetone mixtures provide a more vigorous extraction medium and this method is used widely [11–14] for the separation of haem and globin. Acidified methylethylketone is another solvent suitable for extraction of haems and, unlike acetone, is immiscible with water. This allows direct extraction of haem into the organic phase leaving the undenatured protein behind in the aqueous medium [15]. Preparation of haem from blood has been known since the nineteenth century and is classically achieved by direct addition of blood to salt-saturated glacial acetic acid at 100°C to produce precipitation of crystalline haemin [16, 17]. For bulk preparations, use of acetone/acetic acid with 2% strontium chloride is recommended [18, 19] with yields of up to 80% being reported by this method.

Isolation of *c*-type cytochromes and other haems covalently bound to their proteins is best achieved as the whole pigment protein complex. Extraction of the protein in aqueous medium is possible, followed by separation by ion-exchange chromatography [20] to yield a relatively pure solution of cytochrome. Separation of haem from its covalently attached apoprotion in *c*-type cytochromes is rather difficult, and has only satisfactorily been achieved by the use of silver sulphate in hot acetic acid [21]. This is a reflection of the natural stability of cytochrome *c in vitro*, and in the preparation of colorants is actually advantageous.

In terms of the preparation of material for food use, the suitability of any of the methods so far described may be limited. However, the existence of a logical rationale for the extraction of haem from tissues, and the large body of scientific reference available must surely augur well for such techniques being developed in the near future.

Absorbance. We have already seen that the binding of oxygen to haemoglobin *in vivo* produces a shift in the observed absorption spectrum of the pigment. The binding of a number of alternative ligands is also possible *in vitro*. Molecules such as carbon monoxide, nitrous oxide, hydroxide and cyanide are all capable of forming ligand bonds to the central iron atom in place of oxygen. The overall effect of their presence on the spectrum of haem is dependent on the electronic nature of the ligand itself. Ligands capable of donating electrons to the delocalized system of the porphyrin ring will, in general, produce a shift to shorter wavelengths in the absorbance maximum of the α peak. Conversely, addition of electron withdrawing ligands should produce a shift to longer wavelengths [10].

The advantages of using such ligands to bind to haemoglobin are several. Firstly, irreversible binding of such a ligand may lead to increased stability of haemoglobin itself, and secondly it leads to the formation of a single absorbing form of the parent molecule. Thus the rather broad spectrum of haemoglobin described *in vivo* is transformed to a single absorbing peak at 540 nm by the binding of cyanide. Other ligands known to have similar effects—carbon monoxide, nitrous oxide and hydroxide groups have been characterized in the laboratory, and have some commercial potential. Thirdly, it may be possible to modify the wavelength of absorption by the use of different ligands to produce a group of pigments of varying absorption characteristics for different purposes. The haemin-anion complexes could lend themselves to this use as they are relatively easily prepared [22] and complexation with several different anions is possible. Table 4.3 summarizes the effects of several known ligand substitutions on the absorbance characteristics of haems.

The absorption properties of free haems without their associated apoproteins are well documented [23] and that of free protohaem is included in Table 4.3 for comparison. The spectra of free haems in aqueous solution is known to be dependent on several factors that affect their aggregation [24].

Table 4.3 Effects of ligand substituents on the absorbance characteristics of haems [10, 23]

Name	5th and 6th ligands	Solvent	λmax	EmM
Haem	H_2O, H_2O	pH 7.0 to 12.0 buffer	570–580	5.5–6.5
Haematin	OH, H_2O	pH 10 buffer	600	4.5
		0.1 M NaOH	610	4.6
			385	58.4
Haemin	Cl, H_2O	Ether	638	
		Acetic acid	635	
		Glacial acetic acid	510	90
		Ethanol	400	
Cyanmet-haemoglobin	CN, H_2O	buffer	540	44
Protohaemochrome	Pyridine, pyridine	0.2 M NaOH	557	34.4
		25% pyridine	525	17.5
Cytochrome *b*-haemoglobin-myoglobin-haemochromes	As protohaemochrome			
Cytochrome *c*-haemochrome	Pyridine, pyridine	0.2 M NaOH	550	29.1
		25% pyridine	522	18.6

These factors include the concentration of the haem itself, the pH, ionic strength, and temperature of the solution and the presence in it of several cations. In the case of protoferrihaem, over 95% of the molecules may exist as dimers in aqueous solution, and at pHs greater than 11.0 even higher aggregates may form. The dimeric structure has been shown to be joined by an oxygen molecule bridging the two iron atoms of the haems, and results in a pronounced decrease in the intensity of the Soret band. There is also an increase in the visible region of the spectrum of a component at around 360 nm. Although this may have relatively little effect on the perceived colour of haems in aqueous solution, it is nevertheless important to remember when using absorption spectra to characterize or quantify haems; indeed, this is one of the reasons why absorption of free haems is usually used only in conjunction with other methods for this purpose.

Stability. The relative stability of *b*- and *c*-type haemoproteins has already been discussed with reference to their extraction and purification procedures. The effect of ligand binding in increasing the stability of haems has also been mentioned. It is a fact of nature that haemoglobin, whose function it is to transport oxygen, is itself susceptible to oxidative breakdown. Paradoxically, oxygen itself protects the haem from oxidative attack, which appears to occur only in that fraction of molecules that at any one moment do not have bound oxygen [25]. For this reason, the use of ligands that are capable of binding irreversibly to the haem molecule, protects it from oxidation, as would the binding of oxygen itself were it not in dynamic equilibrium with its unbound form.

This principle underlies the relative lack of stability in crude blood or tissue extracts, where in addition the presence of endogenous proteolytic enzymes may lead to the breakdown of the haemoprotein itself. Progressive breakdown of apoprotein is probably the single most important cause of loss of spectral characteristics in ageing blood or tissue preparations. This problem may be overcome either by purification of the haemoprotein or by simple extraction of the haem moiety itself from the sample, by the methods already described. Several adaptations of these methods have been made for stabilization of haems for commercial use, and these are discussed in section 4.2.3.2.

Quantitative analysis. The property of haems in producing single sharp absorption bands on binding of certain ligands, is used to effect in their quantitative assay. The extinction coefficients given in Table 4.3 for the various haem-ligand complexes may thus be used for quantitative analysis. One ligand complex has, however, proved particularly useful for qualitative and quantitative analysis of haems from a variety of sources. That is the use of their dipyridine complexes [23] known as a dipyridine ferrohaemochrome or pyridine haemochromogen. Sodium dithionite ($Na_2S_2O_4$) is used to reduce the haem to its Ferro (FeII) form, while sodium hydroxide is required to disrupt the interaction of the protein with the fifth ligand position of iron in haemoproteins. The dipyridine complex formed under these circumstances results in a reproducible spectrum characteristic of the individual haem compound. Table 4.3 also gives the absorption maxima and extinction coefficients of several examples of these.

As the pyridine haemochromogen method for quantification of haems is simple, cheap and suitable for haems from almost any source, it is rarely necessary to use more modern techniques for quantification. Such techniques do, however, exist and have been used mainly in the structural determination and analysis of natural products on a milligram scale. High-performance liquid chromatography (HPLC) techniques have been developed for the separation of porphyrins both as the naturally occurring free acids and as their methyl ester derivatives [26]. Despite the central importance of these techniques to clinical and biochemical studies of porphyrins, it seems likely that, in situations where microscale analysis is not required, the older (and less expensive) techniques will prevail.

4.2.3.2 Commercial exploitation. Some of the physical properties and chemical characteristics, capable of being used for the commercial exploitation of haems, have already been alluded to. It seems pertinent here to review the processes and patents already in existence as commercial propositions and the scientific basis of their use.

Several patents covering the use of haem compounds exist [27], and may be considered to fall into three main classes: (i) those dealing with preparations of unmodified whole blood or tissue products; (ii) several concerned with stabilization by binding of different ligands to crude samples; and (iii) a few dealing with the more complicated question of purification. Some processes under patent for the preparation of apoprotein from haemoprotein complexes have been reported to be suitable also for recovery of haem but this is often not the case, for example following decolorization with peroxide or active oxygen, which causes oxidative cleavage of the haem macrocycle.

Patents covering the use of whole blood or tissue slurries are generally rather limited in application, although both can be spray-dried for direct use in foodstuffs. Concentration of haem from blood or red blood cells is possible by a simple process of dehydration using alcohols. Several haem derivatives are described including methods for ozonolysis, and formation of carbon monoxide-, nitrous-oxide-, hydroxy- and carboxy-adducts, all of which serve to stabilize the haem content of the source without extraction from it.

Simple methods for extraction have also been covered, *e.g.* release of haem from haemoglobin by acid or alkali treatment, enzyme hydrolysis or heating. Solvent extraction is rarely used, presumably because of the expense and the need to remove solvents prior to addition to food. One process that uses imidazole to promote separation of haem and globin is presumably similar to the action of pyridine in occupying the fifth and sixth ligand position of the haem iron and thus disrupting its association with the apoprotein.

It is generally only possible to discern from a patent whether the product intended for use is in the form of a haemoprotein or free haem derivative, by examination of the processes described within it. While this may be an interesting academic pastime, it seems reasonable to expect that prospective users of the products should have available very clear information on this, and on the nature of the bound ligands. Accurate dissemination of such information would surely do much to allay current fears of 'artificial' derivatives of these natural pigments.

Nature's own haem derivatives—the bile pigments bilirubin and biliverdin—are not without their commercial uses. Obtained from gall stones and bezoar (hair balls), bilirubin has been used for centuries in Chinese medicine and is reputed to have aphrodisiac properties. Demand for these products in modern China now greatly outstrips the availability from natural sources, and the opportunity for commercialization of a 'nature alike' pigment is now open.

4.3 Phycobilins

4.3.1 Introduction

Phycobilins are deeply coloured, fluorescent, water-soluble pigment protein complexes. They are characteristic proteins of blue-green, red and cryptomonad algae, and represent major biochemical constituents of the organisms in which they are found. The phycobilins can be classified on the basis of their spectral characteristics into three major groups: phycoerythrins (PEs), phycocyanins (PCs), and allophycocyanins (APCs). The phycoerythrins are red in colour with a bright orange fluorescence, while the phycocyanins and allophycocyanins are blue and fluoresce red. The range of colours found within these groups of pigments is large, dependent on the source of the biliprotein and the medium in which it is isolated. These pigments thus provide a variety of natural colours, distributed ubiquitously within the algal kingdom, and with the potential for future commercial development. This section describes the known chemical and biochemical properties of the biliproteins and aims to assess their future commercial and industrial potential.

4.3.1.1 Structure and nomenclature. All biliproteins are based on a structure consisting of a linear tetrapyrrole or bilin, covalently bound to an apoprotein. The structure of these algal bilins is very similar to that of mammalian bile pigments, and their nomenclature follows the same pattern, as shown in Figure 4.5a [1]. The chromophore of the blue phycocyanin and allophycocyanins is the same in both groups of pigments, and is called phycocyanobilin. Its structure is shown in Figure 4.5b. The chromophore of the red phycoerythrins differs from phycocyanobilin in two respects: phycoerythrobilin has two extra atoms of hydrogen, forming a 15,24-dihydrobilin, and also has an ethyl rather than a vinyl group at position 18 of the bilin structure (Figure 4.5c). The chromophores released by methanol

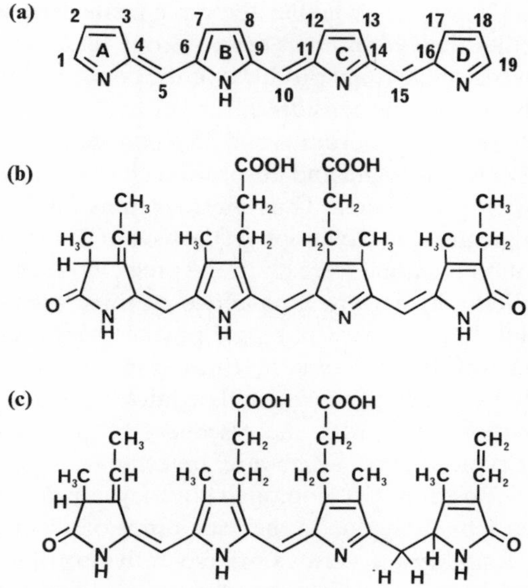

(a) 1 2 3 A 4 N 5 6 7 8 B 9 N H 10 11 12 13 C 14 N 15 16 17 18 D N 19

(b) COOH COOH CH3 CH2 CH2 CH3 H3C CH H3C CH2 H2C CH3 H3C CH2 H O N H N H N N H O

(c) COOH COOH CH3 CH2 CH2 CH2 H3C CH H3C CH2 H2C CH3 H3C CH2 H O N H N H N H H H N H O

Figure 4.5 The structures of: (b) phycocyanobilin, (c) phycoerythrobilin and (a) their common bilin nucleus. The numbering of the bilin structure is according to the IUB-IUPAC recommendations [1].

hydrolysis of the pure biliproteins are spectrally distinct to those produced by HCL cleavage of the proteins [28]. Evidence from nuclear magnetic resonance and mass spectroscopic studies of the pigments obtained by methanolysis led to the structures described in Figure 4.5 [29]. The phycocyanobilin chromophore released by methanolysis is identical to that excreted by *Cyanidium caldarium* following exogenous administration of ALA [30]. It is also known that bilin cleaved by methanol can be re-incorporated into phycocyanin in *C. caldarium*, providing good evidence that its structure represents that of the true biliprotein chromophore [31]. Some similar evidence exists in the case of phycoerythrobilin. When red algae are eaten by the marine gastropod mollusc *Aplysia* their phycoerythrin chromophores are released by digestion and excreted as free-bile pigments. The structure of this pigment is reported to be identical to that of phycoerythrobilin obtained by methanol hydrolysis [32].

The simplest interpretation of the known data is that the two structures shown represent the only prosthetic groups of the algal biliproteins—the wide spectral differences observed among the biliproteins being a result of their different protein environments. Thus phycocyanobilin is the only chromophore found in C-phycocyanin, cryptomonad phycocyanin and allophycocyanin, and only phycoerythrobilin is obtained from C-, R-, B- and cryptomonad phycoerythrins. R-phycocyanin remains the exception to this rule and releases both phycoerythrobilin and phycocyanobilin on methanol cleavage [33, 34]. Other chromophores have been reported from time to time by several authors [28, 35–39], whether these pigments arise as artifacts of extraction, or whether they are present *in vivo* is not yet clear. They do not, however, represent major cleavage products of the common biliproteins and for that reason they will not be considered further here.

Historically, individual phycocyanins and phycoerythrins have been classified according to their algal origin and absorption characteristics (see Bennet and Siegelman [29] for a review). Thus phycocyanins and phycoerythrins isolated from blue-green algae (Cyanophyta), which both have single prominent visible absorption maxima, were given the prefix 'C-', while phycobilins from red algae (Rhodophyta), which have two or three such maxima were prefixed by 'R-' [40]. Later a 'new' red algal phycoerythrin was found with two major visible absorption maxima and was given the prefix 'B-', as the original source was a red alga of the order Bangiales. A final addition to this system of classification came with the discovery of phycobilins in cryptomonad algae (Cryptophyta), which are unicellular flagellate algae of indefinite taxonomic position. Cryptomonad phycocyanin and phycoerythrin have been classified separately due to the wide range of absorption maxima of the phycobilins found in the various representatives of this family. Allophycocyanin (APC) is rather simpler to classify, as there are no spectral differences between allophycocyanins isolated from red or blue-green algae, and it has not yet been characterized in cryptomonad species.

This nomenclature is still fairly widely used, although it is cumbersome, and its usefulness is limited by several exceptions to the taxonomic rules. The eight classes of phycobilins described by this system of classification are listed in Table 4.4, along with their characteristic absorption and fluorescence maxima.

The great diversity of absorption maxima observed for the biliproteins in general is a function of the different apoproteins to which the chromophores are attached, and their physico-chemical environment within the organism. The attachment of the bilin chromophore to its apoprotein is particularly stable and is achieved by means of one or two covalent thioether bonds. The first of these is between a cysteine amino-acid residue of the apoprotein and an unusual ethylidene group at position 3 of the bilin. A second covalent linkage is also possible, between another cysteine residue and the vinyl group at position 18 of the bilin chromophore [41]. The particular properties that this covalent attachment confers on the stability of the biliproteins will be discussed in section 4.3.3.1.

The native forms of many algal bilins consist of several bilin chromophores attached to one of two apoproteins, to produce distinct biliprotein subunits designated α and β. These subunits are present in the organism in different aggregated forms, as part of the architecture of the light harvesting apparatus of photosynthesis. This aspect of bilin structure will now be discussed.

Table 4.4 Classification of phycobilins and their characteristic absorption and fluorescence maxima [29, 65]. NR = not recorded

Phycobiliprotein	Absorption λmax (nm)	Fluorescence emission λmax (nm)
C-phycoerythrin	280 308 380 565	577
R-phycoerythrin	278 308 370 497 538	578
	278 308 370 497 556	
B-phycoerythrin	278 307 370 497 545	578
	278 307 370 497 563	
Cryptomonad phycoerythrin	275 310 370 544	NR
	275 310 370 556	580
	275 310 370 568	NR
C-phycocyanin	280 360 620	654
	280 360 615	647
R-phycocyanin	275 355 522 610	637
	275 355 533 615	637 565
Cryptomonad phycocyanin	270 350 583 625 643	600
	270 350 588 625 630	NR
	270 350 588 625 615	637
Allophycocyanin	280 350 598 629 650	663

4.3.2 Phycobilins in nature

4.3.2.1 Function and occurrence. Phycocyanin and phycoerythrin function as major light-harvesting pigments of the photosynthetic algae. In these organisms, phycobilins exist alongside chlorophyll and act in concert with it to collect and transform light energy, by means of photosynthesis, into the chemical energy of the cell. The broad absorption maxima of phycobilins in their native forms allow algae to trap light energy at wavelengths over a large proportion of the spectrum of sunlight. The absorption spectra of phycocyanin and phycoerythrin effectively fills the optical gap left by the absorbance of chlorophyll.

Table 4.5 shows the occurrence of phycocyanins and phycoerythrins among

Table 4.5 Distribution of biliproteins in some algae of the classes Cyanophycae, Rhodophycae and Cryptophycae [49, 75, 80]

Species containing predominantly phycoerythrin, with trace amounts of phycocyanin	Species containing both phycoerythrin and phycocyanin	Species containing phycocyanin, with no phycoerythrin
Cyanophyceae (contain C-PE)	***Cyanophyceae*** (contain C-PE & C-PC)	***Cyanophyceae*** (contain C-PC)
Hydrocoleum spp.	*Aphanocapsa* spp.	*Anabaena cylindrica*
Oscillatoria rubescens	*Aphanothece sacrum*	*Anabaena variabilis*
Phormidium autumnale	*Calothrix membranacea*	*Anabaena 6411*
Phormidium ectocarpii	*Dermocarpa violacea*	*Anabaenopsis* spp.
Phormidium fragile	*Fremyella diplosiphon*	*Anacystis nidulans* (=*Synechococcus 6301*)
Phormidium persicinum	*Glaucosphaera vacuolata*	*Arthrospira maxima*
Phormidium uncinatum	*Schizothrix alpicola*	*Aphanizomenon flos-aquae*
Pseudanabeana spp.	*Tolypothrix distorta tenuis*	*Calothrix scopulorum*
Rhodophyceae (contain B-PE)	***Rhodophyceae*** (contain B-PE & R-PC)	*Chlorogloea fritschii*
Rhodella violacea maculata	*Porphyridium aerugineum*	*Lyngbya lagerheimii*
Cryptophyceae (contain cryptomonad PE)	*Porphyridium cruentum*	*Mastigocladus laminosus*
Cryptomonas ovata maculata	*Porphyridium marinum* (=*purpureum*)	*Nostoc punctiforme*
Hemiselmis rufescens	***Cryptophyceae***	*Oscillatoria amoena*
Plagioselmis prolonga	Unknown	*Oscillatoria animalis*
Rhodomonas lens		*Oscillatoria aghardii*
		Oscillatoria chalybea
		Oscillatoria formosa
		Oscillatoria minima
		Oscillatoria subbrevis
		Oscillatoria tenuis
		Phormidium faveolarum
		Phormidium luridum
		Plectonema boryanum
		Plectonema calothricoides
		Spirulina platensis
		Rhodophyceae (contains C-PC)
		Cyanidum caldarium
		Cryptophyceae (contain cryptomonad PC)
		Chroomonas spp.
		Cryptomonas cyanomagna
		Hemiselmis virescens

the three algal classes in which they occur. It illustrates both the variation between species, and also the possibility of identifying species that predominantly synthesize individual pigments.

4.3.2.2 Biosynthesis. Plant bilins are synthesized by ring cleavage of haem to form a linear tetrapyrrole, in exact analogy to the formation of bile pigments in mammalian systems following haem breakdown [42]. The biosynthesis of chlorophyll (as discussed in chapter 3) occurs in parallel to that of bile pigments, so that in organisms using both chlorophyll and biliproteins for photosynthesis, a single branched pathway is responsible for the biosynthesis of both pigments. This pathway is outlined in Figure 4.6.

As has been described previously for mammalian haem synthesis, the biosynthetic pathway of haem in plants has the substituted amino acid 5-aminolaevulinate (ALA) as its first committed intermediate. However, in contrast to the mammalian pathway plants, algae and several photosynthetic bacteria are now known to produce ALA from the intact 5-carbon skeleton of glutamate, rather than by the condensation of glycine and succinyl co-enzyme A [43]. This process is presently the subject of intense interest, as the formation of ALA is likely to be a key control point in the biosynthesis of both chlorophyll and the bilins in these photosynthetic organisms. In particular, it appears that light control of pigment synthesis is likely to be exerted at this point, although the mechanisms by which this occurs are not yet known.

The next phase in the biosynthesis of phycobilins is the formation of haem. The process by which this occurs is essentially identical to that known for the biosynthesis of haem in mammalian tissues (see section 4.2.2.2), and will not be described further here. It is, in fact, remarkable that the process of tetrapyrrole biosynthesis is so strictly conserved over the range of organisms in which it occurs.

The formation of phycocyanobilin and phycoerythrobilin following the ring cleavage of haem, is believed to occur through the intermediates biliverdin IX α and phytochromobilin [31] (see Figure 4.6) with the two algal bilin chromophores being produced by the action of one of two possible reductions of the phytochromobilin precursor. This pathway is particularly attractive as it provides a universal means by which the major algal bilins may be synthesized, it also incorporates the chromophore of phytochrome, the higher plant bile pigment responsible for light perception (described in chapter 3), as an intermediate. Indeed, it is proposed that the general pathway outlined here for the synthesis of algal bilins exists throughout the plant kingdom for the biosynthesis of phytochrome.

The final step in the biosynthesis of all bile pigments is likely to be the attachment of the bilin chromophore to its apoprotein [31]. Although little is known about the enzymology of this process, it is clear that in most organisms strict mechanisms exist for the co-ordinate regulation of apoprotein and chromophore synthesis. As a result, the accumulation of free chromophore

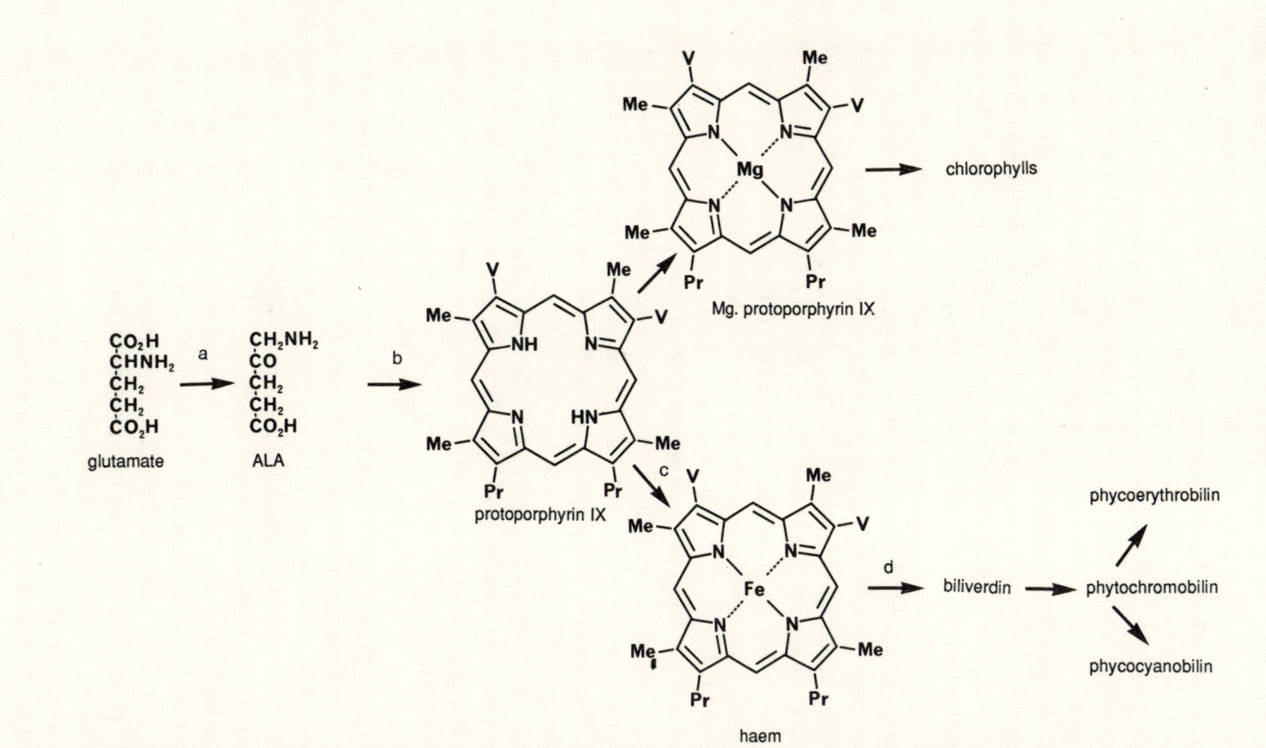

Figure 4.6 The branched biosynthetic pathway for the formation of haem, bilins and chlorophylls from glutamate. The enzymes responsible for the biosynthetic steps shown are: (a) glutamic acid tRNA ligase, glutamyl tRNA dehydrogenase, and glutamate 1-semialdehyde aminotransferase, (b) as described in Figure 4.2, (c) ferrochelatase, (d) as described in Figure 4.3. V = $CHCH_2$, Me = CH_3, Pr = $CH_2CH_2CO_2H$.

within an organism occurs only when the normal metabolic state of the organism is altered. This will be discussed in Section 4.3.2.3.

The converse situation in which free apoprotein is produced without concomitant synthesis of the appropriate chromophore, is not known to occur under any conditions. It would appear from this that formation of the chromophore is not limiting in biliprotein synthesis [44]. This is supported by results recently obtained using cloned genes for phycocyanin apoproteins. The genes for the α and β subunits of phycocyanin from the blue green alga *Synechococcus* spp. PCC7002 were cloned [45] and re-introduced into *Synechococcus* at higher multiplicity. This resulted in over expression of the apoproteins and the accumulation of corresponding greater amounts of phycocyanin [46]. The results suggest that cells are able to synthesize sufficient free bilin to accommodate the increased demand for pigment, and this opens up the possibility of genetic manipulation of algal strains for increased pigment production.

In addition to the major plant and algal bilins already described, there have, over several years, been reports of unusual bilin-like pigments from other sources. A blue bile-pigment based on the parent structure of uroporphyrin has been described in two bacteria—*Clostridium tetanomorphum* and *Propionibacterium shermanii* [47]. This pigment has visible absorption maxima at 369 and 644 nm. However, such pigments are not major products of the organisms concerned and will not be considered further here.

4.3.2.3 Metabolism. This is divided into: (i) chromatic adaptation; (ii) sun to shade adaptation; (iii) effectors of bilin synthesis; and (iv) pigment mutants.

Chromatic adaptation. One of the characteristic properties of many algal species is their ability to produce different proportions of individual light absorbing pigments in response to the different qualities of light in which the organism is growing. This photoregulated response of pigment synthesis is known as chromatic adaptation. Typically red light at wavelengths above 600 nm promotes the production of phycocyanin and allophycocyanin but not phycoerythrin, while green light of 500 to 600 nm promotes formation of phycoerythrin alone [48). In nature, this allows the algae to make the most efficient use of light at the available wavelengths, with the minimum expenditure of metabolic energy in synthesizing pigments.

This characteristic has been known for some time [49], and has been analysed in detail in several different species. In the filamentous blue-green alga *Tolypothrix tenuis*, for example, the production of bilins is known to be specifically driven towards phycoerythrin by light of 541 nm wavelength and towards phycocyanin by illumination at 641 nm [50]. It is thus possible to maximize production of individual bile pigments by using the organisms natural mechanisms for adapting to light quality. It is of particular interest that the genetic basis of photoregulation is beginning to be unravelled [51] as this may offer further possibilities for the future manipulation of pigment synthesis in algal species.

Two other metabolic influences on the overall level of bilins are important. The accumulation of any metabolite is dependent on its net rate of biosynthesis and degradation within the cell. Although specific degradation of phycocyanin is known to occur in several species in response to nitrogen starvation [52], so far as is known adaptation to changes in the wavelength and intensity of light does not involve specific degradative processes [53] and decreases in individual bilin concentrations within the cell are usually considered to be a result of dilution following cell division. This is consistent with studies on exponentially growing cells, which report the remarkable stability of phycobilins *in vivo* [53, 54].

In many algal species when cells are grown under conditions in which nitrogen supply is non-limiting, phycobiliproteins comprise a significant proportion of the total cellular protein. Moving such cells to conditions of limiting nitrogen results in the specific degradation of these pigments. This has been shown to occur in several cyanobacterial species [51, 55]. A similar effect is also known following depletion of sulphur compounds from the growth medium of cyanobacteria [56], although this seems surprising as C-phycocyanin has a rather low ratio of sulphur to nitrogen compared with many proteins. These characteristics may, however, be used to increase pigment synthesis under certain conditions, as it has been observed that administration of sulphate following a period of starvation leads to a rapid and preferential increase in C-phycocyanin content prior to resumed cell growth.

Sun to shade adaptation. Just as algae are able to adapt to the quality of light available for photosynthesis, synthesis of phycobiliproteins is also affected by light intensity. Increasing the light intensity causes an overall decrease in phycobiliprotein concentration, as part of an adaptation mechanism in which photosynthesis is maintained at its maximum rate using the minimum pigment necessary for light harvesting and energy transduction. Reductions of up to three-fold in phycobiliproteins were reported in the unicellular rhodophyte *Porphyridium cruentum* [57] in response to increasing light intensity. Similar reductions were observed in the blue-green alga *Anacystis nidulans* [58] in which it was concluded that light intensity was one of the primary factors affecting pigment synthesis as a whole [59].

Effectors of bilin synthesis. Addition of several substrates and intermediates of tetrapyrrole metabolism is known to induce bilin chromophore synthesis [31]. Addition of aminolaevulinic acid, as a committed intermediate in bilin synthesis, is found *not* to increase overall levels of biliprotein, but instead causes the excretion of the free bilin chromophore. The addition of excess glutamate would be expected to have a similar although less-marked effect as it is a substrate for several metabolic processes other than porphyrin synthesis. It appears that the co-ordinate control of chromophore and apoprotein synthesis prevents increased biliprotein formation in response to increased levels of bilin alone; this suggests that efforts to specifically increase biliprotein content would be more logically aimed at promoting apoprotein formation, as the rate of bilin biosynthesis may respond to the availability of apobiliprotein [44].

Pigment mutants. One means by which pigment synthesis is controlled very strictly, if artificially, is by the use of certain mutations in which biosynthesis of an individual pigment is blocked. In the case of the unicellular red alga *Cyanidium caldarium*, several such mutants exist [60]. Like the wild-type cells, these mutants are capable of growing heterotrophically in the presence of glucose in the dark. On transfer of dark grown cells to the light, they synthesize pigments characteristic of their specific mutation as summarized in Table 4.6. The existence of these mutants demonstrates the possibility of using the independent functions of the branched biosynthetic pathway to produce individual pigments.

4.3.2.4 Physical properties. Cellular localization, absorbance and fluorescence, and stability of the phycobilins in nature will be discussed here.

Cellular localization. In red and blue-green algae phycobilins participate in the overall process of photosynthesis in the form of large multimeric aggregates called phycobilisomes. These structures are present in the chloroplast of the algae, and appear as discrete morphological features on the outer face of the thylakoid membrane of the chloroplast, when viewed under an electron microscope. The size, composition, structure and function of the phycobilisomes from several organisms have now been studied [61–63], and much is known about the mechanisms by which light energy is absorbed and transferred within their structure. In cryptomonad algae phycobilisomes are absent, instead, the biliproteins are aggregated within the intrathylakoid spaces of the chloroplasts. This has a similar effect of producing closely packed pigments, and thereby enhancing their energy transfer efficiency.

The phycobilisomes of red and blue-green algae vary in their composition according to several factors. These include the organism from which they are isolated and the metabolic and spectral conditions in which the organism was grown. A model structure for a phycobilisome has been proposed; it allows for the variation in pigments found in different organisms and accounts for the known characteristics of energy transfer between pigments within the photosynthetic apparatus (Figure 4.7).

The basic phycobilisome building block is a monomer containing two dissimilar apoprotein subunits α and β. The α subunit contains one chromophore per apoprotein, and the β subunit has two [64]. The model proposed consists of a core composed primarily of allophycocyanin, present as trimers of $\alpha\beta$ subunits. As allophycocyanin is thought to be present in trace amounts in all species [65], it is probable that this core is a universal structure on which all phycobilisomes are built. The core is surrounded by several rod structures composed of hexamers of phycocyanin and phycoerythrin $\alpha\beta$ subunits. The

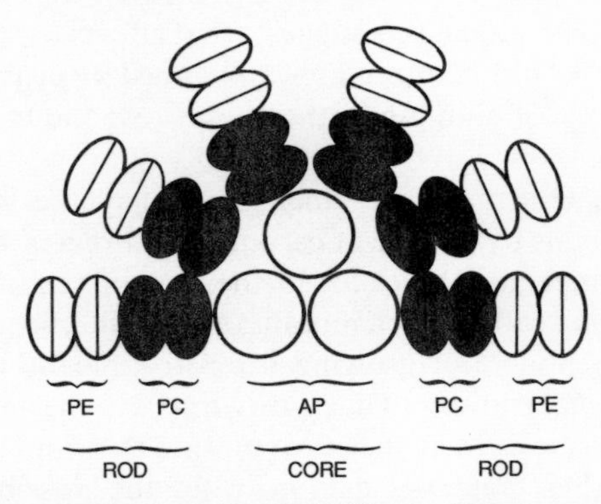

Figure 4.7 Schematic representation of a phycobilisome. The core structure consists of allophycocyanin (AP) associated with specific linker polypeptides. The rods contain phycocyanin (PC) and phycoerythrin (PE) hexamers, and their linker polypeptides.

whole structure is stabilized by the presence of associated linker peptides, which also serve to modulate the absorption characteristics of the pigments within the phycobilisome. The model illustrated is specifically proposed for the alga *Synechocystis* 6701 [66], but it is possible to adapt it for organisms containing different proportions of phycocyanin and phycoerythrin by substituting the pigments in their different proportions in the rod segments of the structure.

The overlapping absorption and emission spectra produced by the biliproteins in the environment of the phycobilisome, result in a one-way flow of light energy from pigment to pigment in the following order:

$$\text{Phycoerythrin} \rightarrow \text{Phycocyanin} \rightarrow \text{Allophycocyanin} \rightarrow \text{Chlorophyll}$$

The overall efficiency of this process is over 90% [67], making the phycobilisome one of the most efficient structures for transducing light energy in nature.

Absorption and fluorescence. The spectrum of an intact phycobilisome is a composite representation of its component chromophores and the individual associations and energy coupling of the overall structure. Representative absorption and fluorescence spectra of phycocyanin and phycoerythrin contained within the intact phycobilisomes of *Fremyella diplosiphon* [61] are shown in Figure 4.8. The diversity of spectra obtained from the photosynthetic apparatus of an organism is a reflection of the different environmental factors affecting pigment accumulation, as described in section 4.3.2.3. Definitive spectra of individual biliproteins are only obtained following isolation from *in vivo* structures, and these are described in section 4.3.3.1.

Stability. The stability of phycobilins *in vivo* has already been mentioned,

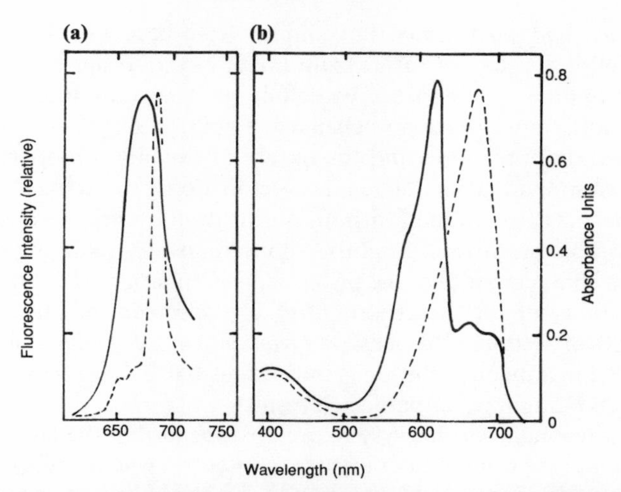

Figure 4.8 Fluorescence emission (a) and absorbance (b) spectra of intact phycobilisomes from *Fremyella diplosiphon* grown under differing light conditions. Cells grown under red light (----) contain phycocyanin and allophycocyanin (λ max 620 nm and 650 nm respectively), while cells grown under green light (——) also contain phycoerythrin (λ max 565 nm).

with regard to the degradation and turnover of pigment during growth under different metabolic conditions. Among the characteristics of the algae that use phycobilins as light-harvesting pigments, is the ability of phycobilisomes to function under a variety of environmental stresses. Thus algae that are classified as thermophilic (high-temperature tolerant), acidophilic (acid-pH tolerant), halophilic (high-salt tolerant) or psychrophilic (low-temperature tolerant) might be expected to show different characteristics with regard to the stability of their pigments. Published data on the phycocyanins of several blue-green and red algae [68] indicate that resistance to denaturation varies from one species to another. Using denaturation by urea as a model system, greatest stability to denaturation was shown in three thermophilic organisms tested—*Mastigocladus laminosus*, I-30-m and NZ-DB2-m, and the acid-tolerant unicellular red alga *Cyanidium caldarium.* The stability of phycocyanins from the halophilic organism *Coccochloris elabens* was also reported as being enhanced compared with those of organisms growing at room temperatures and lower. The authors suggest that the increased stability of the different phycocyanins is produced directly by differences in the primary amino-acid structure of the proteins.

4.3.3 Free phycobilins

4.3.3.1 Physical properties. Extraction and purification, absorbance and fluorescence, stability and quantitative analysis will be considered here.

Extraction and purification. Phycobilins have been extracted from algae and purified in a range of forms from intact phycobilisomes to the protein-free chromophore. These forms, by definition, have very different properties of absorption, fluorescence and chemical stability, thus it is useful to define the purification procedures and the nature of the bilin products before discussing their individual charactersitics. Before doing this it may be noted that phycobilins may be obtained without purification by simple freeze-drying of algal cells. This produces a brightly coloured powder retaining many of the absorption characteristics of the phycobilins in their native form and without loss of stability of the intact structure. The spectrum of phycocyanin and chlorophyll in freeze-dried cells of *Cyanidiuum caldarium* that have been resuspended in aqueous solution is indistinguishable from that of whole cells *in vivo* (J.D. Houghton, unpublished result).

The disadvantages of this system are obvious in that the colours obtained are mixtures of the total cellular pigment contents—usually bilins, chlorophyll and carotenoids. One possible solution to this could be by use of the pigment mutant GGB (see Table 4.6). This mutant is capable of growing heterotrophically in the dark using glucose as a carbon source at rates comparable with those of wild-type cells. On transfer to light, the mutant synthesizes phycocyanin without any detectable synthesis of chlorophyll *a*, and the cells acquire a pale-blue coloration. The amount of phycocyanin accumulated by the mutant is over 80% of that achieved by wild-type cells [69].

Traditionally the problem of purity has been approached by successive separation of phycobilins from the associated cellular pigments and proteins. This process is aided by the fact that phycobilins are very readily soluble in water. Following cell breakage it is therefore possible to separate the phycobilins from the lipid soluble chlorophylls and carotenoids by high-speed centrifugation of the homogenate. This process yields brightly coloured solutions of soluble protein, of which 40% may be phycobilin. In this form, the phycobilins are probably present as aggregates, derived from partial breakdown of the phycobilisome structure (Table 4.7).

Table 4.6 Pigmentation of *Cyanidium caldarium* mutants transferred to light following heterotrophic growth [69, 75][a]

Cell type	Ch 1*a*	PC	APC
Wild type	+(100)	+(100)	+
III-D-2	+(145)	+(140)	+
III-C	+(55)	−(0)	−
GGB	−(0)	+(83)	+
GGB-Y	−(0)	−(0)	−

[a] Numbers in parentheses represent percentage of pigment produced in mutant compared with wild-type cells.

Table 4.7 Extinction coefficients and stable assembly forms of algal biliproteins [23, 80]

	Stable assembly form	λmax	E × 10³ l/mol/cm/ chromophore	Solvent
Allophycocyanin	$(\alpha\beta)_3$	662.5	32.2	8M urea
C-phycocyanin	$(\alpha\beta)_3$ $(\alpha\beta)_6$	662.5	35.5	8M urea
R-phycocyanin	$(\alpha\beta)_3$	662	36.2	8M urea
Cryptomonad phycocyanin	?	—	—	—
Phycocyanobilin	—	670	37.9	5% HCl/MeOH
Phycocyanobilin	—	600	18.9	Neutral $CHCl_3$
C-phycoerythrin	$(\alpha\beta)_6$	550	43.5	8M urea
B-phycoerythrin	$(\alpha\beta)_6\gamma$	550	42.9	8M urea
R-phycoerythrin	$(\alpha\beta)_6\gamma$	567	—	8M urea
Cryptomonad phycoerythrin	$\alpha\beta$	—	—	—
Phycoerythrobilin	—	556	—	HCl/MeOH
Phycoerythrobilin	—	576	—	Acid $CHCl_3$

Further purification of the supernatant by ammonium sulphate precipitation and ion-exchange chromatography yields almost pure phycobilin in the form of linked $\alpha\beta$ subunits [70]. Separation of the subunits themselves is possible by denaturation with urea [71], while purification of the protein-free chromophore may be achieved by refluxing in boiling methanol for 16 h [72]. Each of these purification steps is associated with a loss of overall yield and would require justification in terms of the separation from contaminating coloured proteins or a requirement for absolute purity.

An alternative approach to that of purifying phycocyanobilin from the native biliprotein, in situations where absolute purity is required, would be to use the ability of certain organisms to synthesize free chromophore. Thus in *Cyanidium caldarium* addition of 5-aminolaevulinic acid to the growth medium causes excretion of phycocyanobilin as an almost pure product. Harvesting of the whole cells yields a clear-blue solution from which phycocyanobilin may be separated from any contaminating porphyrins by simple chloroform extraction of the supernatant [73].

Absorption and fluorescence. The visible spectral properties and fluorescence of free biliproteins are dependent on their attached apoproteins and the stable isolation state that is characteristic of the organism in which they are found. The spectral characteristics of phycocyanins and phycoerythrins from three classes of algae are shown in Figure 4.9. The spectra of the individual biliproteins themselves are dependent on the conformational state of the apoproteins, for example, with the transition from aggregate (hexamer) to trimer to monomer [74]. Fluorescence is even more sensitive to dissociation of pigment aggregates as, for example, in the observed shift of fluorescence

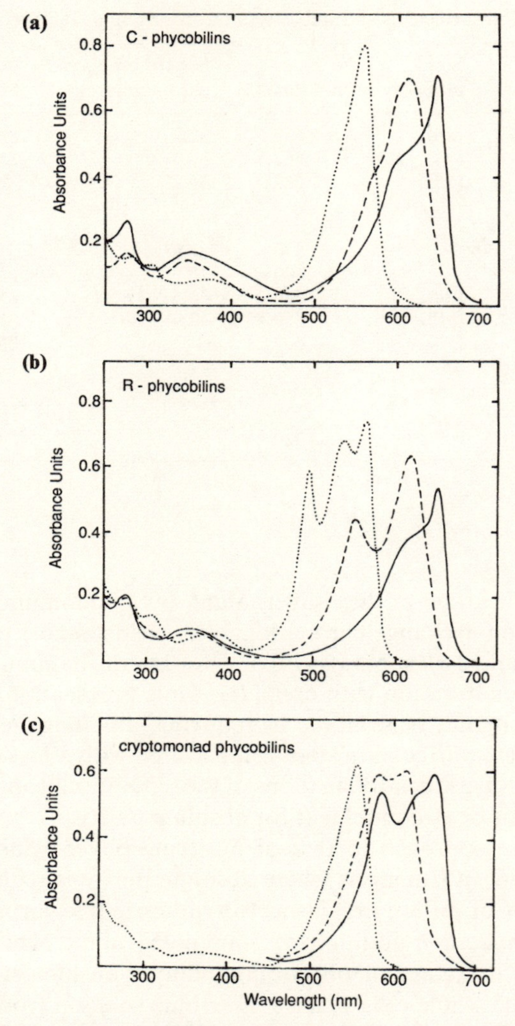

Figure 4.9 Absorbance spectra of isolated biliproteins from three classes of algae, taken in pH 7 phosphate buffer. The spectra show phycoerythrins (····), phycocyanins (----), and allophycocyanins (——) from representative species of the (a) cyanophyta, (b) rhodophyta and (c) cryptophyta.

absorption following dissociation of intact phycobilisomes [75] as shown in Figure 4.10.

Complete unfolding or denaturation of the apoprotein causes a substantial decrease in the visible light absorbance [74] and completely abolishes the visible fluorescence [76]. The spectral shifts associated with these changes are

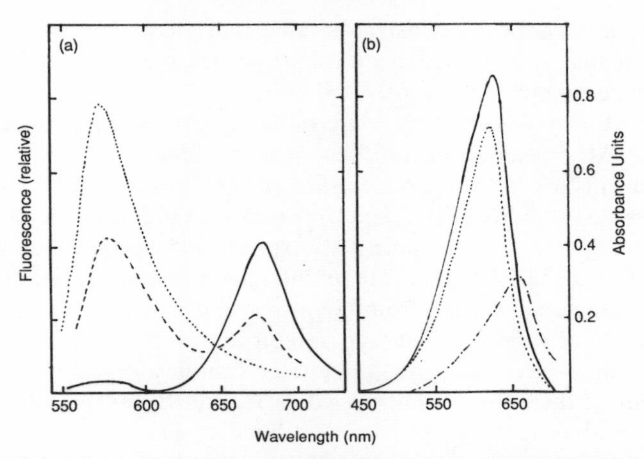

Figure 4.10 Fluorescence emission (a) and absorbance (b) spectra of phycobilins from two species of *Porphyridium*. The spectra show the shifts occurring following dissociation of aggregates (——) to form partially (- - - -) and totally (· · · ·) dissociated products and, finally, on denaturation of the subunits (— · — · —).

considerable and are shown for phycocyanin from *Porphyridium luridum* in Figure 4.10.

More subtle effects are also known. In dilute solution, the extinction coefficient decreases at the absorption maxima, and absorbance at a given concentration varies with ionic strength and pH [77–79]. Thus, factors affecting aggregation also affect absorbance, and it is important when making comparisons between different pigments to ensure that the physical conditions in which they are measured are comparable. Table 4.7 is a summary of the stable aggregation states and known extinction coefficients for phycobilins, and gives for comparison the known values for free phycocyanobilin and phycoerythrobilin.

Stability. The stability of phycobilins *in vitro* is a notional concept, as the stable state of the biliprotein varies from simple $\alpha\beta$ monomers to distinct hexameric aggregates. In the case of C-phycoerythrin and C-phycocyanin, this aggregation behaviour appears to vary according to the species of origin [80], and is dependent on concentration, pH, ionic strength and the presence of denaturants and detergents. In contrast, native forms of R- and B-phycoerythrins exist as stable $\alpha\beta$ hexamers, with some evidence for irreversible dissociation to form $\alpha\beta$ monomers [81]. R-phycocyanin appears to exist in two well-defined aggregation states as $\alpha\beta$ trimers and hexamers. The transition between the two forms is regulated by pH, with values of 6.5 or above favouring the trimer and values below 6.5 the hexamer. C-phycocyanin follows a similar aggregation pattern, but tends to dissociate at low protein concentrations to form $\alpha\beta$ monomers [82] as shown in Table 4.7.

Phycocyanin has been extensively studied in terms of aggregation phenomena in particular and physico-chemical properties in general, and a review on this subject has appeared recently [74].

Phycoerythrins as a class are more stable towards denaturation than phycocyanins. All phycocyanins are known to be denatured in 4M urea while phycoerythrins are not [82]. Among the phycoerythrins, the R and B types show increased stability over C and Cryptomonad pigments, denaturing only slowly in 8M urea, and retaining absorption and fluorescence maxima at temperatures up to 70°C [82]. These differences in conformational stability have been suggested to arise from the presence of covalent disulphide bonds between cysteine residues of the separate protein subunits [82). Thus R- and C-phycocyanins have only enough cysteine residues to provide for covalent attachment of their chromophores, while R- and B-phycoerythrins contain sufficient residues to allow intra-subunit bonding to take place. Dissociation of biliproteins into subunits using mercurial or sulphydryl reagents may be mediated through interaction with sulphydryl groups on the proteins [83–85], while the action of denaturants, such as urea and guanidine hydrochloride, and detergents may be assumed to be through hydrophobic interactions with the proteins. It is interesting to note that renaturation of biliproteins can be facilitated by simple removal of denaturants [86], for example, by dialysis. Although this reflects an inherent stability of the pigment protein complex of biliproteins, there is an implication that renaturation is not complete as only partial restoration of spectral properties is observed [87].

The stability of the free chromophores of phycocyanobilin and phycoerythrobilin is rather less well studied. It is thought that the free phycobilin structures are stabilized by intermolecular hydrogen bonds to form a semicycle structure [88, 89]. This structure is soluble in acid, methanol and chloroform but is both acid-labile and subject to oxidative degradation.

Quantitative analysis. In many instances, the most convenient method for quantitative analysis of biliproteins is by use of their UV-visible absorption spectra and known extinction coefficients. The accuracy of this method is, however, dependent on a knowledge of the physical state of the phycobilin with regard to the presence of apoprotein and the state of aggregation.

For absolute accuracy, it may sometimes be desirable to use modern methods of high-pressure liquid chromatography (HPLC) for separation and quantification of bilins. The subject of bile pigment separation by HPLC has been reviewed recently by McKavanagh and Billing [90]. Although this specifically deals with bile pigments derived from the breakdown of haem in mammalian tissues, the principles of separation are identical to those employed in dealing with phycobilins, and the methods may be adapted accordingly.

In general, it is more convenient to quantify bile pigments by HPLC as protein-free chromophores either in the free-acid form or with the propionic acid substituents esterified to form the corresponding bilin dimethyl ester.

Separation of unesterified bile pigments is achieved by reverse-phase HPLC using a hydrophilic column support and a hydrophobic mobile phase. Esterified bilins have been separated by both normal and reverse-phase systems with success, and this method of sample preparation would seem to confer several advantages in terms of stability and ease of handling. A method for the HPLC separation of algal bilin methyl esters has also recently been published [91].

One advantage of using HPLC as a tool for the quantification of bile pigments is that, in addition to information on the amount of pigment present, the HPLC provides details of the nature and quantity of contaminating pigments, and may in the future prove to be of considerable use as a tool in quality control of pigment production.

4.3.3.2 Commercial exploitation. The commercial exploitation of biliproteins is a rapidly advancing field. The range and intensity of colours available, their relative abundance in the organisms in which they are found, and the inherent stability of their pigment-protein complexes, make them ideal candidates for future development. Current interest is roughly divided into biotechnology and applications in industry.

Biotechnology. Although there is a long tradition in certain cultures of harvesting algae for food use, the single most promising area for further advances lies in the use of modern biotechnological methods for large-scale culture and extraction of microalgae [92–94].

This field of technology has advanced considerably in recent years following pioneering work on the extraction of β-carotene from the halophilic green alga *Dunaliella salina*. Other microalgae are now beginning to be exploited for mass cultivation and harvesting of fine chemicals, the enhanced value of which justifies the technology required for optimum algal growth. To date, the algae most commonly cultured among the Rhodophyta and Cyanophyta are respectively from the genera *Porphyridium* and *Spirulina*. It is fair to say that a great many other species of algae could prove suitable for mass culture given the basic knowledge of their physiology and biochemistry and the financial incentive to develop appropriate technology. This financial incentive is becoming increasingly more apparent in the use of microalgae for production of vitamins, pigments, polysaccharides and pharmaceutical compounds [95].

Current technology for mass culture is commonly based on open pond plants in hot areas where algae with halophilic or alkali-tolerant properties can grow in near monoculture all year round. This system represents relatively low-cost technology and uses natural sunlight to promote photosynthetic growth. Its use is also limited to those organisms whose growth conditions preclude major contamination by indigenous species, and this has proved to be a major limiting factor. Several other methods for large-scale culture are available and, although their use represents increased expense in equipment and running costs, the production of high-value metabolites from individual algae could well offset these. Closed systems using polyethylene tubes to incubate algae under natural sunlight represent an advance on techniques for photosynthetic growth [96], as have several variations on this theme such as the use of covered troughs and vertical glass columns. Such enclosed photobioreactors will shortly be available for industrial systems using microalgae [92] and should considerably enhance the number of species for which it is feasible to consider large-scale culture.

An alternative approach is to use already existing fermentation technology to grow those algal species capable of heterotrophic growth using an external carbon source. Again, the investment in equipment and substrates may be justified by production of novel high-value products. It is in this area in particular that the use of genetically engineered algae might be envisaged, as the economic rewards of unique product synthesis would be great.

So far as pigment synthesis from microalgae is concerned, the technology existing for the growth and extraction of the blue-green alga *Spirulina platensis* is typical and may be used as a general example. The natural occurrence of this alga is in saltwater ponds and inland lakes with saline and alkaline water; as such, it is capable of colonizing extreme environments that are unsuitable for many other organisms. In the wild, *S. platensis* grows at 27°C at pHs from 7.2 to 9.0 and at salt concentrations of over 30 g/1, under which conditions it is found almost as a monoculture [97]. The high amino-acid content of *Spirulina* and digestibility to humans, has led to its being used as a food supplement from earliest times among indigenous tribes in Africa and Mexico where *Spirulina* is found. Its use is fashionable now as a health food, and the extraction of C-phycocyanin from *Spirulina* is a recent addition to the commercial potential of the organism. Mass cultivation is generally in covered ponds or closed reactors, and in such conditions of temperature, light and nutritional status as are suitable for maximum cell growth, pigment synthesis *etc*. [97, 98]. Harvesting may be achieved by use of a vibrating screen or filtration and the cells are then sun- or spray-dried.

The situation for the unicellular red alga *Porphyridium cruentum* is similar [99]. It is generally grown in artificial seawater at temperatures of 21°C and with an added nitrogen source such as potassium nitrate (KNO_3). The alga can be grown in ponds or closed tubular reactors and, in addition has been immobilized on glass beads for use as a biocatalyst in the production of excreted polysaccharides [92, 99, 100]. Other products of interest from this alga include the dietary supplement arachidonic acid and R-phycoerythrin.

The processes described for the culture of *Spirulina* and *Porphyridium* are similar to those required for any number of algal species, given the biochemical knowledge to optimize their culture conditions and the economic incentive for doing so. It is thus reasonable to expect that, should the future demand for pigments justify it, a range of algal organisms might be employed to produce not only the C-phycocyanin and R-phycoerythrin currently available but also the many other spectral variants of the pigments occurring in nature.

The economic potential of any individual alga is a combination of the input in its cultivation and the value of its products. In this respect, the concept of obtaining several products from a single organism must appear attractive. In the case of *Porphyridium* this possibility has been assessed and costed [99, 101], with the conclusion that the gross value of a number of products would repay any additional cost of their individual separation. The removal of pigments from algal biomass may, in fact, be a requisite for its proposed use, for example as animal feed, and in this case the pigment itself becomes a by-product of the feed production process, further enhancing the commercial potential of the process.

The potential of several other microalgae remains to be investigated. It seems highly likely that among the organisms listed in Table 4.5, are some whose economic possibilities are such that efficient production of individual pigments would be possible, and the time is now ripe to capitalize on this.

Applications in industry. Although in its infancy, the development of bilins as food colorants has already received much interest [27, 102–105]. Several patents already exist for the extraction, stabilization purifications and use of phycocyanin from *Spirulina* spp. and *Aphanothece nidulans* [27]. Niohan Siber Hegner Ltd. (Tokyo) and Dainippan Ink and Chemicals Inc. (Tokyo) both market phycocyanin as a food colorant. It has been used in chewing gums, and is suggested as an additive in frozen confections, soft drinks, dairy products, sweets and ice-creams [102, 106]. No patents have been filed to date on the use of phycoerythrin as a food colorant, but the potential for the development of a natural red pigment is thought to be great because of concern about the safety of the alternatives currently available. It is likely that this will be an area of rapid development in the immediate future.

In addition to the use of phycobilins in the food industry, several other high-value uses have recently emerged. The fluorescent properties of phycobilins are being employed as novel tracers in biochemical research (see Glazer and Stryer [107] for review). Biliproteins are conjugated with molecules that confer biological specificity to produce a new class of fluorescent probes—the phycofluors. The high absorbance coefficients and intense fluorescence emission of phycobiliproteins make them highly sensitive fluorescent labels. These properties are almost unchanged on conjugation with other molecules through acylation of specific amino-acid residues of the protein.

The phycobiliproteins may be conjugated to specific linker molecules, allowing formation of several derivatives [108]. One of the most widely used methods involves formation of a derivative with a monoclonal antibody for use in fluorescent immunoassay [109]. Phycobiliprotein conjugates have also been used for fluorescence-activated cell sorting and fluorescence microscopy and are promising tools in several new areas of research [107]. R-phycoerythrin conjugates are already commercially available from the Sigma

Chemical Co. (Poole), and Vector Laboratories (Peterborough), and represent an important high-value outlet for phycoerythrin production from algae.

4.3.3.3 Comparison with synthetic colours. Several comparisons have been made of the colours of the phycobiliproteins with the synthetic red and blue pigments available commercially. Phycocyanin has been reported as being 'between blue colour no. 1 (brilliant blue) and blue colour no. 2 (indigo carmine)' [106]. However, as already discussed, the exact spectral properties of all the phycobiliproteins are dependent on their algal source, state of purification and physicochemical environment. These spectra have been presented throughout the chapter in relation to each particular state of the pigment. It seems pertinent here to present for accurate comparison with the phycocyanins and phycoerythrins the spectra of several synthetic blue and red colours [110] (Figure 4.11, Table 4.8).

4.3.3.4 Future prospects. In one sense, the field of haem and bilin colorants as a whole is one of future promise rather than current practice. It is, however, clear that sufficient scientific know-how and commercial incentives exist for this particular area of colorant use to undergo a rapid transformation from theory into practice. Several current trends have already been alluded to, but it seems likely that future expansion will be based on advances in the following areas:

1. Increasing acceptance of 'natural' and 'nature-alike' pigments and their derivatives as food colorants. Public awareness of these pigments is growing, and should be met by the provision of accurate information on the nature of the product offered. Clear analysis and definition of the structures of natural pigment derivatives used within the industry would provide the background to better public understanding and acceptance.
2. The production of customized pigments either by chemical or biological means. Simple chemical transformations of natural products could be used

Table 4.8 Absorbance characteristics of some synthetic red and blue food colorants

Name	λmax (nm)	E 1%	EmM
Ponceau 4R (E124)	505	431	71
Erythrosine (E127)	526	1154	—
Red 2G	528	620	122
Patent blue V (E131)	635	2000	172
Indigo carmine (E132)	610	489	105
Brilliant blue (FCF)	629	1637	207

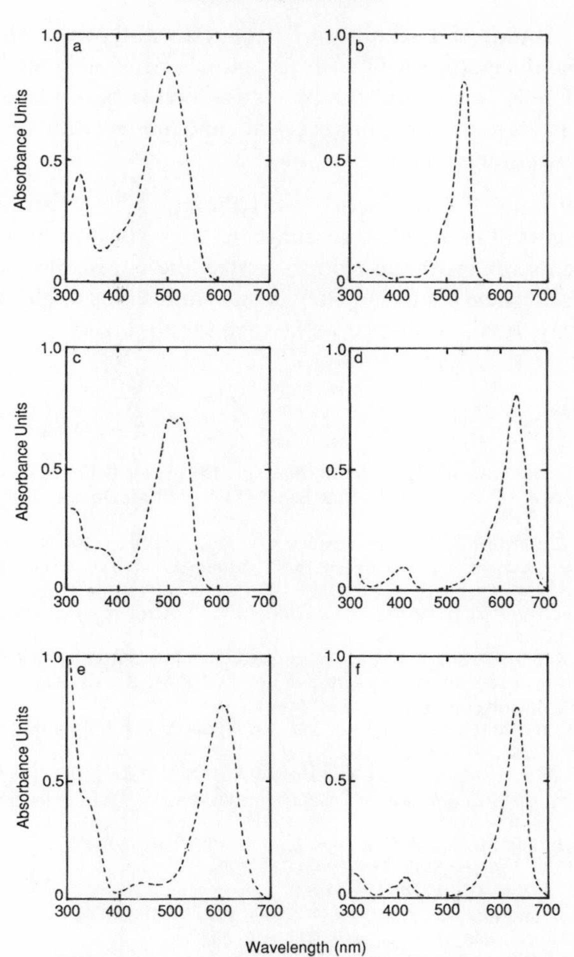

Figure 4.11 The absorbance spectra of several synthetic blue and red food colorants: (a) ponceau 4R (E124); (b) erythrosine (E127); (c) red 2G; (d) patent blue V (E131); (e) indigo carmine (E132); and (f) brilliant blue FCF.

to produce predictable spectral shifts desirable for specific applications. Similarly, the great diversity of spectral forms of individual pigments available in nature could be better used.

3. Development of new technology for the growth of a wider range of micro-organisms on a commercial scale would allow the bulk production of a great variety of natural pigments not presently available.

4. Similarly, the greater use of the techniques of molecular biology in enhancing the production of natural pigments within an organism could

increase the commercial viability of production of some otherwise minor pigments. The development of such techniques could also lead to the transformation of cells to perform the biological synthesis of novel compounds. This subject is currently one of great academic interest and has considerable future promise commercially.

The overall aim of future work must therefore be to produce naturally derived pigments of high public acceptability, to satisfy the growing demands of replacements for synthetic colours across the entire spectrum. With the co-operation of science and industry, haems and bilins could offer just such an opportunity. It is a challenge well worth taking up.

References

1. Merrit, J.E. and Loening, K.L. *Eur. J. Biochem.* **108** (1980) 1–30.
2. Battersby, A.R., Fookes, C.J.R., Matchan, G.W.F. and McDonald, E. *Nature* **285** (1980) 17–21.
3. Stryer, L. *Biochemistry*, W.H. Freeman & Co., San Francisco (1975).
4. Kikuchi, G., Kumar, A., Talmage, P. and Shemin, D. *J. Biol. Chem.* **233** (1958) 1214–1219.
5. Bogorad, L. *The Porphyrins* Vol. 6 (D. Dolphin, Ed.), Academic Press Inc., London (1979) pp.125–172.
6. Lemberg, R. and Barrett, J. *Cytochromes*, Academic Press, London (1973).
7. Brown, S.B. and Troxler, R.F. in *Bilirubin* Vol. 2 (K.P.M. Heirwegh and S.B. Brown, Eds), CRC Press, Boca Raton, Florida (1982) pp.11–38.
8. Schmid, R. and McDonagh, A.F. in *The Porphyrins* Vol. 6 (D. Dolphin, Ed.) Academic Press, New York (1979) pp.257–292.
9. Hendry, G.A.F., Houghton, J.D. and Brown, S.B. *New Phytol.* **107** (1987) 255–302.
10. Smith, K.M. (Ed.) *Porphyrins and Metalloporphyrins* (1975), Elsevier Scientific, Amsterdam.
11. Anson, M.L. and Mirsky, A.E. *J. Gen. Physiol.* **13** (1930) 469–472.
12. Drabkin, D.L. *J. Biol. Chem.* **158** (1945) 721–722.
13. Lewis, U.J. *J. Biol. Chem.* **206** (1954) 109.
14. Sumner, J.B. and Dounce, A.L. *J. Biol. Chem.* **127** (1938) 439–447.
15. Teale, F.W.J. *Biochim. Biophys. Acta* **35** (1959) 543.
16. Shalfejeff, M. *Chem. Ber.* **18** (1885) 232–233.
17. Fischer, H. *Org. Syn.* **3** (1955) 442–445.
18. Labbe, R.F. and Nishida, G. *Biochim. Biophys Acta* **26** (1957) 437.
19. Chu, T.C. and Chu, E.J. *J. Biol. Chem.* **212** (1955) 1–7.
20. Dixon, H.B.F. and Thompson, C.M. *Biochem. J.* **107** (1968) 427–431.
21. Morrison, M. and Stotz, E. *J. Biol. Chem.* **213** (1955) 373–378.
22. Hamsik, A. and Hofman, M. *Hoppe-Seylers, Z. Physiol. Chem.* **305** (1956) 143–144.
23. Dawson, R.M.C., Elliot, D.C., Elliott, W.H. and Jones, K.M. *Data for Biochemical Research* 3rd edn. Clarendon Press, Oxford (1986) pp.213–231.
24. Brown, S.B., Hatzikonstantinou, H. and Herries, D.G. *Int. J. Biochem.* **12** (1980) 701–707.
25. Perutz, M.F. *Trends in Biochem. Sci.* **14** (1989) 42–44.
26. Rossi, E. and Curnow, D.H. in *HPLC of Small Molecules: A Practical Approach* (C.K. Lim, Ed.), IRL Press, Oxford (1986) pp.260–304.
27. Francis, F.J. *Handbook of Food Colorant Patents*, Food and Nutrition Press Inc., Westport, Connecticut (1986).
28. O'Carra, P., Murphy, R.F. and Killilea, S.D. *Biochem. J.* **187** (1980) 303–309.
29. Bennet, A. and Siegelman, H.W. in *The Porphyrins*, Vol. 6 (D. Dolphin, Ed.), Academic Press, New York (1979) pp.493–520.

30. Troxler, R.F. and Bogorad, L. *Plant Physiol.* **41** (1966) 491–499.
31. Brown, S.B., Houghton, J.D. and Vernon, D.I. *J. Photochem. Photobiol. B. Biology* **5** (1990) 3–23.
32. Rudiger, W., O'Carra, P. and O'hEocha, C. *Nature* **215** (1967) 1477–1478.
33. O'Carra, P. and O'hEocha, C. *Phytochem.* **5** (1966) 993–997.
34. Chapman, D.J., Cole, W.J. and Siegelman, H.W. *Biochem. J.* **105** (1967) 903–905.
35. Chapman, D.J., Cole, W.J. and Siegelman, H.W. *Am. J. Bot.* **55** (1968) 315–316.
36. Bryant, D.A., Glazer, A.N. and Eiserling, F.A. *Arch. Microbiol.* **110** (1976) 61–75.
37. Morschel, E. and Wehrmeyer, W. *Arch. Microbiol.* **113** (1977) 83–89.
38. MacColl, R., Berns, D.S. and Gibbons, O. *Arch. Biochem. Biophys.* **177** (1976) 265–275.
39. Glazer, A.N. and Cohen-Bazire, G. *Arch. Microbiol.* **104** (1975) 29–32.
40. Svedberg, T. and Katsurai, I.T., *J. Am. Chem. Soc.* **51** (1929) 3573–3583.
41. Lagarias, J.C., Klotz, A.V., Dallas, J.L., Glazer, A.N., Bishop, J.E., O'Connell, J.F. and Rapoport, H. *J. Biol. Chem.* **263** (1988) 12977–12985.
42. Brown, S.B., Houghton, J.D. and Wilks, A. in *Biosynthesis of Heme and Chlorophylls* (H.A. Dailey, Ed.), McGraw Hill, New York (1990) pp.543–575.
43. Kannangara, C.G., Gough, S.P., Bruyant, P., Hoober, J.K., Kahn, A. and von Wettstein, D. *Trends in Biochem. Sci.* **13** (1988) 139–143.
44. Beale, S.I. and Weinstein, J.D. in *Biosynthesis of Heme and Chlorophylls* (H.A. Dailey, Ed.), McGraw Hill, New York (1990) pp.287–391.
45. de Lorimer, R., Bryant, D.A., Porter, R.D., Liu, W-Y., Jay, E. and Stevens, S.E. Jr. *Proc. Natl. Acad. Sci. (USA)* **81** (1984) 7946–7950.
46. de Lorimer, R., Wang, Y-J. and Yeh, M-L. *Proc. Third Ann. Penn. State Symp., Plant Physiol.* (1988) pp.332–326.
47. Brumm, P.J., Fried, J. and Friedmann, H.C. *Proc. Natl. Acad. Sci. (USA)* **80** (1983) 3943–3947.
48. Bogorad, L. *Ann. Rev. Plant Physiol.* **26** (1975) 369–401.
49. Hattori, A. and Fujita, Y. *J. Biochem.* **46** (1959) 521–524.
50. Fujita, Y. and Hattori, A. *Plant and Cell Physiol.* **1** (1960) 293–303.
51. Tandeau de Marsac, N., Mazel, D., Damerval, T., Guglielmi, G., Capuano, V. and Houmard, J. *Photosyn. Res.* **18** (1988) 99–132.
52. Wood, N. B. and Haselkorn. in *Limited Proteolysis in Microorganisms* (G.N. Cohen and H. Holzer, Eds), US DHEW Publication no (NIH) Bethesda, MD (1979) pp.79–1591.
53. Bennet, A. and Bogorad, L. *J. Cell. Biol.* **58** (1973) 419–435.
54. Tandeau de Marsac, N. *J. Bacteriol.* **130** (1977) 82–91.
55. Canto de Loura, I., Dubacq, J.P. and Thomas, J.C. *Plant Physiol.* **83** (1987) 838–843.
56. Schmidt A., Erdle, I. and Kost, H-P. *Z. Naturforsch.* **37c** (1982) 870–876.
57. Brody, M. and Emerson, R. *Am. J. Bot.* **46** (1959) 433–440.
58. Myers, J. and Kratz, W.A. *J. Gen. Physiol.* **39** (1955) 11–22.
59. Jones, L.W. and Myers, J. *J. Phycol.* **1** (1965) 7–14.
60. Nichols, K.E. and Bogorad, L. *Botan. Gaz.* **124** (1962) 85–93.
61. Gantt, E. *Int. Rev. Cytol.* **66** (1980) 45–80.
62. Gantt, E. *Ann. Rev. Plant Physiol.* **32** (1981) 327–347.
63. Grossman, A.R., Lemaux, P.G. and Conley, P.B. *Photochem. Photobiol.* **44** (1986) 827–837.
64. Brown, A.S., Offner, G.D., Ehhart, M.M. and Troxler, R.F. *J. Biol. Chem.* **254** (1979) 7803–7811.
65. Chapman, D.J. in *The Biology of Blue Green Algae* (N.G. Carr and B.A. Whitton, Eds), Univ. California Press, Berkeley (1973) pp.162–185.
66. Glazer, A.N. *Annu. Rev. Biochem. Biophys.* **14** (1985) 47–77.
67. Porter, G., Tredwell, C.J., Searle, G.F.W. and Barber, J. *Biochim. Biophys. Acta.* **501** (1978) 532–545.
68. Chen, C-H. and Berns, D.S. *Biophys. Chem.* **8** (1978) 203–213.
69. Troxler, R.F. in *Chemistry and Physiology of Bile Pigments* (P.D. Berk and M.I. Berlin, Eds), Dept. Health Education and Welfare, US, pp.431–454.
70. Brown, A.S. and Troxler, R.F. *Biochem. J.* **163** (1977) 571–581.
71. Glazer, A.N. and Fang, S. *J. Biol. Chem.* **248** (1973) 659–662.
72. Troxler, R.F., Kelly, P. and Brown, S.B. *Biochem. J.* **172** (1978) 569–576.

73. Brown, S.B., Holroyd, J.A., Troxler, R.F. and Offner, G.D. *Biochem. J.* **194** (1981) 137–147.
74. Berns, D.S. and MacColl, R. *Chem. Rev.* **89** (1989) 807–825.
75. Gantt, E. and Lipschultz, C.A. *Biochim. Biophys. Acta* **292** (1973) 858–861.
76. Rudiger, W. in *Pigments in Plants* (F-C. Czygan, Ed.), Akademie-Verlag, Berlin (1981) 314–351.
77. MacColl, R., Lee, J.J., Berns, D.S. *Biochem. J.* **122** (1971) 421–426.
78. Neufeld, G.J. and Riggs, A.F. *Biochim. Biophys. Acta* **181** (1969) 234–243.
79. Davis, L.C., Radke, G.A., Guikema, J.A. *J. Liq. Chromatogr.* **9** (1986) 1277–1295.
80. Bennett, A. and Bogorad, L. *Biochemistry* **10** (1971) 3625–3634.
81. van der Velde, H.H. *Biochim. Biophys. Acta* **303** (1973) 246–257.
82. O'Carra, P. and O'hEocha, C. in *Chemistry and Biochemistry of Plant Pigments* 2nd edn. Vol. 2 (T.W. Goodwin, Ed.), Academic Press, London (1976) pp.328–376.
83. Fujimori, E. and Pecci, J. *Biochemistry*, **5** (1966) 3500–3508.
84. Fujimori, E. and Pecci, J. *Biochim. Biophys. Acta* **207** (1970) 259–261.
85. Pecci, J. and Fujimori, E. *Biochim. Biophys. Acta* **154** (1968) 332–341.
86. Murphy, R.F. and O'Carra, P. *Biochim. Biophys. Acta* **214** (1970) 371–373.
87. Boucher, L.J., Crespi, H.L. and Katz, J.J. *Biochemistry*, **5** (1966) 3796–3802.
88. Sheldrick, W.S. *J. Chem. Soc. Perkin Trans.* **2** (1976) 1457–1461.
89. Falk, H. and Grubmayr, K. *Angew Chem. Int. Ed.* **16** (1977) 470–471.
90. McKavanagh, S. and Billing, B.H. in *HPLC of Small Molecules: A Practical Approach* (C.K. Lim, Ed.), IRL Press, Oxford (1986) 305–323.
91. Dawson, L., Houghton, J.D., Vernon, D.I. and Brown, S.B. *Mol. Asp. Med.* **11** (1989) 133–134.
92. Parkinson, G., Shota, U., Hunter, D. and Sandler,N. *Chem. Eng.* **94** (1987) 19–23.
93. Borowitzka, M.A. and Borowitzka, L.J. *Micro-algal Biotechnology*, Cambridge University Press, Cambridge (1988).
94. Cohen, Z. in *CRC Handbook of Microalgal Mass Culture* (A. Richmond, Ed.), CRC Press, Boca Raton (1986) pp.421–454.
95. Borowitzka, M.A. in *Microalgal Biotechnology* (M.A. Borowitzka, and L.J. Borowitzka, Eds), Cambridge University Press (1988) pp.173–175.
96. Lee, Y.K. *Trends in Biotechnol.* **4** (1986) 186–189.
97. Richmond, A. in *Microalgal Biotechnology* (M.A. Borowitzka and L.J. Borowitzka, Eds), Cambridge University Press, Cambridge (1988) pp.85–97.
98. Tomasseli, L., Giovannetti, L., Sacchi, A. and Bocci, F. in *Algal Biotechnology* (T. Stadler *et al.*, Eds), Elsevier Applied Science, London (1988), pp.305–314.
99. Vonshak, A. in *Microalgal Biotechnology* (M.A. Borowitzka and L.J. Borowitzka, Eds), Cambridge University Press, Cambridge (1988) pp.130–133.
100. Robinson, P.K., Mak, A.L. and Trevan, M.D. *Process Biochem.* **21** (1986) 122–127.
101. Thepenier, C., Chaumont, D. and Gudin, C. in *Algal Biotechnology* (T. Stadler *et al.*, Eds), Elsevier Applied Science, London (1988) pp.413–421.
102. Francis, F.J. *Food Technol.* April (1987) 62–68.
103. Taylor, A.J. in *Developments in Food Colours—2* (J. Walford, Ed.), Elsevier Applied Science, London (1984) pp.159–206.
104. Wadds, G. in *Developments in Food Colours—2* (J. Walford, Ed.), Elsevier Applied Science, London (1984) pp.23–74.
105. Cohen, Z. in *CRC Handbook of Microalgal Mass Culture* (A. Richmond, Ed.), CRC Press Inc. Boca Raton, Florida (1986) 421–454.
106. Jacobson, G.K. and Jolly, S.O. in *Biotechnology* Vol. 7b (H-J. Rehm and G. Reed, Eds), VCH Publishers, Weinheim (1989).
107. Glazer, A.N. and Stryer, L. *Trends in Biochem. Sci.* **9** (1984) 423–427.
108. Vernon, T.O., Glazer, A.N. and Stryer, L. *J. Cell. Biol.* **93** (1983) 981–986.
109. Kronick, M.N. *J. Immunol. Meth.* **92** (1986) 1–13.
110. Walford, J. (Ed.) *Developments in Food Colours—2*, Elsevier Applied Science, London (1984).
111. Hendry, G.A.F., Houghton, J.D. and Jones, O.T.G. *Biochem. J.* **194** (1981) 743–751.

5 Carotenoids

G. BRITTON

5.1 Summary

The carotenoids are very widespread natural pigments in plants and animals, so they provide the natural yellow, orange or red colours of many foods as well as being used extensively as non-toxic natural or nature-identical colorants. This chapter discusses all aspects of carotenoids, especially in relation to their presence or use in food. Topics discussed include the distribution of carotenoids, particularly in plants and animals, the natural functions and commercial uses and applications of carotenoids, the pathways and regulation of carotenoid biosynthesis, the absorption and transport of carotenoids and their metabolism, especially into vitamin A. Examples are given of the use of natural carotenoid extracts (*e.g.* annatto) and synthetic carotenoids (*e.g.* β-carotene, canthaxanthin, 8′-apo-β-caroten-8′-al) as colorants in food, feed, cosmetics and medical and health products. The properties and stability of carotenoids in the free form and in oil-based and water-dispersible formulations are assessed. All practical aspects of the handling, isolation and analysis of carotenoids are outlined, with emphasis on their purification and identification by chromatography, including HPLC, and their characterization and quantitative determination by physicochemical methods such as UV-visible and NMR spectroscopy and mass spectrometry. Future prospects for the production and use of carotenoids as colorants are assessed, and substantial advances in the biological and biotechnological production of carotenoids by microalgae or bacteria are predicted.

5.2 Introduction: structures and nomenclature

Carotenoid hydrocarbons collectively are called *carotenes*. Derivatives that contain oxygen functions (most commonly hydroxy, keto, epoxy, methoxy or carboxylic acid groups) are called *xanthophylls*. Some carotenoids are acyclic (*e.g.* lycopene) but more common are those that contain a six-membered (or occasionally five-membered) ring at one end or both ends of the molecule. The structure and numbering of the parent acyclic and bicyclic carotenoids, lycopene and β-carotene, are illustrated in Figure 5.1.

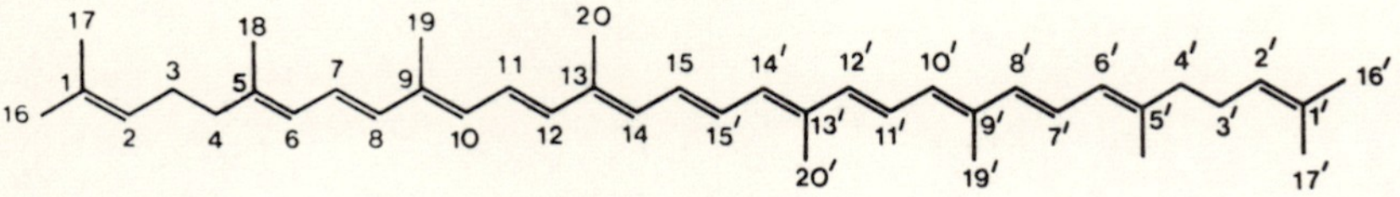

Lycopene

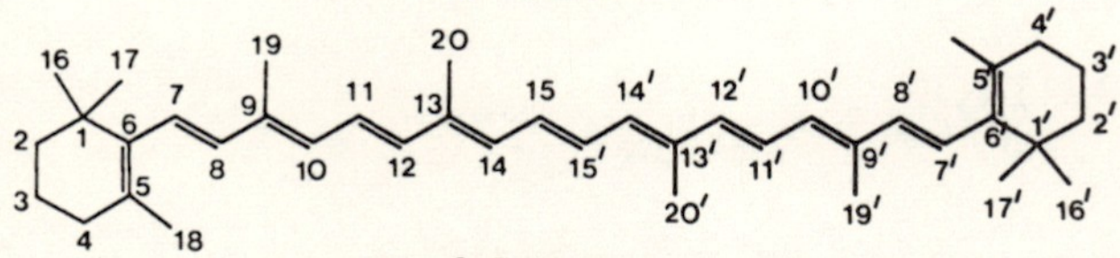

ß-Carotene

Figure 5.1 Structures and numbering scheme for the parent acyclic and bicyclic carotenoids, lycopene and β-carotene.

Traditionally, natural carotenoids have been given trivial names that are usually derived from the name of the biological source from which they were first isolated. In recent years, however, a semi-systematic nomenclature that conveys structural information has been devised. According to this scheme, the carotenoid molecule is considered in two halves. Each individual compound is then named as a derivative of the parent carotene, as specified by the Greek letters that describe its two end groups, with conventional prefixes and suffixes being used to indicate changes in hydrogenation level and the presence of substituent groups. The seven end groups that have been recognized are illustrated in Figure 5.2.

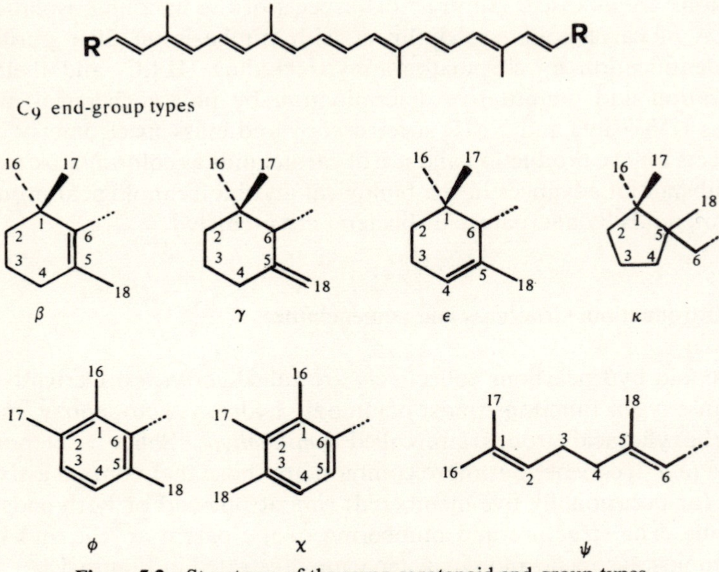

C_9 end-group types

β γ ϵ κ

ϕ χ ψ

Figure 5.2 Structures of the seven carotenoid end-group types.

Compounds from which an end group has been removed from the normal C_{40} structure are known as *apo-carotenoids*; some of these are important food colorants. In keeping with current practice, trivial names will be used in this article for common carotenoids, but the semi-systematic names of these are given in Table 5.1 (see also Appendix for the structures of individual carotenoids mentioned in the text). Rules for application of the semi-systematic nomenclature are given in the monograph by Isler [1], which is also a vast source of information about the chemistry, biochemistry, properties and functions of carotenoids.

Because of the extensive double-bond system in the molecule, any carotenoid can, theoretically, exist in many geometrical isomeric forms (*Z-E*

Table 5.1 Semi-systematic names of carotenoids mentioned in the text (stereochemistry is not specified)

Trivial name	Semi-systematic name
Actinioerythrol	3,3′-dihydroxy-2,2′-dinor-β,β-carotene-4,4′-dione
Antheraxanthin	5,6-epoxy-5,6-dihydro-β,β-carotene-3,3′-diol
Astaxanthin	3,3′-dihydroxy-β,β-carotene-4,4′-dione
Bixin	methyl hydrogen 9′-*cis*-6,6′-diapocarotene-6,6′-dioate
Canthaxanthin	β,β-carotene-4,4′-dione
Capsanthin	3,3′-dihydroxy-β,κ-caroten-6′-one
Capsorubin	3,3′-dihydroxy-κ,κ-carotene-6,6′-dione
α-Carotene	β,ε-carotene
β-Carotene	β,β-carotene
γ-Carotene	β,ψ-carotene
δ-Carotene	ε,ψ-carotene
ε-Carotene	ε,ε-carotene
ζ-Carotene	7,8,7′,8′-tetrahydro-ψ,ψ-carotene
Citranaxanthin	5′,6′-dihydro-5′-apo-18′-nor-β-caroten-6′-one
Crocetin	8,8′-diapocarotene-8,8′-dioic acid
α-Cryptoxanthin	β,ε-caroten-3-ol
β-Cryptoxanthin	β,β-caroten-3-ol
Isocryptoxanthin	β,β-caroten-4-ol
Isozeaxanthin	β,β-carotene-4,4′-diol
Lactucaxanthin	ε,ε-carotene-3,3′-diol
Lutein	β,ε-carotene-3,3′-diol
Lutein-5,6-epoxide	5,6-epoxy-5,6-dihydro-β,ε-carotene-3,3′-diol
Lycopene	ψ,ψ-carotene
Neoxanthin	5′,6′-epoxy-6,7-didehydro-5,6,5′,6′-tetrahydro-β,β-carotene-3,5,3′-triol
Neurosporene	7,8-dihydro-ψ,ψ-carotene
Norbixin	6,6′-diapocarotene-6,6′-dioic acid
Phytoene	7,8,11,12,7′,8′,11′,12′-octahydro-ψ,ψ-carotene
Phytofluene	7,8,11,12,7′,8′-hexahydro-ψ,ψ-carotene
Violaxanthin	5,6,5′,6′-diepoxy-5,6,5′,6′-tetrahydro-β,β-carotene-3,3′-diol
Violerythrin	2,2′-dinor-β,β-carotene-3,4,3′,4′-tetrone
α-Zeacarotene	7′,8′-dihydro-ε,ψ-carotene
β-Zeacarotene	7′,8′-dihydro-β,ψ-carotene
Zeaxanthin	β,β-carotene-3,3′-diol

isomers). Most carotenoids occur naturally in the all-*E* (all-*trans*) form, but *Z*-isomers (*cis*-isomers) are frequently present in small amounts, and are easily formed as artifacts from the all-*E* isomer.

5.3 Distribution and natural functions

5.3.1 Distribution

Carotenoids are often considered to be plant pigments alone but they also occur widely in bacteria, fungi, algae and in animals, especially birds, fish and invertebrates. Two volumes of Goodwin [2a,b] provide extensive details of the distribution of carotenoids in plants and animals, respectively.

5.3.1.1 Plants. In plants, the carotenoids occur universally in the chloroplasts of green tissues, but their colour is masked by the chlorophylls. The leaves of virtually all species contain the same main carotenoids, that is β-carotene (usually 25 to 30% of the total), lutein (around 45%), violaxanthin (15%) and neoxanthin (15%). Small amounts of α-carotene, α- and β-cryptoxanthin, zeaxanthin, antheraxanthin and lutein-5,6-epoxide are also frequently present, and lactucaxanthin is a major xanthophyll in a few species, notably lettuce (*Lactuca sativa*).

The quantitative carotenoid compositions of most leaves are similar, although some changes are seen in plants under stress.

Carotenoids are also widely distributed in non-photosynthetic tissues of plants and are responsible for the yellow, orange and red colours of many flowers and fruit. They are usually located in chromoplasts, and the xanthophylls are frequently present as complex mixtures of fatty acyl esters.

Several distinctive carotenoid patterns have been recognized in fruit:

- the normal collection of chloroplast carotenoids
- large amounts of lycopene and its hydroxy-derivatives (*e.g.* tomato)
- large amounts of β-carotene and its hydroxy-derivatives (*e.g.* peach)
- collections of 5,6- or 5,8-epoxycarotenoids (*e.g.* carambola)
- large amounts of apocarotenoids (*e.g. Citrus* spp.)
- some specific or unusual carotenoids (*e.g.* capsanthin in red pepper, *Capsicum annuum*).

Although carotenoids are not widely found in roots, there are some well-known examples, for example carrot and sweet potato, where large amounts of carotenoids, mainly carotenes, are present. Similarly, some seeds are coloured by carotenoids, for example zeaxanthin in maize. One particular example, *Bixa orellana* (annatto) contains very large amounts of the apocarotenoid bixin in its seed coat (up to 10% dry weight).

Most yellow flowers are coloured by carotenoids; here xanthophyll epoxides are common. Natural extracts of some flowers, such as marigold (*Tagetes erecta*) contain large amounts of lutein and are used for coloration.

5.3.1.2 Animals. In animals, although carotenoids are found in some birds and fish, they are particularly associated with invertebrate animals. In birds, carotenoids colour some yellow or red feathers but they are also important for skin colour in chickens and especially in egg yolk. In fish, important examples are the flesh of salmon and trout (astaxanthin and canthaxanthin). In many marine invertebrate animals (*e.g.* shrimps, crabs and lobsters), astaxanthin and related carotenoids may be present in large amounts, often as carotenoprotein complexes, which are green, purple or blue in the living animal but are denatured to reveal the red carotenoid colour when the animal is cooked.

5.3.2 Natural functions

The carotenoids owe their distinctive properties and functions to the presence of a long chromophore of conjugated double bonds, which gives them their light-absorbing (colour) properties and renders the molecules extremely susceptible to oxidative degradation.

In flowers, fruits and many animals, the function of carotenoids is simply to provide colour. In green plant tissues, they play important roles in photosynthesis [3–6]. The carotenoids are located, in the chloroplast thylakoid membranes, in the pigment-protein complexes of photosystems 1 and 2, where they serve as accessory light-harvesting pigments (mainly the xanthophylls). They are particularly important as protectants against photo-oxidation by singlet oxygen, $^1O_2$. When more light energy is absorbed by the light-harvesting chlorophylls than can be used to drive photosynthesis, some excited chlorophyll molecules will undergo intersystem crossing to give the slightly lower energy but longer-lived triplet excited state, ^{3}CHL. The energy of the ^{3}CHL can be transferred to oxygen to give singlet oxygen, $^1O_2$, which is a highly reactive species that can rapidly cause damage to lipids, membranes and tissues. Carotenoids, particularly β-carotene in the chloroplast pigment-protein complexes, give protection by quenching the energy of ^{3}CHL and thus preventing the formation of $^1O_2$, and they can also quench the energy of $^1O_2$ if any should be formed. Carotenoids can similarly protect non-photosynthetic organisms or tissues against photo-oxidation by $^1O_2$ since this can also be produced in the presence of other suitable photosensitizers. Thus, in humans, some conditions such as erythropoietic protoporphyria, in which patients are extremely sensitive to light because haem synthesis is abnormal, lead to the accumulation of free porphyrins in the skin. The free porphyrin molecules can act as photosensitizers and cause the production, in the skin, of $^1O_2$, which causes tissue damage, inflammation and so on. β-Carotene affords substantial protection against this [7].

5.4 Biosynthesis

Carotenoids are biosynthesized in higher plants, algae, fungi and bacteria. Animals cannot biosynthesize them *de novo*, although many are able to metabolize and modify structurally some carotenoids that they ingest.

Many review articles on carotenoid biosynthesis are available, and should be consulted for details of the pathways, reactions, enzymes and regulation [8–14].

The carotenoids are isoprenoid compounds and are biosynthesized from acetyl coenzyme-A via mevalonic acid as a branch of the great isoprenoid or terpenoid pathway (Figure 5.3). The main stages of carotenoid biosynthesis are outlined in Figure 5.4. The early stages of the pathway are common to the biosynthesis of all isoprenoid compounds. The first step that is specific to carotenoids is the formation of the first C_{40} hydrocarbon, phytoene, from two molecules of geranylgeranyl diphosphate (GGDP) via prephytoene diphosphate (PPDP). In plants, the phytoene that is formed appears to be the 15*Z* isomer, although the 15*E* form is made by some bacteria directly. The colourless phytoene (with three conjugated double bonds) then undergoes a series of desaturation reactions (Figure 5.5), each of which introduces a new double bond and increases the chromophore by two conjugated double bonds

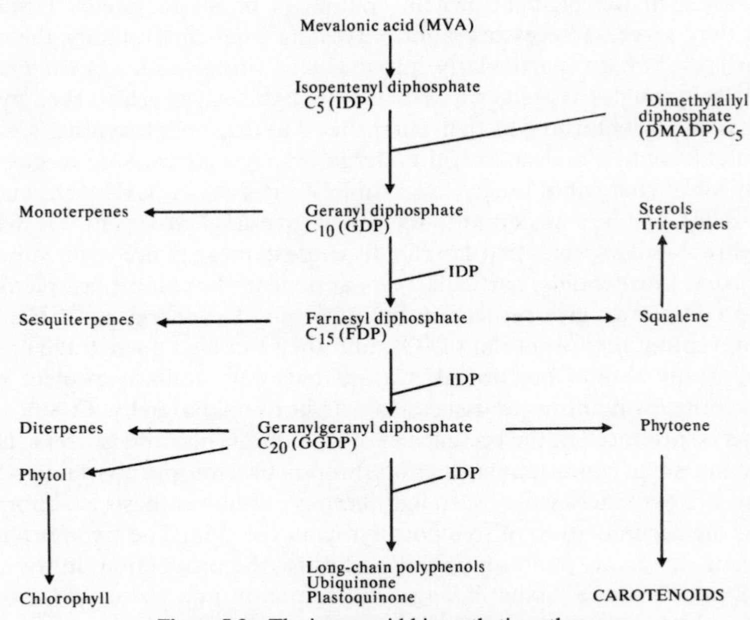

Figure 5.3 The isoprenoid biosynthetic pathway.

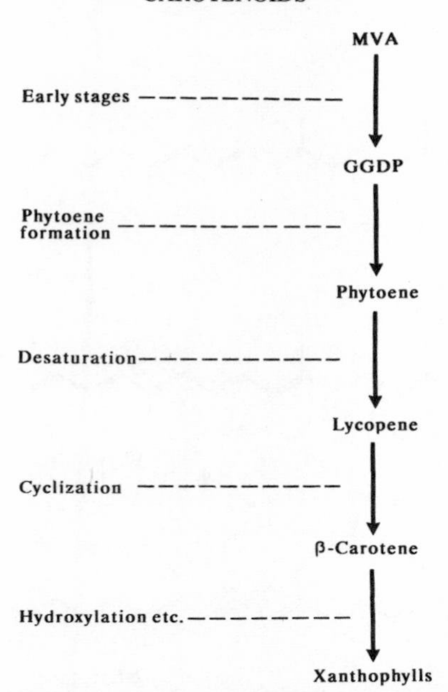

Figure 5.4 Summary of the main stages of carotenoid biosynthesis.

(c.d.b.). The intermediates are phytofluene (5 c.d.b.), ζ-carotene (7 c.d.b.), neurosporene (9 c.d.b.) and the normal final product, lycopene (11 c.d.b.).

Lycopene can then undergo cyclization at one end or both ends of the molecule to give the monocyclic *γ*-carotene and *δ*-carotene and the bicyclic *β*-carotene, *α*-carotene and *ε*-carotene (Figure 5.6). Similar cyclization of the more desaturated end group of neurosporene gives *β*- and *α*-zeacarotenes. The cyclizations occur by the mechanism shown in Figure 5.7, in which alternative losses of H^+ from C-6, C-18 or C-4 give rise to the *β*-, *γ*- or *ε*-rings, respectively.

Oxygen functions are normally introduced as the final steps in the biosynthesis. Thus the common 3-hydroxy-xanthophylls lutein and zeaxanthin are formed by the stereospecific hydroxylation of the corresponding carotenes (*α*- and *β*-carotene) by mixed-function oxidase enzymes (Figure 5.8).

Pathways have been proposed for the formation of most of the structural variations that are found in carotenoids, for example, of several different end groups in plant carotenoids from a 5,6-epoxy-*β*-ring (Figure 5.9). In most cases, however, there is little or no biochemical evidence to support these proposals.

The apocarotenoids, including the apo-*β*-carotenals, crocetin and bixin, that occur in plants are believed to be biosynthesized by excision of fragments

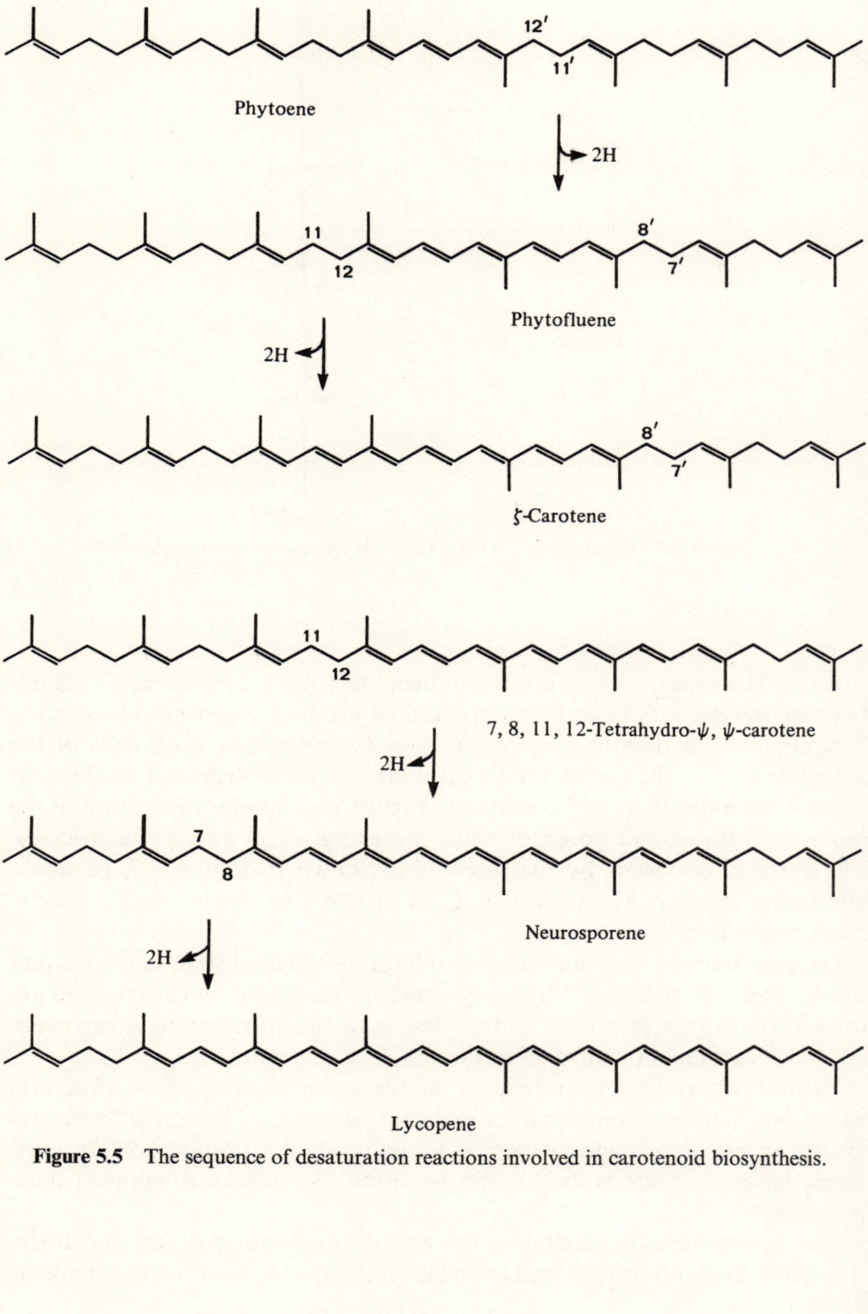

12′
11′
Phytoene
2H
11
12
8′
7′
Phytofluene
2H
8′
7′
ζ-Carotene
11
12
7, 8, 11, 12-Tetrahydro-ψ, ψ-carotene
2H
7
8
Neurosporene
2H
Lycopene

Figure 5.5 The sequence of desaturation reactions involved in carotenoid biosynthesis.

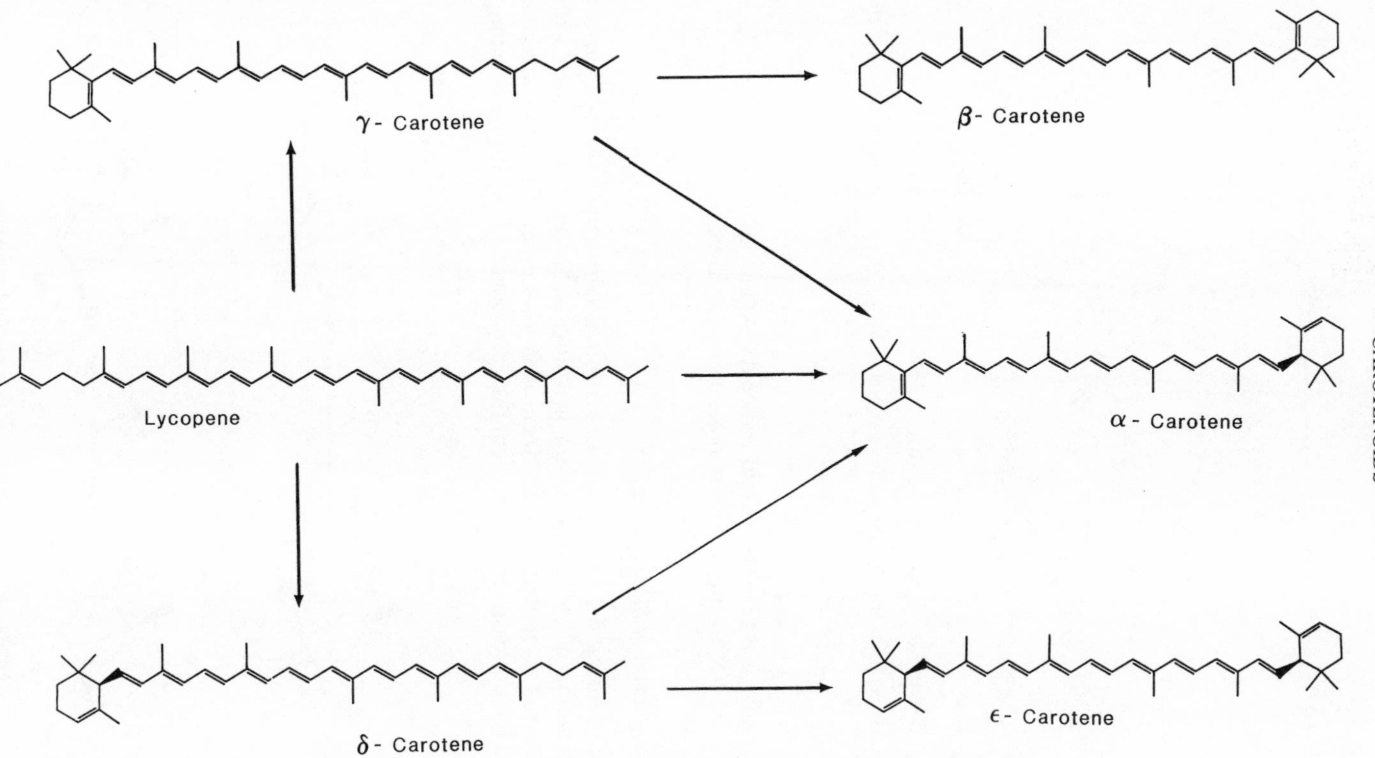

Figure 5.6 The biosynthesis of monocyclic and bicyclic carotenes from lycopene.

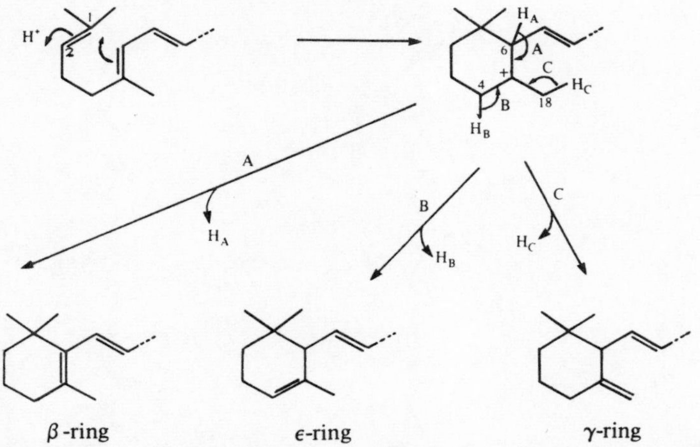

β-ring ε-ring γ-ring

Figure 5.7 Mechanisms for the alternative formation of β-, γ- and ε-rings from the acyclic precursor, neurosporene.

from the ends of a C_{40} carotenoid molecule. Again, there is no direct biochemical evidence to support these reasonable suggestions.

5.4.1 *Regulation of biosynthesis*

In plants, carotenoid biosynthesis occurs universally as an essential part of the construction of chloroplasts, but it is also particularly associated with chromoplast development as fruit ripen or flowers open. Thus, in *Capsicum* or tomato fruit, the carotenoid content increases several fold during ripening.

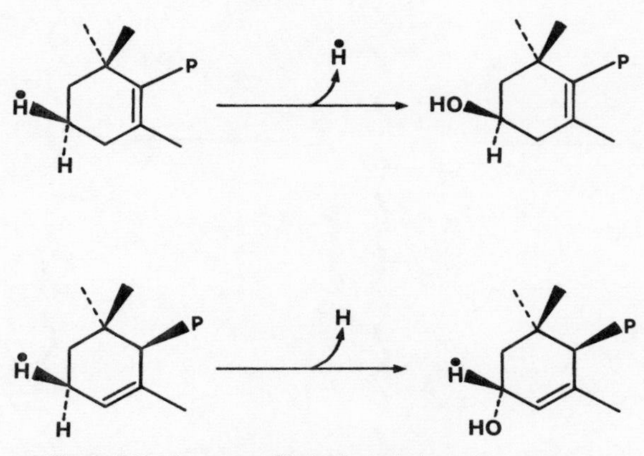

Figure 5.8 Stereospecific hydroxylation of carotenes.

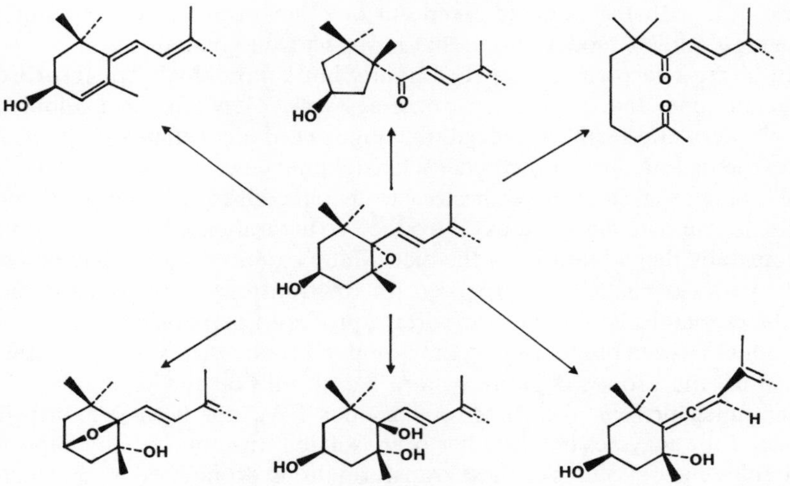

Figure 5.9 Scheme for the formation of different end-groups of plant carotenoids.

Carotenoid contents and compositions in plants are affected by environmental and nutritional conditions, and can be altered drastically by treatment with a range of chemicals, especially amines. The genetics of carotenoid biosynthesis in tomatoes has been studied intensively and many strains with different carotenoid compositions have been produced.

Rapid progress is now being made on the molecular genetics of carotenoid biosynthesis, particularly in phototrophic bacteria but also in other organisms [12]. Some fungi (*e.g. Blakeslea trispora* and *Phycomyces blakesleeanus*), bacteria (*e.g. Flavobacterium* spp.) and microalgae (*e.g. Dunaliella bardawil*) are capable of very high levels of carotenoid synthesis, which is likely to be amenable to genetic manipulation for biotechnological production.

5.5 Absorption, transport and metabolism

There is very wide variation in the ability of different animal species to absorb and metabolize carotenoids [2, 15]. Humans are generally good and indiscriminate absorbers of all carotenoids that are offered in the diet. Many other mammals, for example the cat, rat, mouse and the sheep, are very poor absorbers and metabolizers. Cattle efficiently absorb β-carotene but not xanthophylls; β-carotene colours the fat and may accumulate in high concentration in the corpus luteum, and it is considered to be implicated in maintaining fertility in cattle. Although most work on the absorption and metabolism of carotenoids has been performed on rats and mice, neither these nor, with the possible exception of other primates, other mammals provide a reliable model system that is applicable to humans.

In birds, β-carotene is absorbed poorly but xanthophylls are absorbed in large amounts and transported to the egg yolk. They may also colour skin (in chicken) and feathers; indeed, flamingos need a constant supply of oxo-carotenoids to maintain their characteristic pink colour.

Although they cannot synthesize carotenoids, birds, fish and invertebrate animals, can, in many cases, introduce structural modifications into the carotenoids they obtain from the diet. Many oxidative and reductive metabolic pathways have been proposed for the transformations of carotenoids in these organisms, the most important processes perhaps being those that introduce oxygen functions into the β-ring to produce astaxanthin. In almost all cases, the proposed pathways are based only on the structures of the compounds present and direct conversions have not been demonstrated. None of the enzymes involved has been isolated. In commercially important examples such as salmon, these transformations do not occur; satisfactory pink-fleshed salmon are only obtained when astaxanthin itself is supplied in the diet [16].

Dietary carotenoids are absorbed in the gut along with other lipids. A most important process, following absorption, is the conversion of β-carotene and other suitable molecules, that is those having one unsubstituted β-ring, into vitamin A (retinol) [17]. This conversion occurs mainly in the small intestine, although there are reports of its occurrence also in other tissues, for example liver and kidney. The accepted mechanism of conversion is central cleavage, which requires a 15,15′-dioxygenase enzyme and can provide two molecules of retinal from one molecule of β-carotene. In addition to central cleavage, it is likely that excentric cleavage by oxidation of other double bonds also occurs to give apocarotenals of different chain lengths (C_{22} to C_{30}), which undergo further oxidation to give retinal (one molecule from each molecule of β-carotene). The retinal that is produced by either of these processes is then reduced enzymically to retinol, which is transported by retinol-binding protein to its required sites of action or to the liver for storage as acyl esters. The formation of vitamin A from the parent carotenoid is regulated so that excessive amounts of the highly toxic vitamin are not produced.

Carotenoids that are absorbed intact are also transported on blood lipoproteins, and can be deposited in tissues [18, 19]. β-Carotene and other carotenes are associated with the low-density lipoprotein, the various xanthophylls in the human diet are transported on the high-density lipoprotein fraction [20]. When excessive amounts of carotenoids are supplied they may be deposited in many tissues and can accumulate in concentrations sufficient to cause yellow-orange coloration of the skin, especially that of the hands and feet. Although this normally happens only in individuals who ingest a large amount of pure carotenoid, carotenodermia is also sometimes seen in people who eat large amounts of, for example, carrots and oranges which have a high carotenoid content. Coloration by carotenoids is generally harmless and reversible, although there are cases where the intake of large amounts of canthaxanthin (200 mg/day) in the form of oral 'sun-tanning' capsules not only imparts the sought-after bronzed (although somewhat orange) appearance of the skin, but has resulted in the appearance of crystals of canthaxanthin in the eye. There is, fortunately, no risk of this happening with the much smaller amounts (microgram quantities) of canthaxanthin that are ingested as food colorants. The absorption of β-carotene and other carotenoids from plant food sources is described as 'poor to fair', while that of pure β-carotene is rather better. Absorption is facilitated by the presence of fats and agents such as lecithin, which aid emulsification. Virtually nothing is known about the degradative metabolism of carotenoids in humans.

Carotenoids can persist in the human body for a considerable time. β-Carotene has a half-life of a few days [21] but canthaxanthin, if large amounts have been taken, can persist for 6 to 12 months. The β-carotene status and general dietary carotenoid profile of an individual can be monitored rapidly by HPLC analysis of a small sample of blood or serum.

5.6 Natural and synthetic carotenoids as colorants

Carotenoids and carotenoid-containing extracts have been widely used for centuries as colorants in food. The coloration may be introduced by direct addition to the foodstuff or indirectly by feeding to animals (*e.g.* chickens and fish), which thus become suitably coloured ready for use as food. All aspects of the use of carotenoids as colorants are covered in detail in Bauernfeind [22].

5.6.1 Natural carotenoids and extracts

The natural extracts first used were those of annatto, carrot, palm oil, saffron, tomato and paprika. The powdered, dried plant materials and extracts of them have been used for many years. These are not, however, pure carotenoids or simply mixtures of carotenoids but generally contain large amounts and large numbers of other, mainly unidentified substances.

5.6.1.1 Paprika. One of the oldest and most important such extracts is that of paprika (*Capsicum annuum*), which is produced as a dry powder or an oil extract or oleoresin. These products provide hot and spicy flavour as well as colour, and their use is thus limited to suitably savoury products. The main carotenoids present are capsanthin and capsorubin (largely as acyl esters) but there are many other minor components.

5.6.1.2 Annatto. The use of annatto also has a long history. Annatto is obtained as extracts of the red-brown resinous coating of the seeds of *Bixa orellana*, a tree that grows abundantly in the tropics. The major pigment present is the apocarotenoid bixin (9-*cis*) and this methyl ester is the main component in oil-based preparations. Hydrolysis (saponification) liberates the dicarboxylic acid, norbixin, salts of which provide the water-soluble annatto preparations. The seedcoat contains a high concentration of bixin together with small amounts of many unidentified but probably related compounds. The biosynthesis of bixin has not been elucidated but is presumed to follow the normal C_{40} pathway to lycopene (normal C_{40} phytoene, phytofluene and ζ-carotene are present). Many preparations of annatto are available with different hues (usually pinkish) and are used to colour a wide range of food products.

5.6.1.3 Saffron. One of the earliest additives, saffron is the powdered dried flowers of *Crocus sativus*. The main constituent is crocin, the bis-gentiobioside (and other glycosides) of the diapocarotenedioic acid, crocetin. Saffron is used particularly to impart a pure yellow colour to rice and other foods. It is also used as a spice.

5.6.1.4 Tomato. Extracts of tomato (*Lycopersicon esculentum*) have a high lycopene content (80 to 90% of total carotenoid). Their use as colorants is limited because of the strong tomato flavour.

5.6.1.5 Xanthophyll pastes. These usually consist of concentrated extracts of leaves (*e.g.* nettles, alfalfa and broccoli), and are green rather than yellow unless saponification has been used. Many 'xanthophyll' pastes contain as much as 30% carotene. Other xanthophyll extracts are obtained from flowers. Marigold extracts (*Tagetes erecta*), or the dried, crushed flowers themselves, are very rich in lutein and are used to pigment egg-yolk and chickens.

5.6.1.6 Palm oil. The red oil obtained from the fruit of the tropical palm tree *Elaeis guineensis* contains 500 mg of carotenoid, mainly β-carotene, per kilogram of oil.

5.6.2 Synthetic carotenoids

Following the successful commercial synthesis of vitamin A, a similar procedure was used for the production of β-carotene, which was introduced onto the market in 1954. Several companies now manufacture β-carotene by processes based on Grignard reactions, enol ether condensations, ethynylation, Wittig and sulphone reactions; the total annual output now exceeding 500 tonnes. Application of the same synthetic methods subsequently led to the commercial production of 8′-apo-β-caroten-8′-al, 8′-apo-β-caroten-8′-oic acid ethyl or methyl ester and citranaxanthin, which are also widely used. More recently canthaxanthin has entered production, being made originally by oxidation of β-carotene, later by direct synthesis. Commercial production of the natural optical isomers of zeaxanthin (3*R*, 3′*R*) and astaxanthin (3*S*, 3′*S*) is now under development [23].

The commercial synthetic products are crystalline and of high purity and they are identical in all respects to the natural substances. They are marketed as crystals, as micronized suspensions in oil and as water-dispersible formulations.

5.7 General properties and stability

Carotenoids crystallize in a variety of forms, and the crystals vary in colour from orange-red to violet and almost black, depending on their shape and size. The melting points are high, usually in the range 130 to 220°C. Even the crystals are very sensitive to oxidative decomposition when exposed to air, and they must be stored in an inert atmosphere or under vacuum. The pure carotenoids are greatly stabilized by suspension or solution in vegetable oil, especially in the presence of antioxidants such as α-tocopherol. Peroxidation of unsaturated lipids in the oil can, however, cause the rapid oxidative destruction of carotenoids. Carotenoids *in situ* are also very susceptible to enzymic degradation by lipoxygenases when tissues are disrupted.

Carotenoid crystals tend to have only poor solubility. They are insoluble in water, only slightly soluble in vegetable oils and very soluble only in chlorinated solvents (*e.g.* chloroform and dichloromethane). The crystals are generally very slow to dissolve, although rates of solution increase with heating.

5.8 General procedures for carotenoid work

Full details of recommended methods for working with carotenoids, including some worked examples, are given in several articles [24–27]. The conjugated polyene chromophore of carotenoids renders them sensitive to oxygen, light and heat. Other structural features are easily modified by acid or alkali. Stringent precautions must be observed, therefore, if losses of material or unwanted structural changes are to be avoided. Speed of manipulation is important. All procedures that inevitably introduce risks of oxidation, isomerization, and so on, must be carried out as rapidly as possible. The sensitivity of modern analytical methods has greatly increased the possibility of detecting artifacts produced during extraction and purification. The strong colour of carotenoids can mask the presence of substantial amounts of colourless contaminants that are not detected during the normal spectrophotometric assay of carotenoids.

The need to protect carotenoids against oxidation cannot be overemphasized. The presence of traces of oxygen in stored samples (even at deep-freeze temperatures) and of peroxides in solvents (especially diethyl ether) or of any oxidizing agents even in extracts containing carotenoids can rapidly lead to bleaching, or to the formation of artefacts such as epoxides or apocarotenals. The presence of unsaturated lipids and metal ions can greatly enhance the oxidative breakdown, especially if lipoxygenase enzymes are also present. Carotenoids, or extracts containing them, should always be stored in the complete absence of oxygen, either *in vacuo* or in an inert atmosphere (Ar or N_2).

In commercial, water-dispersible formulations and in the blue carotenoproteins, carotenoid molecules are generally stabilized by protein. When plant tissues are disrupted, however, even the carotenoids that are located in the pigment-protein complexes of the thylakoids can undergo rapid photochemical or enzymic oxidation; β-carotene is particularly susceptible.

Changes in the geometrical isomeric composition of a carotenoid can easily occur during isolation and purification. To minimize such changes, exposure of carotenoids to heat or light, especially direct sunlight, should be avoided if possible. In the presence of chlorophyll or other potential sensitizers, photoisomerization, presumably via the carotenoid triplet state, occurs very rapidly and appreciable amounts of artifactual geometrical isomers can be produced. Carotenoids should be shielded from light during chromatography. To avoid excessive heating, solvents with low boiling points should be used whenever possible, since these can subsequently be removed at low temperature.

Almost all carotenoids are susceptible to decomposition, dehydration or isomerization if subjected to acid conditions. Carotenoid 5,6-epoxides undergo particularly facile isomerization to the corresponding furanoid 5,8-epoxides; many plant tissues are sufficiently acid to bring about this isomerization during extraction, but it can usually be prevented by the inclusion of a neutralizing agent such as sodium bicarbonate ($NaHCO_3$) during extraction. Acidic adsorbents, especially silica gel and silicic acid, can cause isomerization during chromatography and should be avoided, as should acidic solvents, notably chloroform, which usually contains traces of hydrochloric acid (HCl). Acidic reagents and strong acids should not be used in rooms where carotenoids are being handled.

Most carotenoids are stable to alkali, so saponification is used routinely to hydrolyze carotenoid esters or to destroy contaminating chlorophylls or oils. Some carotenoids, however, notably those containing the 3-hydroxy-4-oxo-β-ring as in astaxanthin, are altered by treatment with even weak alkali. Saponification must be avoided if it is suspected that any such compounds may be present or, of course, if carotenoid esters are to be isolated.

5.9 Extraction and purification

Carotenoids should be extracted from tissues as rapidly as possible, in order to minimize oxidative or enzymic degradation. Extraction normally employs a water-miscible organic solvent, usually acetone, methanol or ethanol; chloroform-methanol is not recommended for carotenoids because of the possible presence of HCl. Extraction from dried tissue is usually more efficient if the tissue is first treated with a little water. Efficient extraction from plant material usually requires mechanical tissue disruption. On the laboratory scale, the usual procedure involves direct homogenization of the tissue with the solvent in a suitable electric blender. For small samples, grinding the tissue in acetone with clean sand (pestle and mortar) may be used, but quantitative extraction by this means is not easy to achieve. On a commercial scale, prolonged steeping in solvent, with suitable agitation, is often used. Although cold solvent should be used if possible, extraction may sometimes be so much more efficient with hot solvent that the advantages of rapid extraction outweigh the possible deleterious effects of heat. After filtration, the solid debris is re-extracted with fresh solvent until no more colour is recovered (usually twice or three times). For large-scale commercial preparations, the solvent extracts are concentrated directly. In analytical work, the carotenoid-containing lipid extract is transferred to ether, washed two to three times with water and evaporated to dryness. The extract may then be saponified, if required, in 6 to 10% ethanolic potassium hydroxide (KOH) at room temperature, in the dark, under N_2, overnight. All traces of acetone must be removed before saponification, to avoid the possibility of producing artifacts by facile aldol condensation between apocarotenals and acetone, or introducing contaminants by the polymerization of acetone.

Column chromatography, thin-layer chromatography (TLC) and high-performance liquid chromatography (HPLC) are widely used with carotenoids, but GLC is not suitable because of their instability. A rapid preliminary examination by TLC gives an indication of the number and variety of carotenoids present and allows the best separation and purification procedure for the extract to be determined. Column chromatography is used mainly for large-scale purifications or, in analytical work, to separate an extract into fractions that contain groups of compounds of similar polarity. TLC may then be used to isolate and purify the individual carotenoids. A general strategy that will usually yield a pure carotenoid is to use column chromatography on alumina followed by TLC successively on silica gel, MgO-kieselgur G and silica again, but with a different solvent. Samples for analysis by mass spectrometry (MS) and nuclear magnetic resonance (NMR) need further treatment to bring them to the rigorous state of purity required. HPLC is now the method of choice for routine qualitative and quantitative analysis of carotenoid compositions. It should be noted, however, that identification by HPLC, even combined with UV-visible absorption spectra, does not constitute rigorous characterization. Alumina and silica are widely used for column chromatography and TLC. Other materials, particularly basic inorganic substances such as magnesium oxide (MgO) and calcium hydroxide ($Ca(OH)_2$), are also extremely valuable for carotenoid purification.

Separation on alumina and silica depends upon polarity; the carotenes are very weakly adsorbed, the xanthophylls more strongly, depending on the functional groups present. In contrast, separation on MgO, $Ca(OH)_2$ and so on is determined by the number and arrangement of double bonds in the molecule. Carotenoids with the most extensive conjugated polyene system are most strongly adsorbed on MgO (*e.g.* lycopene > neurosporene > ζ-carotene). Acyclic carotenoids are much more strongly adsorbed than cyclic ones that have the same number of double bonds (*e.g.* lycopene > γ-carotene > β-carotene), and β-ring compounds are more strongly held than the corresponding ε-ring isomers (*e.g.* β-carotene > α-carotene; zeaxanthin > lutein). Carotenoid-5,6-epoxides are much less strongly adsorbed than the corresponding 5,8-epoxides. Other polar groups, even hydroxy groups, are generally not a strong influence; lycopene is much more strongly adsorbed than the diol zeaxanthin.

Other basic adsorbents, especially $Ca(OH)_2$ and zinc carbonate ($ZnCO_3$) are very useful for separating geometrical isomers that may otherwise be difficult or impossible to resolve.

The instability of some carotenoids can cause problems even with the commonly used adsorbents. Thus, unless neutralized, silica gel or silicic acid can be sufficiently acidic to cause isomerization of 5,6-epoxycarotenoids. On alumina, some carotenoids, especially those containing a 3-hydroxy-4-oxo-β-ring (*e.g.* astaxanthin) undergo irreversible oxidation to the corresponding 2,3-didehydro (diosphenol) derivatives, which are virtually impossible to elute. Indiscriminate use of activated alumina (grade O or I) can cause geometrical isomerization. If used without dilution with a filter aid (celite) or binder (kieselgur G), MgO is sufficiently basic to cause polymerization of acetone in the solvent, and to catalyze aldol condensation between acetone and carotenoid aldehydes.

The usual adsorbent for column chromatography is alumina (deactivated to Grade III by addition of 6% water), although silica can also be used. The column is eluted with solvents of increasing polarity. Suitable solvents and the carotenoid groups eluted by each are shown in Table 5.2.

For TLC on silica, ether-petrol or acetone-petrol mixtures are very suitable for carotenoids, and the solvent composition that was used to elute the fraction from an alumina column usually gives effective separation. Other mixtures, for example of methanol-toluene, propan-2-ol-petrol or ethyl acetate-carbon tetrachloride may be used for the second TLC on silica. The most effective solvents for TLC on MgO-kieselgur G are mixtures of petrol with acetone and toluene, or both. The adsorptive strength of MgO is somewhat variable, but the following solvent mixtures are approximately correct: 4%

Table 5.2 Suitable solvents and the carotenoid groups eluted using column chromatography

Solvent system	Carotenoids separated
Light petroleum (petrol) (or diethyl ether (ether) : petrol (1 : 99)	Carotenes
Ether : petrol (5 : 95)	Carotene epoxides
Ether : petrol (10 : 90)	Xanthophyll esters
Ether : petrol (20 : 80)	Monooxocarotenoids
Ether : petrol (1 : 1)	Dioxocarotenoids, Monohydroxy carotenoids
Ether (or ethanol : ether) (5 : 95)	Dihydroxy—more polar xanthophylls
Ethanol : ether (1 : 4)	Carotenoid glycosides

acetone in petrol for α- and β-carotene, 20 to 25% acetone in petrol for lutein and zeaxanthin and acetone-toluene-petrol (1 : 1 : 4) for lycopene. Elution from MgO is best achieved with acetone, plus some toluene and ethanol for strongly adsorbed carotenoids.

For further details and discussion of the purification of carotenoids, the general references [24–27] should again be consulted.

5.10 High-performance liquid chromatography (HPLC)

HPLC is the method of choice for both qualitative and quantitative analysis of carotenoids. The carotenoids are readily detected by their UV-visible light absorption, and the procedure is extremely powerful when a photodiode array detector is used to permit simultaneous detection and monitoring at any chosen wavelengths and also continuous determination and memorizing of absorption spectra during the chromatography. More detailed treatment of HPLC of carotenoids is given by Rüedi [28] and by Goodwin and Britton [29].

5.10.1 Normal-phase (adsorption) HPLC

Gradient elution is usually employed for complex extracts that contain carotenoids of widely different polarities. Particularly impressive resolution of carotenediols and more polar xanthophylls can be achieved with the following programme:

- 0–8 min: isocratic, 10% solvent A, 90% solvent B
- 8–20 min: linear gradient 10–25% solvent A
- 20–30 min: isocratic 40% solvent A
- 30–40 min: isocratic 70% solvent A,

where solvent A is hexane propan-2-ol (8 : 2) and solvent B is hexane, and the flow rate is 2 ml/min.

This procedure will separate efficiently mixtures of geometrical isomers of the main xanthophylls such as zeaxanthin, lutein, violaxanthin and neoxanthin. The use of propan-2-ol is not so satisfactory, however, for separating less polar carotenoids. For this, mixtures of hexane with ethyl acetate are recommended, but long pre-equilibration is needed to remove residual traces of more polar solvents, especially alcohols.

Bonded nitrile phase columns (*e.g.* Spherisorb S5-CN) [28] permit the very efficient resolution of mixtures of similar carotenoids, including structural isomers (lutein and zeaxanthin), geometrical isomers, 5,6- and 5,8-epoxides and even the 8*R* and 8*S* epimers of the 5,8-epoxides. The best results are achieved by isocratic separation of fractions (*e.g.* 'dihydroxy-fraction') that have been obtained from natural extracts by classical column chromatography or TLC. Combinations of two solvent mixtures are used; solvent A is of constant composition, namely hexane containing 0.1% N-ethyldiisopropylamine, whereas solvent B comprises dichloromethane containing a variable proportion of methanol. Solvent compositions recommended for separating the carotenoids of various polarity groups have been tabulated by Rüedi [28].

5.10.2 Reverse-phase HPLC

Reverse-phase partition chromatography is now most widely used for the routine analysis of carotenoids in natural extracts. There is virtually no risk of decomposition or structural modification, even of the most unstable carotenoids. The stationary phases usually used are those with C_{18}-bonded chains (ODS) and the resolution is somewhat dependent on the extent of endcapping and carbon loading of the stationary phase. The order of elution from reversed-phase columns is not exactly the opposite of that from normal-phase columns, for example lutein is eluted before zeaxanthin in both methods. As well as the overall polarity of the compounds (presence of polar substituent groups) other important factors influence separations. The strongest association with the C_{18} stationary phase is that of an unsubstituted carotenoid end group, so that a diol with both hydroxy groups in one end group will be substantially more strongly retained than one with one hydroxyl in each end group. Acyclic carotenoids generally have longer retention times than cyclic ones containing the same functional groups, and carotenoids with a greater degree of saturation are usually more strongly retained.

Non-aqueous solvent systems (acetonitrile-dichloromethane-methanol) have been widely explored for reversed-phase HPLC of carotenoids [30]. In the author's laboratory, a simple gradient procedure is used as a routine method for screening carotenoid compositions [29]. This procedure (adapted from Wright and Shearer [31]), employs a linear gradient of ethyl acetate (0 to 100%) in acetonitrile-water (9 : 1, containing 0.1% triethylamine) at a flow rate of 1 ml/min over 25 min, and gives good resolution both of polar xanthophylls and of non-polar carotenes, carotene epoxides and xanthophyll acyl esters in the same run. The gradient is easily modified, if required, to improve resolution of selected parts of the chromatogram. A natural esterified carotenoid will usually contain a mixture of molecular species containing different esterifying fatty acids. The presence of esters is therefore usually revealed by a pattern of unusual peaks in the chromatogram in the vicinity of β-carotene. The absorption spectra of xanthophylls are not altered by esterification so the carotenoid component can tentatively be identified.

5.10.3 HPLC of apocarotenoids

The chromatographic behaviour of apocarotenoids on a reversed-phase column can cause some confusion. The retention times are short; apo-β-carotenals, for example, are found in the same area of the chromatogram as the very polar xanthophyll, neoxanthin. The retention times increase with increasing chain length, which is a more important determining feature than the nature of the functional groups present.

5.10.4 Resolution of optical isomers

Several methods are available for the resolution of optical isomers of natural carotenoids, either direct on optically active columns or after formation of diastereoisomeric derivatives.

5.11 UV-visible light absorption spectroscopy

Both the position of the absorption maxima (λ_{max}) and the shape or fine structure of the spectrum are characteristic of the chromophore of the carotenoid molecule.

5.11.1 Position of the absorption maxima

Spectra of apolar carotenoids are usually determined in petrol or hexane, those of the more polar xanthophylls in ethanol. The absorption spectra of most carotenoids exhibit three maxima. Values of λ_{max} are markedly dependent on solvent. The values recorded in petrol, hexane and ethanol are almost identical, but values recorded in acetone are greater by around 4 nm, those recorded in chloroform or benzene greater by 10 to 12 nm, and those in carbon disulphide greater by as much as 35 to 40 nm. When spectra are determined on-line during HPLC, it must be remembered that the values for

λ_{max} in the eluting solvent frequently do not correspond to the published values recorded in a pure solvent.

Also, with gradient HPLC, the solvent composition is changing throughout the chromatography, so compounds which in published tables are given the same maximum values may not give the same maximum values during chromatography.

The λ_{max} values of the main carotenoids are given in Table 5.3. Comprehensive tables giving λ_{max} values for a wider range of carotenoids in a variety of solvents are available in other articles [25, 26, 32].

In any given solvent, λ_{max} values increase as the length of the chromophore increases. Non-conjugated double bonds, for example the C-4,5 double bond of the ε-ring, do not contribute to the chromophore. Extension of the conjugated double bond system into a ring (the C-5,6 double bond of the β-ring) does extend the chromophore but, because the ring double bond is not coplanar with the main polyene chain, the λ_{max} occur at shorter wavelengths than those of the acyclic carotenoid with the same number of conjugated double bonds. Thus, although they are all conjugated undecaenes, the acyclic, monocyclic and bicyclic compounds lycopene, γ-carotene and β-carotene have λ_{max} at 444, 470, 502 nm, at 437, 462, 494 nm and at 425, 450, 478 nm, respectively.

Carbonyl groups, in conjugation with the polyene system, also extend the chromophore. A ring keto-group increases λ_{max} by 10 to 20 nm, so that echinenone and canthaxanthin have λ_{max} at 461 and 478 nm, respectively.

Other substituents, such as hydroxy and methoxy groups do not affect the chromophore. All such substituted carotenoids therefore have λ_{max} virtually identical to those of the parent hydrocarbon with the same chromophore. For example, β-carotene and its hydroxy-derivatives β-cryptoxanthin, zeaxanthin, isocryptoxanthin, and isozeaxanthin, all have virtually identical spectra with λ_{max} at 425, 450, 478 nm.

5.11.2 Spectral fine structure

The overall shape or fine structure of the spectrum is also diagnostic and generally reflects the degree of planarity that the chromophore can achieve. Thus the spectra of acyclic compounds are characterized by sharp maxima and minima (equivalent to persistence). The degree of fine structure decreases a little when the chromophore exceeds nine double bonds. Cyclic carotenoids in which conjugation does not extend into the rings have simple linear polyene chromophores.

When conjugation extends into a β-ring, steric strain causes the ring to adopt a conformation in which the ring double bond is not coplanar with the π-electron system of the polyene system. This effect is even more pronounced in carotenoids that have carbonyl groups in conjugation with the polyene chain. Spectral fine structure therefore decreases in the order lycopene > γ-

Table 5.3 Light absorption maxima and specific absorbance coefficients ($A^{1\%}_{1\,cm}$) of some carotenoids

Carotenoid	λ_{max} (nm)			Solvent[a]	$A^{1\%}_{1\,cm}$ [b]
Actinioerythol	470	496	529	P	
	480	508	538	A	
		518		C	
Antheraxanthin	422	445	472	P	
	422	444	472	E	
	430	456	484	C	
8′-Apo-β-caroten-8′-al		*457*		P	2640
		463		E	
		477		C	
8′-Apo-β-caroten-8′-oic acid ethyl or methyl ester		*445*	470	P	2500
Astaxanthin		468		P	
		478		E	
		480		A	
		485		C	
Bixin	432	*456*	490	P	4200
	433	470	502	C	
Canthaxanthin		*466*		P	2200
		474		E	
		482		C	
Capsanthin	450	475	505	P	
	460	*483*	518	B	2072
Capsorubin	445	479	510	P	
	460	489	523	B	2200
α-Carotene	422	*444*	473	P	2800
	423	444	473	E	
	424	448	476	A	
	433	457	484	C	
β-Carotene	425	*449*	476	P	2592
		450	476	E	2620
	(429)	452	478	A	
	435	*461*	485	C	2396
γ-Carotene	437	*462*	494	P	3100
	440	460	489	E	
	439	461	491	A	
	446	475	509	C	
δ-Carotene	431	*456*	489	P	3290
	440	470	503	C	
ε-Carotene	416	*440*	470	P	3120
	417	440	470	E	
ζ-Carotene	378	400	425	P	
	380	*400*	425	H	2555
	377	399	425	E	
Citranaxanthin		463	495	H	
		463	495	P	2145
		475		E	
Crocetin	400	422	*450*	P	4320
	401	423	447	E	
	413	435	462	C	

continued

Table 5.3 *Continued*

Carotenoid	λ_{max} (nm)			Solvent	$A^{1\%}_{1\,cm}$
α-Cryptoxanthin	421	*445*	475	H	2636
	434	456	485	C	
β-Cryptoxanthin	425	*449*	476	P	2386
	428	450	478	E	
Isocryptoxanthin	425	448	475	P	
	425	450	477	E	
	438	464	489	C	
Isozeaxanthin	427	*450*	475	P	2400
	430	451	478	E	
	435	463	489	C	
Lactucaxanthin		438	468	P	
	419	440	470	E	
	427	450	479	C	
Lutein	421	445	474	P	
	422	*445*	474	E	2550
	435	458	485	C	
Lutein epoxide	420	443	472	P	
	420	442	471	E	
	433	453	483	C	
Lycopene	444	*470*	502	P	3450
	446	472	503	E	
	448	474	505	A	
	458	484	518	C	
Neoxanthin	416	438	467	P	
	415	*439*	467	E	2243
	423	448	476	C	
Neurosporene	414	439	467	P	
	416	*440*	470	H	2918
	416	440	469	E	
	424	451	480	C	
Norbixin	442	474	509	C	
Phytoene	276	*286*	297	P	1250
	276	*286*	297	H	915
Phytofluene	331	*348*	367	P	1350
	331	*347*	366	H	1577
Violaxanthin	416	440	465	P	
	419	*440*	470	E	2550
	426	449	478	C	
Violerythrin		566		A	
α-Zeacarotene	398	*421*	449	H	1850
β-Zeacarotene	406	*428*	454	P	2520
	406	427	454	H	1940
	405	428	455	E	
Zeaxanthin	424	*449*	476	P	2348
	428	*450*	478	E	2540
	430	*452*	479	A	2340
	433	462	493	C	

[a] Solvents: P = light petroleum; A = acetone; C = chloroform; E = ethanol; H = hexane; B = benzene.
[b] The $A^{1\%}_{1\,cm}$ values given are for the wavelength printed in italics.

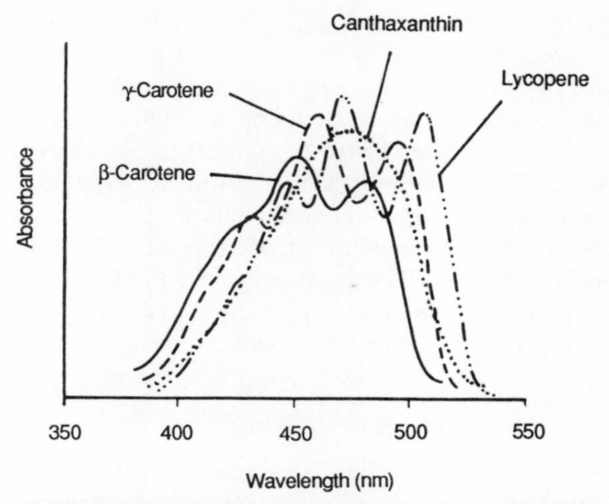

Figure 5.10 Spectral characteristics of common carotenoids.

carotene > β-carotene > canthaxanthin (Figure 5.10). Canthaxanthin shows only a single, rounded, almost symmetrical absorption peak (in ethanol), but a slight degree of fine structure remains if the spectrum is determined in a non-polar solvent such as light petroleum.

5.11.3 Geometrical isomers

For *Z*-isomers, the λ_{max} are generally 1 to 5 nm lower, the spectral fine structure is decreased and a new absorption peak, usually referred to as the '*cis*-peak' appears at a characteristic wavelength in the UV region 142 ± 2 nm below the longest wavelength peak in the main visible absorption region. The intensity of the '*cis*-peak' is greatest when the *Z*-double bond is located at or near the centre of the chromophore.

5.12 Quantitative determination

5.12.1 Spectrophotometry

Spectrophotometric analysis is normally used for the quantitative determination of carotenoids. The amount of carotenoid present is calculated from the equation

$$x = \mathrm{A}y/(\mathrm{A}^{1\%}_{1\,\mathrm{cm}} \times 100) \quad (5.1)$$

where x is the mass of carotenoid (g), y the volume of solution (ml), A the measured absorbance, and $A_{1\,cm}^{1\%}$ is the specific absorbance coefficient, that is, the absorbance of a solution of 1 g of that carotenoid in 100 ml of solution. $A_{1\,cm}^{1\%}$ values for some common carotenoids are included in Table 5.3. More extensive tables of $A_{1\,cm}^{1\%}$ values have been published elsewhere [25, 26, 32]. An arbitrary value of 2500 is often taken when no experimentally determined value has been reported, for an unknown compound, or to give an estimate of the total carotenoid content of an extract. It is difficult to achieve complete solution, especially of crystalline carotenoids even in favourable solvents, so carotenoid contents are easily underestimated. Also there is doubt about the accuracy of some of the published $A_{1\,cm}^{1\%}$ values.

5.12.2 Quantitative determination by HPLC

HPLC provides the most sensitive, accurate and reproducible method for quantitative analysis of carotenoids, particularly when the instrumentation includes automatic integration facilities for measuring peak areas. The relative amounts of each component in the chromatogram can be determined, provided the peak area can be calculated for each component at its λ_{max} by use of a multi-wavelength detector. For the estimation of absolute amounts or concentrations, calibration is necessary and this can be achieved by means of a calibration graph or by use of an internal standard.

5.13 Other physicochemical methods

5.13.1 Mass spectrometry (MS)

Although chemical ionization has been used, most MS work on carotenoids has employed ionization by electron impact. Carotenoids have very low volatility, and samples are usually inserted by means of a direct probe, heated to 200 to 220°C. Almost all carotenoids give good molecular ions. In addition, many fragmentations that are diagnostic of particular structural features have been identified. These have been discussed in detail in several articles [33–38].

5.13.1.1 Fragmentations of the polyene chain. It is characteristic of all carotenoids that they undergo reactions in which the polyene chain is folded and portions of it are then excised. The most intensively studied are the losses of toluene (92 mass units) and *m*-xylene (106 mass units) but similar losses of 79 and 158 mass units are also frequently seen. The abundance ratio of the $[M\text{-}92]^+$ and $[M\text{-}106]^+$ fragment ions can give a good indication of the carbon skeleton of a carotenoid; for example, for acyclic, monocyclic and bicyclic carotenoids the $[M-92]^+ : [M-106]^+$ abundance ratios are in the ranges 0.02 to 0.3 : 1, 0.6 to 1.0 : 1, and 2 to 10 : 1, respectively. Carotenoids with a higher degree of saturation, for example phytoene, ζ-carotene, show an analogous loss of 94 mass units.

5.13.1.2 Functional groups and end groups. The presence of functional groups in a carotenoid is indicated by characteristic fragmentations, for example hydroxy groups, especially if in an allylic position, give rise to strong losses of water (18 m.u.). Acetylation and trimethylsilylation increase the molecular mass by 42 and 73 mass units, respectively, for each hydroxy group reacting. Those single bonds in the carotenoid molecule that are in a position allylic to the main polyene chain and also to an isolated double bond in an end group undergo particularly facile cleavage (*e.g.* C-3,4 of lycopene $[M-69]^+$, C-7,8 of ζ-carotene $[M-137]^+$ and C-11,12 of phytoene $[M-205]^+$). It is usually possible to detect the presence of an ε-ring in a carotenoid by losses of 56 mass units (by a retro-Diels–Alder fragmentation to lose C-1, C-2 and the C-1 geminal methyl groups) and of 123 mass units due to cleavage of the C-6,7 bond (138 mass units for a 3-hydroxy-ε-ring).

Carotenoid epoxides are readily identified. Both 5,6- and 5,8-epoxides show strong loss of 80 mass units (two successive losses for bis-epoxides) and also abundant ions at m/z 165 and 205 (181 and 221 if a 3-hydroxy-group is also present).

5.13.2 Nuclear magnetic resonance (NMR) spectroscopy

Moss and Weedon (1976) discussed the general features of NMR spectra of carotenoids and presented a large amount of tabulated data on the ^1H NMR assignments of many carotenoid end groups, especially the methyl group protons. Englert [39, 40] has provided an extremely valuable collection of figures which illustrate the ^1H and ^{13}C assignments for a wide range of natural and synthetic carotenoid end groups. Englert also described the application in the carotenoid field of some of the sophisticated NMR methods now available. A summary of recent progress in carotenoid NMR, together with tabulated data, is included in the article by Goodwin and Britton [29].

5.13.2.1 End group assignments. The ^1H and ^{13}C chemical shifts for a particular carotenoid end group are the same for any carotenoid that contains that end group. The methyl group signals are usually diagnostic.

5.13.2.2 Olefinic protons and geometrical configuration. In a high-field ^1H-NMR or ^{13}C NMR spectrum, for example at 400 MHz, the signals in the olefinic region are well resolved and can be assigned, their coupling relationships identified and the geometrical configuration of the carotenoid determined.

5.13.3 Circular dichroism (CD) and optical rotatory dispersion (ORD)

An extensive article by Bartlett *et al.* [41] provided ORD data for a wide range of carotenoid half molecules, and formulated an empirical additivity rule by which ORD curves for carotenoids could be predicted. Since then, most work has used CD rather than ORD. Some recent reviews and papers have described general features such as the origin of CD in carotenoids, factors that affect the CD, and empirical rules for the interpretation of CD data [42–45].

In optically active carotenoids, the asymmetric centres are located in the end groups and determine the preferred conformation of the end group. The chiral end groups impose chirality on the main polyene chain and determine the preferred angle of twist about the C-6,7 bond which, in turn, determines the chirality of the twist imposed on the polyene chain. CD is therefore observed in the electronic transitions of the main polyene chain.

Carotenoid CD spectra are strongly dependent on factors such as temperature, concentration (*e.g.* molecular aggregation) and especially geometrical configuration. Great care must therefore be taken to ensure geometrical isomeric purity of a carotenoid sample, and also in interpreting CD data.

5.13.4 Infra-red and Raman spectroscopy

Infra-red spectroscopy does not play a major role in carotenoid analysis, although it can assist in the identification of carbonyl groups and special structural features such as acetylene and allene groups [36].

Resonance Raman spectroscopy has made great advances in recent years [46]. No extensive systematic study of carotenoids has yet been reported, but carotenoids can be detected *in situ* by resonance Raman spectroscopy of intact tissues.

5.14 Commercial uses and applications

Only a brief outline can be given here. For a comprehensive review, the monograph by Bauernfeind [47] is recommended. Shorter, but useful accounts are given by Bauernfeind *et al.* [48] and by Kläui [49]. Natural carotenoids and carotenoid-containing extracts have been used for many years to colour foods. Carotenoids are also added to animal feeds. With the introduction of pure, synthetic, nature-identical carotenoids onto the market and the move away from unnatural synthetic compounds as colorants, the use of carotenoids as colour additives has increased enormously. The use of carotenoids and carotenoid-containing extracts and preparations as health-promoting products is also increasing rapidly and shows great potential for further exploitation.

5.14.1 *Application forms and formulations*

5.14.1.1 Natural extracts. Coloration by natural carotenoids does not normally use the purified pigments. Traditionally, dried and powdered or homogenized plant material was used direct, a practice that is still followed. Now, however, the usual preparations are obtained by extraction with a suitable solvent and removal of the solvent to give a concentrated crude extract. Alternatively, the extract may be prepared with, or reconstituted in, a suitable vegetable oil.

Annatto is the most widely used natural carotenoid extract, especially in dairy and bakery products, and confectionery. Many annatto preparations of widely differing quality are available. Oil-soluble preparations contain the natural bixin, whereas water-soluble forms consist mainly of solutions of norbixin as its potassium salt, as obtained from bixin by saponification. Annatto preparations usually have good stability, but their coloration properties are somewhat pH-sensitive, as would be expected for carboxylic acids.

5.14.1.2 Synthetic carotenoids. The use of crystalline, pure, synthetic carotenoids depends upon the preparation of suitable application forms or formulations that will provide the desired hue and uniform, stable coloration. The dry crystals are rarely used direct because of their poor solubility properties. The usual formulations may be oil-based or water-dispersible. The main oil-dispersible forms consist of solutions or suspensions of micronized carotenoid crystals in a vegetable oil. These preparations are stable and can be stored for long periods, especially if an antioxidant is included.

Pure crystalline carotenoids can be converted into water-dispersible or colloidal preparations. Simple emulsions and colloidal dispersions are generally not very satisfactory because only very low concentrations of carotenoid can be achieved and colour stability is not good. The most successful preparations use emulsions of supersaturated oil solutions or solutions in suitable easily removed organic solvents, and are marketed in the form of beadlets containing surface-active dispersing agents, stabilizing proteins and antioxidants. Products currently on the market contain up to 10% of the carotenoid and dissolve readily in water to give slightly cloudy dispersions. β-Carotene and ethyl 8′-apo-β-caroten-8′-oate give yellow to orange colours (depending on concentration), whereas 8′-apo-β-caroten-8′-al gives yellow to red and canthaxanthin orange-red.

5.14.2 *Uses*

5.14.2.1 Direct use in food coloration. The main use of carotenoids is for the direct coloration of food [18]. Natural extracts are used, but the major market is for synthetic carotenoids in either oil-based or water-dispersible formulations. β-Carotene and 8′-apo-β-caroten-8′-al are the main carotenoids used in oil suspensions and solutions, and can provide yellow-orange and orange-red colours, respectively, depending on concentration. The solubility of canthaxanthin in triglyceride oils is too low for practical application.

The oil-based preparations, especially of β-carotene, are widely used for colouring butter, margarine, cheese, cooking fats, industrial egg products, bakery products, pasta, salad dressings, dairy product substitutes, popcorn, potato products and many others.

Water-dispersible forms of β-carotene, 8′-apo-β-caroten-8′-al and canthaxanthin, and the water-soluble (norbixin) forms of annatto, are used extensively for colouring soft drinks (especially 'orange juice'), ice-cream, desserts, sweets, soups and meat products.

5.14.2.2 Animal feeds. Feeds for cattle, birds and fish will be considered.

Cattle. Natural pasture is very rich in β-carotene and provides sufficient of this compound to fulfil the vitamin A requirement of the animals and also to give a desirable yellow colour to the fat and rich cream colour to cream and butter. Artificial diets must be supplemented with β-carotene to ensure adequate vitamin A levels and to maintain the required yellow colour of dairy products and fat [47].

Birds. Many ornamental birds owe their exotic yellow and red colours to the presence of carotenoids in the feathers. Adequate dietary supplies of carotenoid therefore need to be maintained in captive birds. A well-known example is the flamingo, which needs to be provided with substantial quantities of the oxocarotenoids that the wild birds obtain from their diet of crustaceans, otherwise the characteristic deep pink colour is lost. Of far greater commercial importance, is the poultry industry. Chickens absorb and accumulate xanthophylls rather than carotenes and need substantial supplies of xanthophylls to ensure the required golden-yellow colour of the egg yolk and also the yellow skin colour demanded in many countries. The apocarotenoids and canthaxanthin can be used for yolk coloration, but the most natural colours of yolk and skin are achieved with zeaxanthin. Marigolds, as a source of lutein, are included in the diet in some countries [16, 50–53].

Fish. With the recent rapid development of farming methods for salmon and trout, it is essential that the products should have the same desired pink flesh coloration as the wild fish. This is now attained by the inclusion of astaxanthin (salmon) or canthaxanthin (trout) in the diet for several weeks before harvesting [16, 52–54].

5.14.2.3 Medical and health products. The beneficial role of β-carotene as provitamin A has long been known. β-Carotene is also now used successfully to alleviate the symptoms of light-sensitivity diseases, especially erythropoietic protoporphyria, that are characterized by extreme irritation of the skin when exposed to strong light, because of the formation of singlet oxygen sensitized by the free porphyrins that accumulate in this condition. β-

Carotene, administered at a high level (around 180 mg/day), is deposited in the skin and then quenches the triplet state of the sensitizer and prevents formation of singlet oxygen [19].

The increasing reports that β-carotene may be an important antioxidant that can afford protection against cancer and other diseases [55, 56] is already leading to the appearance of a great number of β-carotene preparations on the health market, including crystalline carotene, extracts of natural sources such as carrots and algae, and even dried algal cells (*Dunaliella*). There seems little doubt that the demand for β-carotene and perhaps other carotenoids as beneficial health products will increase dramatically.

Carotenoids are also used simply for coloration purposes in health products such as pills, capsules and suppositories [57].

5.15 Future prospects

5.15.1 Other carotenoids

Only a very small number of the 600 or so known naturally-occurring carotenoids have been used in food coloration. Some of the other, longer chromophore carotenoids that absorb at longer wavelengths may be worth investigating for the possibility of extending the range of colours or hues available further into the red and purple. Interesting examples include actinioerythrol (cherry-red, from the sea anemone *Actinia equina*) and its oxidized derivative violerythrin (purple-blue).

5.15.2 Potential as protective antioxidants and anti-cancer agents

The nutritional importance of β-carotene as provitamin A has been known for many years. Recently, there have been many reports to suggest that β-carotene could also be important by affording protection against cancer and other diseases, and also against ageing [55, 56, 58]. These reports indicate a physiological role for intact carotenoids independent of their conversion into vitamin A and suggest that carotenoids without provitamin A activity may also be important nutritional factors. As already mentioned, the treatment of light-sensitivity diseases such as erythropoietic protoporphyria by administration of β-carotene has been used successfully for some years [7, 19]. In 1981, a now classic paper by Peto *et al.* [55] discussed the possibility that dietary β-carotene could afford some protection against some forms of cancer. Several studies since then have concluded that individuals whose dietary intake and serum levels of β-carotene are low have higher risk of developing some cancers, for example of the skin, lung and bladder. Experiments with animals as model systems have strongly suggested that the administration of carotene can afford substantial protection against the development and proliferation of induced cancers. It now seems clear that, under appropriate conditions, β-carotene can function as a very effective antioxidant, not only against $^1O_2$ but also against the highly destructive hydroxyl radical $HO^{\bullet}$ that is implicated in many diseases. A protective role of β-carotene against cancer and other diseases is likely to be a consequence of these antioxidant properties. Other work suggests that β-carotene may also have important direct effects on the immunoresponse system [59].

Obviously, there is currently intense research activity in these areas. The general findings seem to indicate the desirability of maintaining or increasing intake of carotenoids. Also, it is likely that other carotenoids will have antioxidant properties similar to those of β-carotene and may thus be equally beneficial dietary components. The increasing use of carotenoids as food additives and in health preparations can therefore be predicted.

5.15.3 Carotenoproteins

Free carotenoids are yellow, orange or red in colour but, in many marine invertebrate animals, they (especially astaxanthin) occur as carotenoprotein complexes, which have a range of colours including green, blue and purple [60–64]. The best-known example of a carotenoprotein is α-crustacyanin, the blue astaxanthin-protein of the carapace of the lobster (*Homarus gammarus*), which has λ_{max} 630 nm in contrast to the 480 nm of free astaxanthin. This 320 kDa protein has been studied intensively. Its protein subunit structure is known and the primary sequences of the apoprotein subunits have been determined. The main interactions between protein and carotenoid occur via the ring oxygen functions of astaxanthin, particularly the C-4 and C-4′ oxo-groups, but the central part of the chromophore is also highly perturbed. Details of the three-dimensional structure of the protein and of the interactions between the protein and carotenoid molecules are likely to be elucidated in the near future.

The carotenoprotein complexes are water soluble and very stable (the colour is stable for years in air at room temperature) and they have, therefore, aroused considerable interest as possible colorants. In addition, when the interactions with protein that lead to the large spectral shift and colour change are understood, it may prove possible to devise formulations that mimic or reproduce these interactions and would therefore provide stable, water soluble, blue, green or purple colorants.

5.15.4 Biotechnology

As outlined above, natural extracts, particularly of fruits, flowers and leaves, have been used for many years as sources of carotenoids as colorants. With increasing general interest in biotechnology, the commercial production of carotenoids by microbial systems is being evaluated [65]. The first sources to be studied were the carotenogenic moulds, *Blakeslea trispora* and *Phycomyces blakesleeanus*. High-yielding mutants and optimized culture conditions have been developed and large-scale production of β-carotene could certainly be achieved. It will be difficult, however, for this product to compete commercially with synthetic β-carotene. A red yeast, *Phaffia rhodozyma*, has been explored as a source of astaxanthin for fish feed (salmon) and has enjoyed some limited success, but it is not an ideal organism to use because the astaxanthin it produces is the wrong optical isomer [(3*R*, 3′*R*)], and extraction of it is difficult.

The major developments in recent years have occurred with microalgae [66], especially *Dunaliella* species (*D. salina* and *D. bardawil*) and, more recently, *Haematococcus* species. Under conditions of stress (high light, high temperature, high salt, mineral stress) these green algae become orange because they produce large amounts of 'secondary carotenoids' outside the chloroplast, *e.g.* in the cell wall or in oil droplets. *Dunaliella* species accumulate β-carotene (up to 10% dry weight), *Haematococcus* accumulates astaxanthin (up to 1 or 2%). The algae are now being cultured on a large scale in vast open ponds in countries with a suitable climate, especially Israel and Australia. As no energy input is needed (except sunlight) and the extreme NaCl concentrations used (up to 4M) with *Dunaliella* prevent contamination with other organisms, production of carotene-rich dried algae or oil-based extracts is cheap and such preparations are competitive. The purification of β-carotene from the extracts, however, increases the costs substantially. The culturing of *Haematococcus* is less favourable and the yields of astaxanthin are much lower, but the high cost of synthesizing optically active astaxanthin makes the feasibility of production by *Haematococcus* worth exploring. In the longer term, the genetic transfer or introduction of the ability to produce the more expensive commercially important carotenoids (*e.g.* astaxanthin or zeaxanthin with the correct stereochemistry) into *Dunaliella* is an attractive prospect.

Some bacteria have been considered for carotenoid production by fermentation, for example strains of *Brevibacterium* (to yield canthaxanthin) and *Flavobacterium* (to yield zeaxanthin), but commercial production has not been established. Very recent molecular genetic work has achieved the introduction of the genes for carotenoid biosynthesis from carotenogenic bacteria into the normally non-carotenogenic *Escherichia coli*. High concentrations of β-carotene and, potentially, of other carotenoids can be obtained in this way, and rapid developments in this area can be expected.

Appendix Structures of the individual carotenoids mentioned in the text

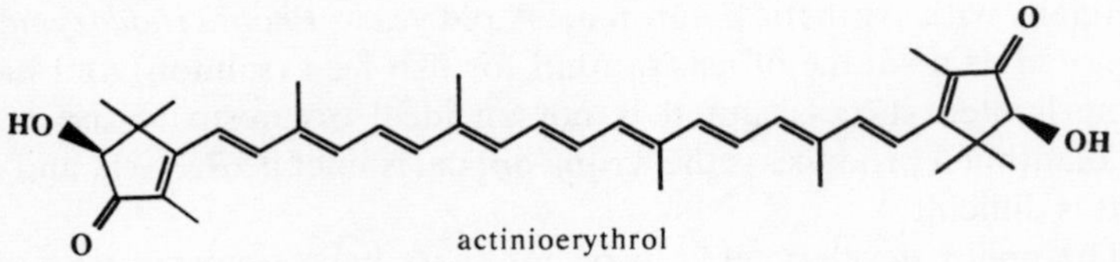

actinioerythrol

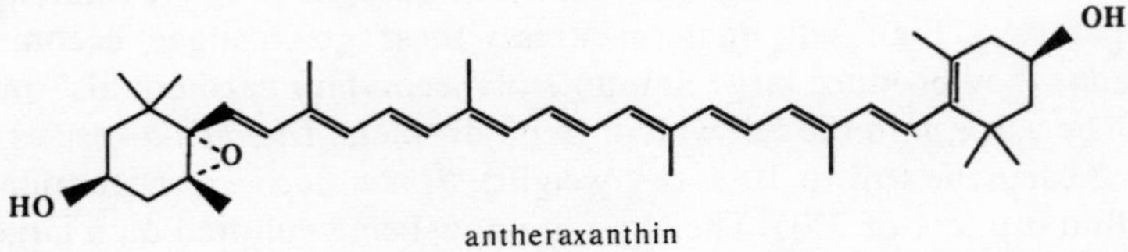

antheraxanthin

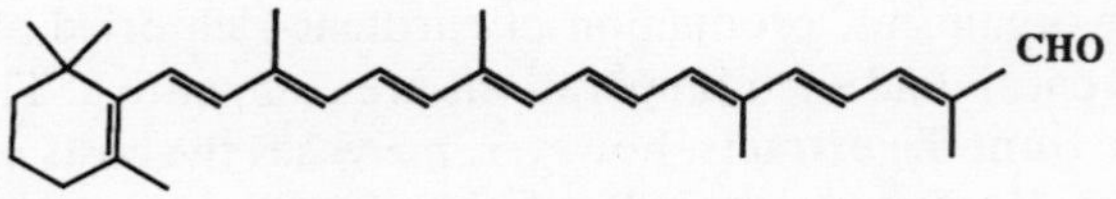

8'-apo-β-caroten-8'-al

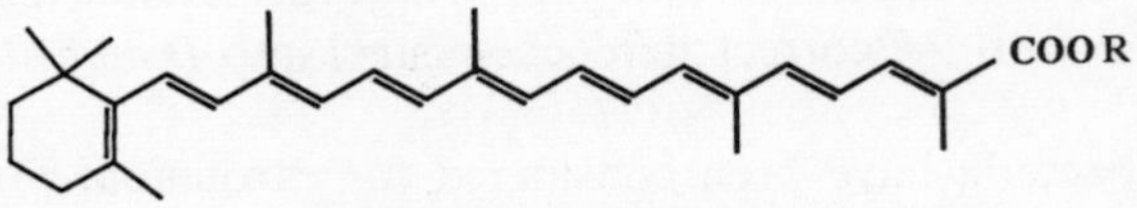

8'-apo-β-caroten-8'-oic acid ester

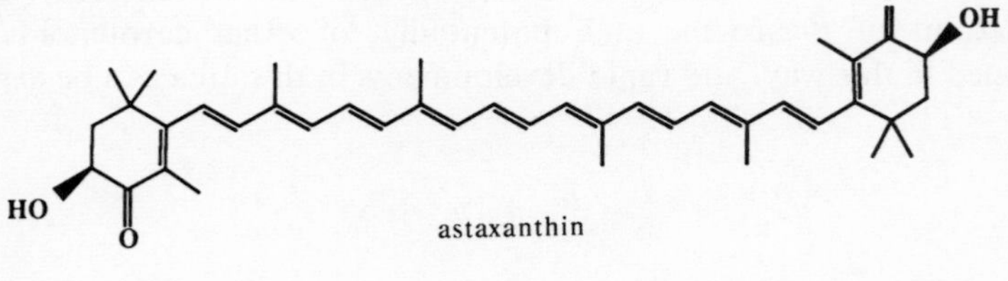

astaxanthin

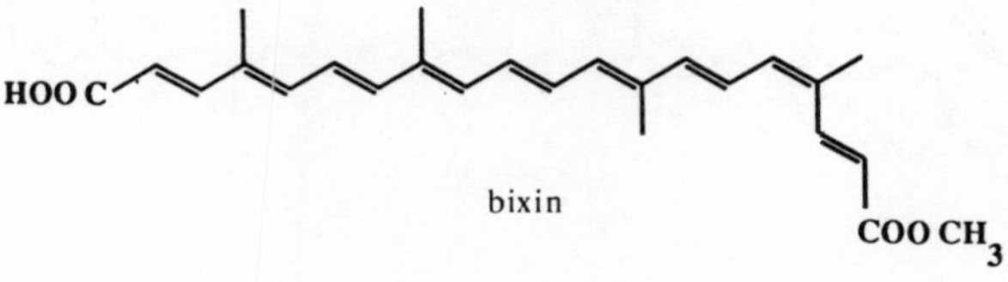

HOOC COOCH$_3$

bixin

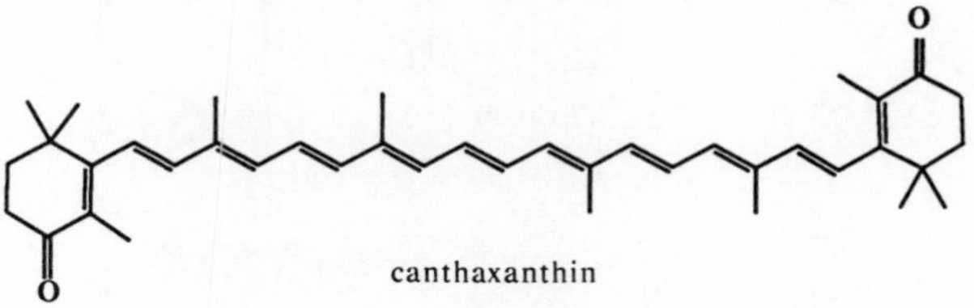

O O

canthaxanthin

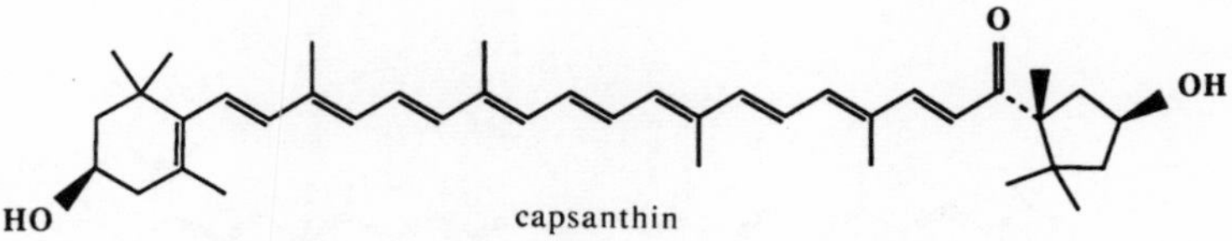

O OH HO

capsanthin

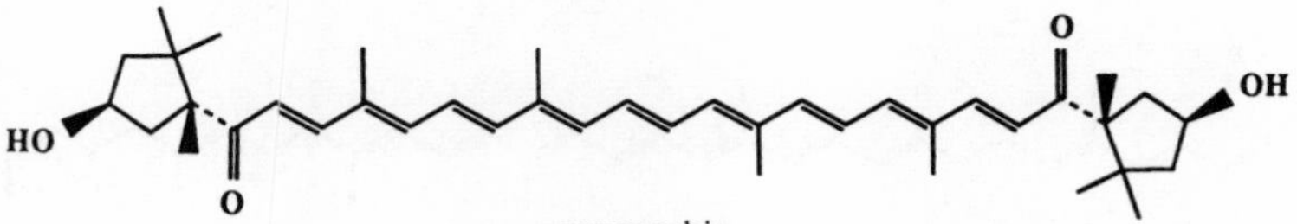

O OH HO O

capsorubin

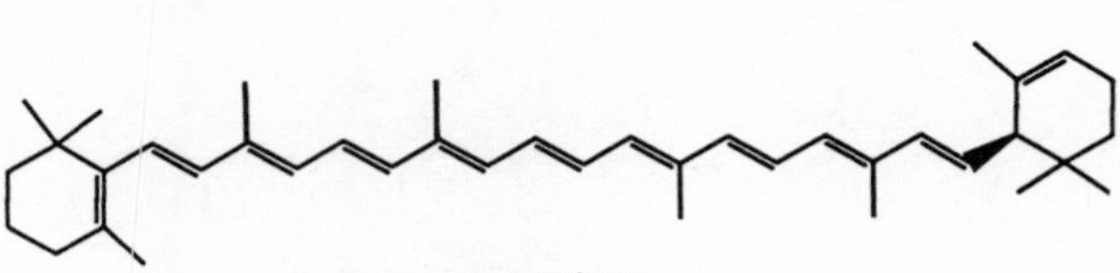

α-carotene

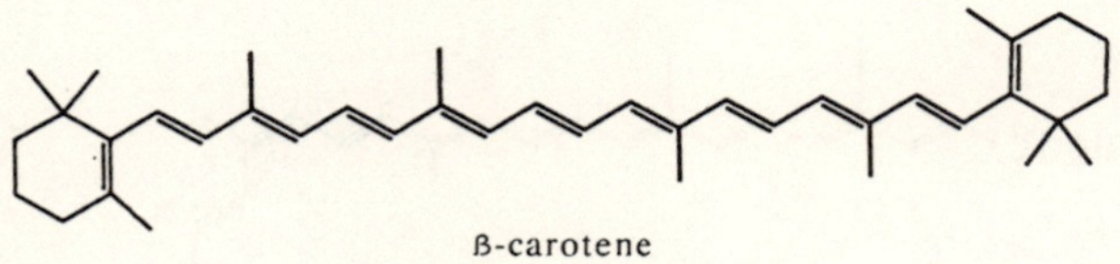

ß-carotene

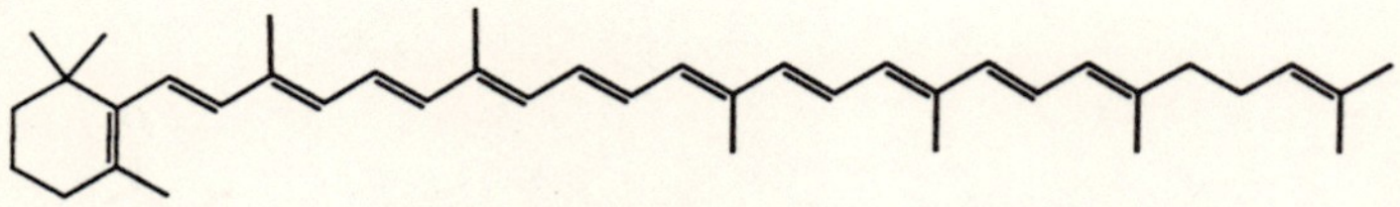

γ-carotene

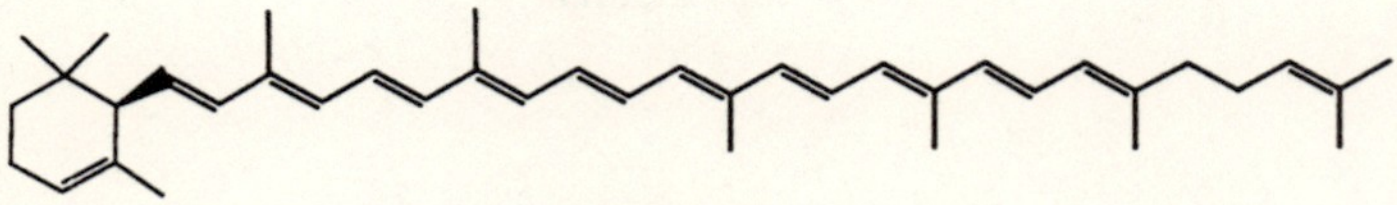

δ-carotene

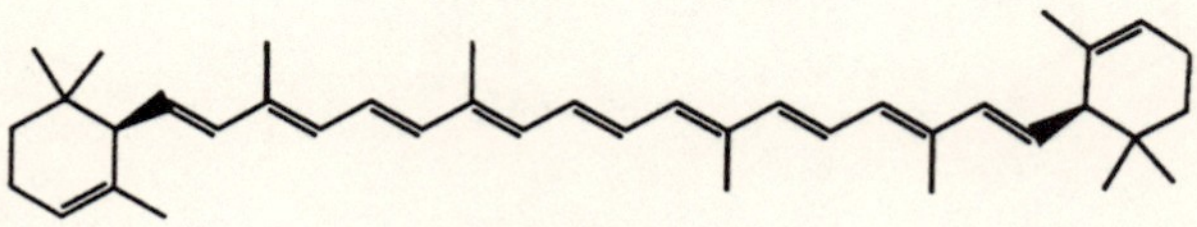

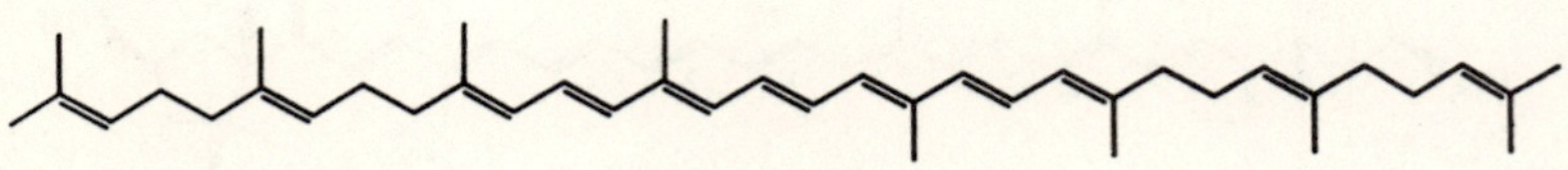

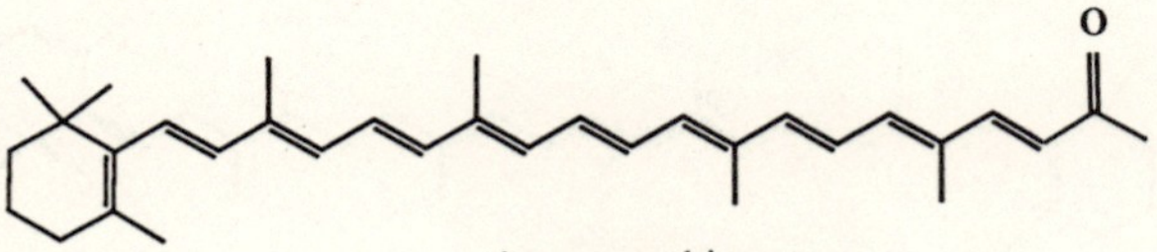

ε-carotene

ζ-carotene

O

citranaxanthin

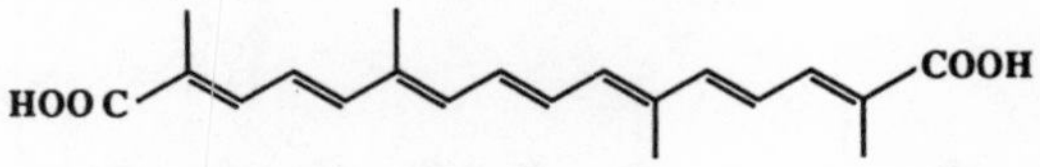

HOOC COOH

crocetin

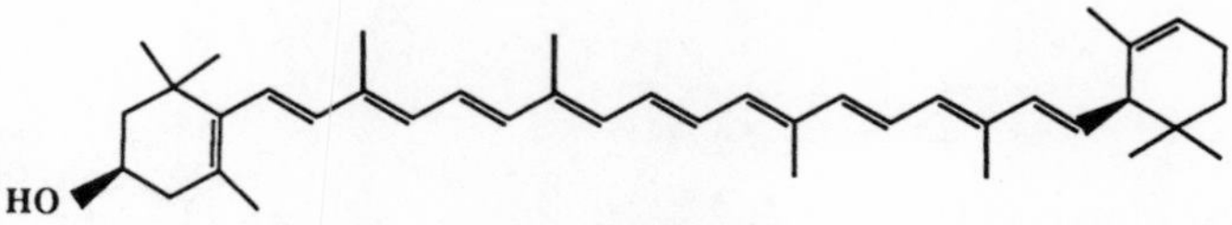

HO

α-cryptoxanthin

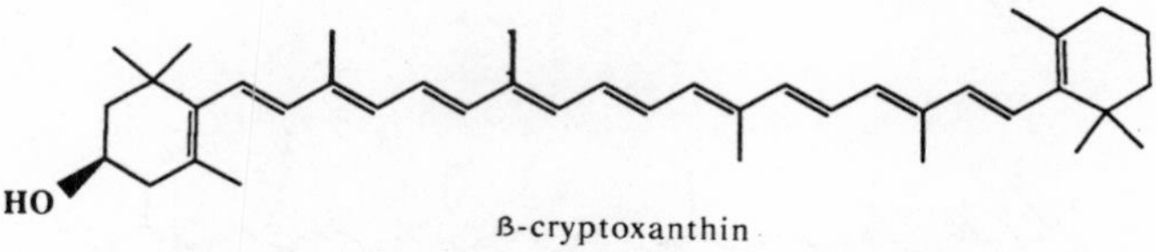

HO

ß-cryptoxanthin

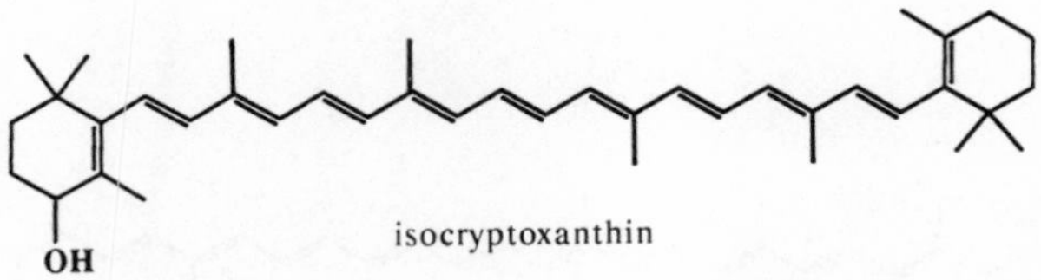

OH

isocryptoxanthin

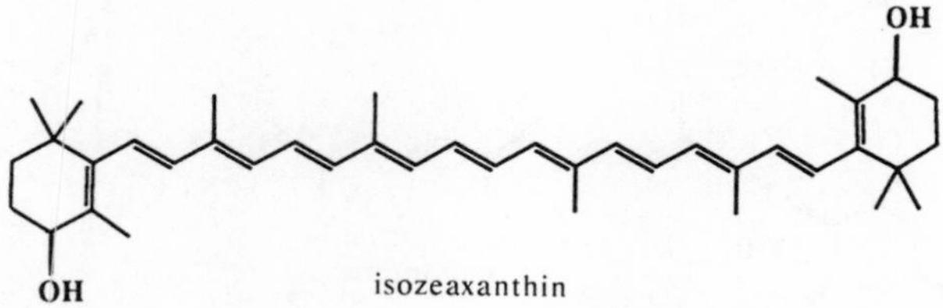

OH

OH

isozeaxanthin

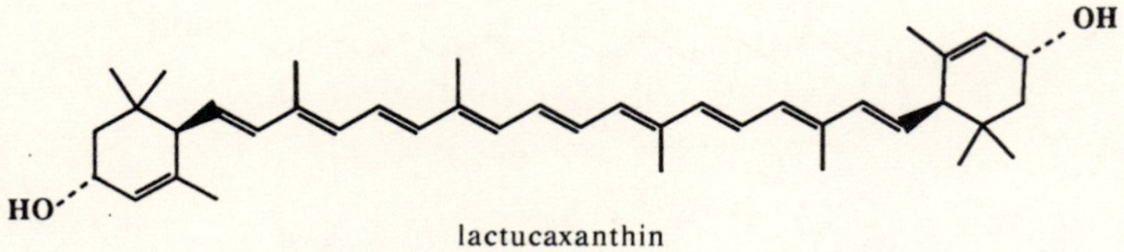

OH
HO
lactucaxanthin

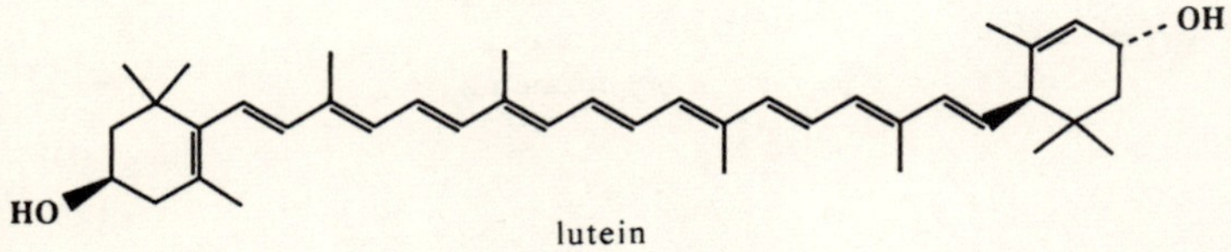

OH
HO
lutein

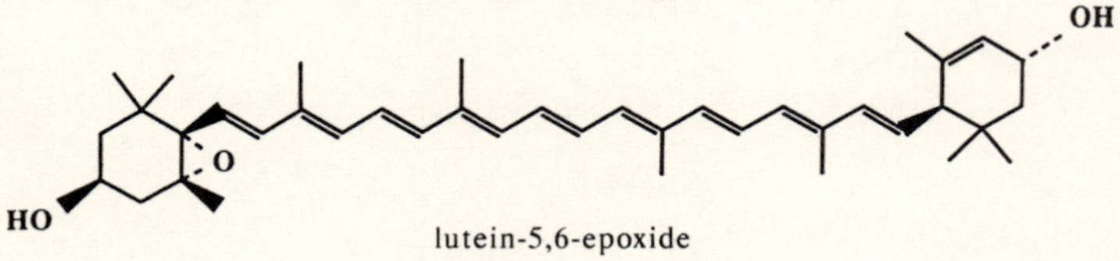

OH
O
HO
lutein-5,6-epoxide

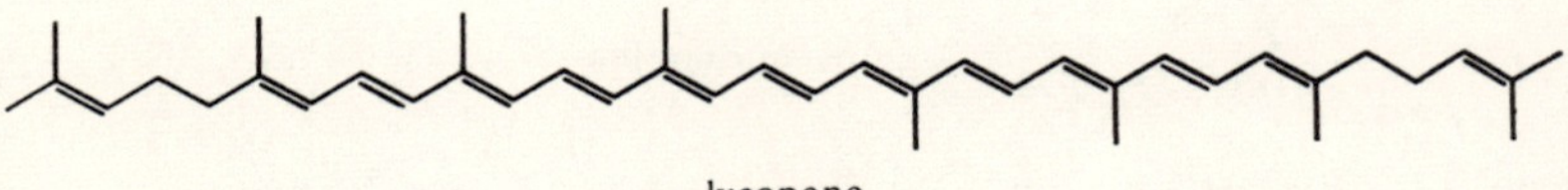

lycopene

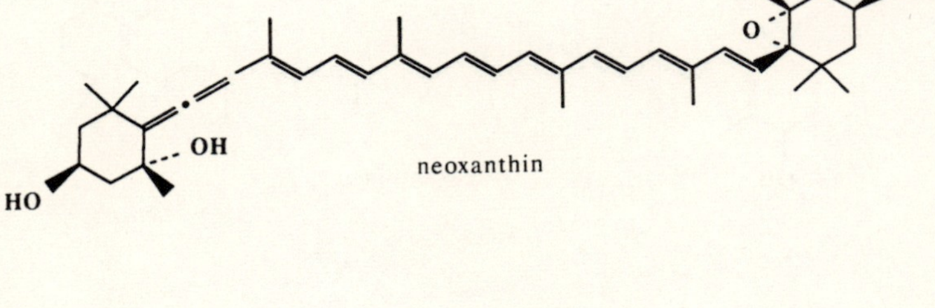

OH
O
OH
HO
neoxanthin

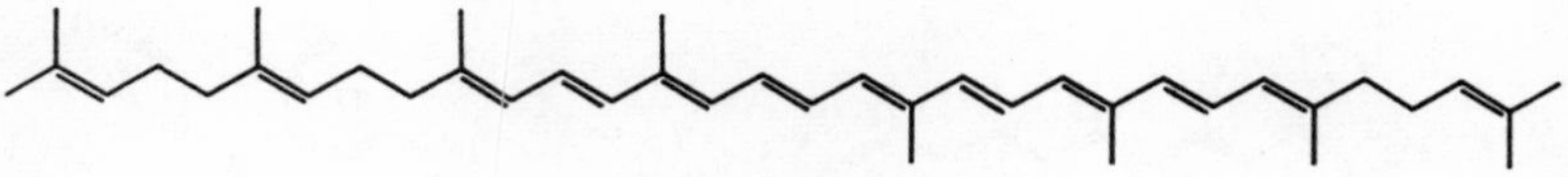

neurosporene

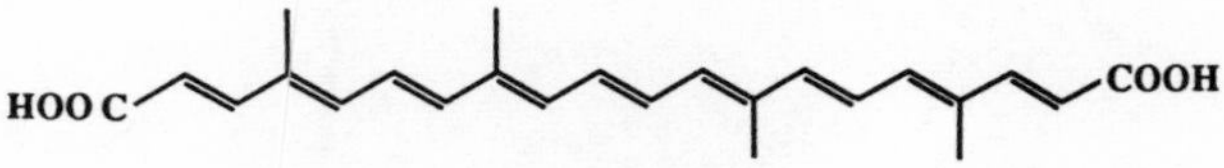

$HOOC$ ... $COOH$

nor-bixin

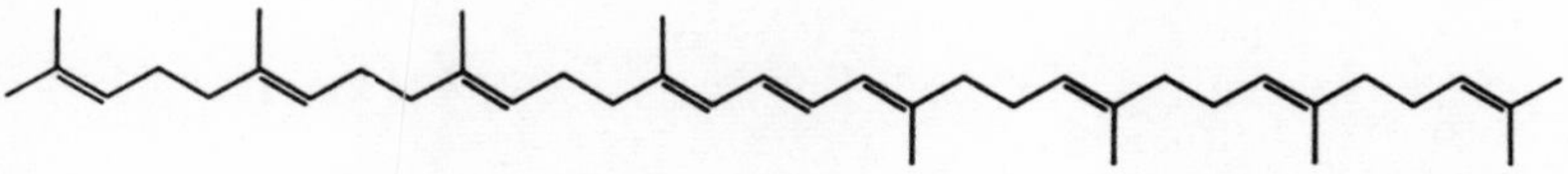

phytoene

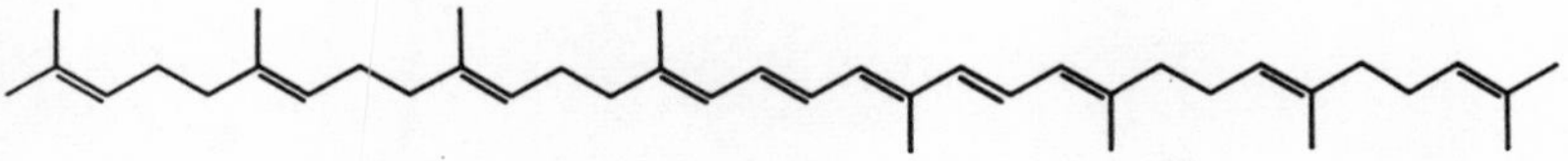

phytofluene

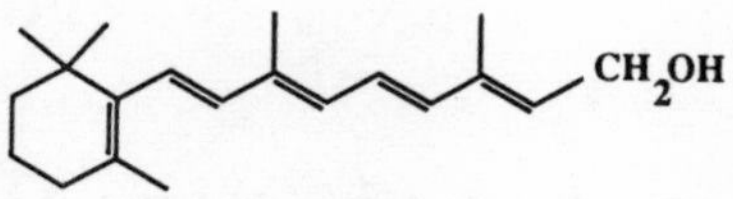

CH_2OH

retinol

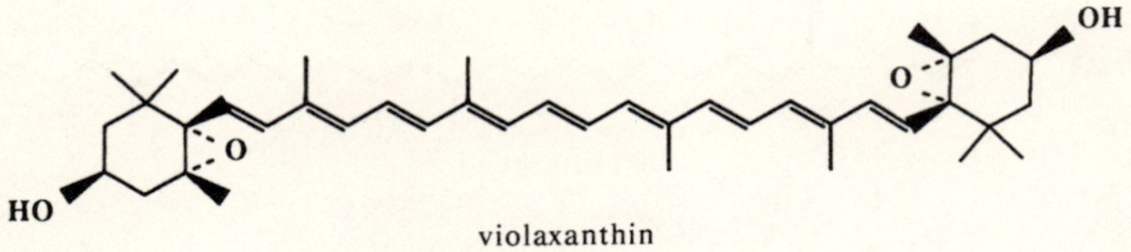

HO O O OH

violaxanthin

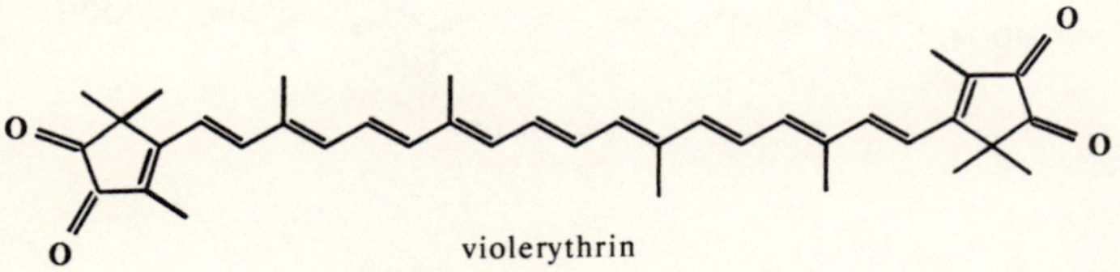

O O O O

violerythrin

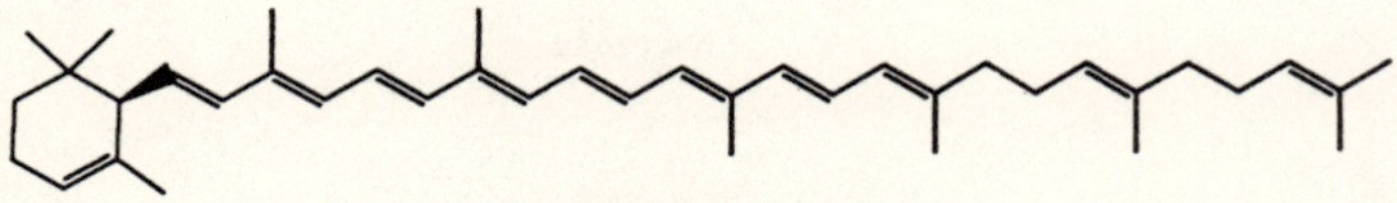

α-zeacarotene

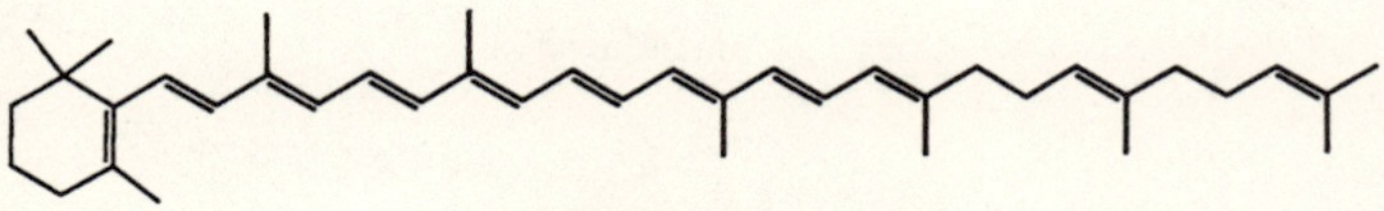

ß-zeacarotene

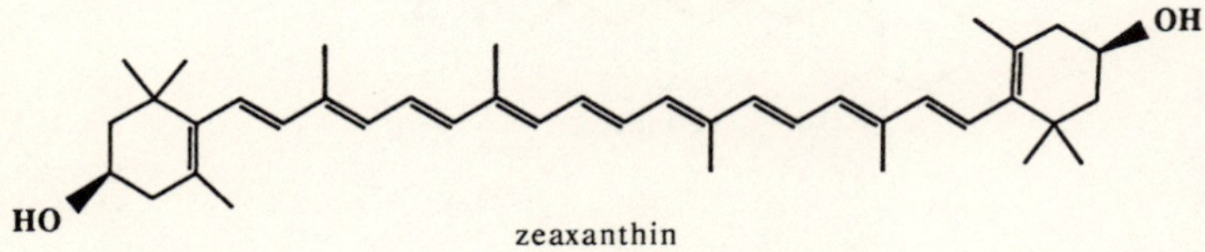

HO OH

zeaxanthin

References

1. Isler, O. *Carotenoids*, Birkhäuser, Basel and Stuttgart (1971).
2a. Goodwin, T.W. *The Biochemistry of the Carotenoids, Vol. 1, Plants*. Chapman and Hall, London (1980).
2b. Goodwin, T.W. *The Biochemistry of the Carotenoids, Vol. 2. Animals*. Chapman and Hall, London (1984).
3. Cogdell, R.J. *Pure Appl. Chem.* **57** (1985) 723–728.
4. Cogdell, R.J. in *Plant Pigments* (T.W. Goodwin, Ed.), Academic Press, London (1988) pp. 183–230.
5. Siefermann-Harms, D. *Biochim. Biophys. Acta* **811** (1985) 325–355.
6. Siefermann-Harms, D. *Physiol. Plantarum* **69** (1987) 561–568.
7. Mathews-Roth, M.M. in *Carotenoid Chemistry and Biochemistry* (G. Britton and T.W. Goodwin, Eds), Pergamon, Oxford (1982) pp. 297–307.
8. Britton, G. in *Chemistry and Biochemistry of Plant Pigments* 2nd edn. (T.W. Goodwin, Ed.), Academic Press, London (1976) pp. 262–327.
9. Britton, G. in *Regulation of Chloroplast Differentiation* (G. Akoyunoglou and H. Senger, Eds), A.R. Liss, New York (1986) pp. 125–134.
10. Britton, G. in *Plant Pigments* (T.W. Goodwin, Ed.), Academic Press, London (1988) pp. 133–182.
11. Britton, G. in *Carotenoids: Chemistry and Biology* (N.I. Krinsky, M.M. Mathews-Roth and R.F. Taylor, Eds), Plenum Press, New York (1990) pp. 167–184.
12. Britton, G. *Pure Appl. Chem.* **63** (1991) 101–108.
13. Bramley, P.M. *Adv. Lipid Res.* **21** (1985) 243–279.
14. Bramley, P.M. and Mackenzie, A. *Curr. Topics Cell Regulat.* **29** (1988) 291–343.
15. Goodwin, T.W. *Ann. Rev. Nutrition* **6** (1986) 273–297.
16. Schiedt, K. in *Carotenoids: Chemistry and Biology* (N.I. Krinsky, M.M. Mathews-Roth and R.F. Taylor, Eds), Plenum Press, New York (1990) pp. 247–268.
17. Olson, J.A. in *Vitamin A Deficiency and its Control* (J.C. Bauernfeind, Ed.), Academic Press, Orlando and London (1986) pp. 19–67.
18. Kläui, H. and Bauernfeind, J.C. in *Carotenoids as Colorants and Vitamin A Precursors* (J.C. Bauernfeind, Ed.), Academic Press, New York (1981) pp. 48–317.
19. Mathews-Roth, M.M. in *Carotenoids as Colorants and Vitamin A Precursors* (J.C. Bauernfeind, Ed.), Academic Press, New York (1981) pp. 755–785.
20. Krinsky, N.I., Cornwell, D.G. and Oncley, J.L. *Arch. Biochem. Biophys.* **73** (1958) 233–246.
21. Dimitrov, N.V. and Ullrey, D.E. in *Carotenoids: Chemistry and Biology* (N.I. Krinsky, M.M. Mathews-Roth and R.F. Taylor, Eds), Plenum Press, New York (1990) pp. 269–277.
22. Bauernfeind, J.C. *Carotenoids as Colorants and Vitamin A Precursors*. Academic Press, New York (1981).
23. Bernhard, K. in *Carotenoids: Chemistry and Biology* (N.I. Krinsky, M.M. Mathews-Roth and R.F. Taylor, Eds), Plenum Press, New York (1990) pp. 337–363.
24. Britton, G. and Goodwin, T.W. *Methods Enzymol.* **18C** (1971) 654–701.
25. Davies, B.H. in *Chemistry and Biochemistry of Plant Pigments, 2nd edn.* (T.W. Goodwin, Ed.), Vol. 2, Academic Press, London (1976) pp. 38–165.
26. Britton, G. *Methods Enzymol.* **111** (1985) 113–149.
27. Britton, G. in *Methods in Plant Biochemistry, Vol. 7 Terpenoids* (D.V. Banthorpe and B.V. Charlwood, Eds), Academic Press, London (1991) (In press).
28. Rüedi, P. *Pure Appl. Chem.* **57** (1985) 793–800.
29. Goodwin, T.W. and Britton, G. in *Plant Pigments* (T.W. Goodwin, Ed.), Academic Press, London and San Diego (1988) pp. 61–132.
30. Nelis, H.J.C.F. and De Leenheer, A.P. *Anal. Chem.* **55** (1983) 270–275.
31. Wright, S.W. and Shearer, J.D. *J. Chromatogr.* **294** (1984) 281–295.
32. De Ritter, E. in *Carotenoids as Colorants and Vitamin A Precursors* (J.C. Bauernfeind, Ed.), Academic Press, New York (1981) pp. 883–923.
33. Vetter, W., Englert, G., Rigassi, N. and Schwieter, U. in *Carotenoids* (O. Isler, Ed.), Birkhäuser, Basel and Stuttgart (1971) pp. 189–266.
34. Johannes, B., Brzezinka, H. and Budzikiewicz, H. *Org. Mass Spectrom.* **9** (1974) 1095–1113.

35. Johannes, B., Brzezinka, H. and Budzikiewicz, H. *Z. Naturforsch.* **34B** (1979) 300–305.
36. Moss, G.P. and Weedon, B.C.L. in *Chemistry and Biochemistry of Plant Pigments*, 2nd edn. (T.W. Goodwin, Ed.), Vol. 1, Academic Press, London (1976) pp. 149–224.
37. Enzell, C.R., Wahlberg, I. and Ryhage, R. *Mass Spectrom. Rev.* **3** (1984) 395–438.
38. Enzell, C.R. and Wahlberg, I. *Biochem. Appl. Mass Spectrom. (Suppl. 1)* (1980) 407–438.
39. Englert, G. in *Carotenoid Chemistry and Biochemistry* (G. Britton and T.W. Goodwin, Eds), Pergamon, Oxford (1982) pp. 107–134.
40. Englert, G. *Pure Appl. Chem.* **57** (1985) 801–821.
41. Bartlett, L., Klyne, W., Mose, W.P., Scopes, P.M., Galasko, G., Mallams, A.K., Weedon, B.C.L., Szabolcs, J. and Toth, G. *J. Chem. Soc. C* (1969) 2527–2544.
42. Noack, K. and Thomson, A.J. *Helv. Chim. Acta* **62** (1979) 1902–1921.
43. Sturzenegger, V., Buchecker, R. and Wagniere, G. *Helv. Chim. Acta* **53** (1980) 1074–1092.
44. Buchecker, R., Marti, U. and Eugster, C.H. *Helv. Chim. Acta* **65** (1982) 896–912.
45. Noack, K. in *Carotenoid Chemistry and Biochemistry* (G. Britton and T.W. Goodwin, Eds), Pergamon, Oxford (1982) pp. 135–153.
46. Merlin, J.C. *Pure Appl. Chem.* **57** (1985) 785–792.
47. Bauernfeind, J.C., Adams, C.R. and Marusich, W.L. in *Carotenoids as Colorants and Vitamin A Precursors* (J.C. Bauernfeind, Ed.), Academic Press, New York (1981) pp. 564–743.
48. Bauernfeind, J.C., Brubacher, G.B., Kläui, H.M. and Marusich, W.L. in *Carotenoids* (O. Isler, Ed.), Birkhäuser, Basel (1971) pp. 743–770.
49. Kläui, H. in *Carotenoid Chemistry and Biochemistry* (G. Britton and T.W. Goodwin, Eds), Pergamon, Oxford (1982) pp. 309–328.
50. Brush, A.H. in *Carotenoids as Colorants and Vitamin A Precursors* (J.C. Bauernfeind, Ed.), Academic Press, New York (1981) pp. 539–562.
51. Marusich, W.L. and Bauernfeind, J.C. in *Carotenoids as Colorants and Vitamin A Precursors* (J.C. Bauernfeind, Ed.), Academic Press, New York (1981) pp. 320–462.
52. Schiedt, K., Leuenberger, F.J., Vecchi, M. and Glinz, E. *Pure Appl. Chem.* **57** (1985) 685–692.
53. Schiedt, K., Bischof, S. and Glinz, E. *Pure Appl. Chem.* **63** (1991) 89–100.
54. Simpson, K.L., Katayama, T. and Chichester, C.O. in *Carotenoids as Colorants and Vitamin A Precursors* (J.C. Bauernfeind, Ed.), Academic Press, New York (1981) pp. 463–538.
55. Peto, R., Doll, R., Buckley, J.D. and Sporn, M.B. *Nature* **290** (1981) 201–208.
56. Santamaria, L. and Bianchi, A. *Preventive Medicine* **18** (1989) 603–623.
57. Münzel, K. in *Carotenoids as Colorants and Vitamin A Precursors* (J.C. Bauernfeind, Ed.), Academic Press, New York (1981) pp. 745–754.
58. Krinsky, N.I. in *Carotenoids: Chemistry and Biology* (N.I. Krinsky, M.M. Mathews-Roth and R.F. Taylor, Eds), Plenum Press, New York (1990) pp. 279–291.
59. Bendich, A. in *Carotenoids: Chemistry and Biology* (N.I. Krinsky, M.M. Mathews-Roth and R.F. Taylor, Eds), Plenum Press, New York (1990) pp. 323–335.
60. Britton, G., Armitt, G.M., Lau, S.Y.M., Patel, A.K. and Shone, C.C. in *Carotenoid Chemistry and Biochemistry* (G. Britton and T.W. Goodwin, Eds), Pergamon, Oxford (1982) pp. 237–251.
61. Findlay, J.B.C., Pappin, D.J.C., Brett, M. and Zagalsky, P.F. in *Carotenoids: Chemistry and Biology* (N.I. Krinsky, M.M. Mathews-Roth and R.F. Taylor, Eds), Plenum Press, New York (1990) pp. 75–104.
62. Zagalsky, P.F. *Pure Appl. Chem.* **47** (1976) 103–120.
63. Zagalsky, P.F. *Oceanis* **9** (1983) 73–90.
64. Zagalsky, P.F., Eliopoulos, E.E. and Findlay, J.B.C. *Comp. Biochem. Physiol.* **97B** (1990) 1–18.
65. Nonomura, A.M. in *Carotenoids: Chemistry and Biology* (N.I. Krinsky, M.M. Mathews-Roth and R.F. Taylor, Eds), Plenum Press, New York (1990) pp. 365–375.
66. Ben-Amotz, A. and Avron, M. *Ann. Rev. Microbiol.* **37** (1983) 95–119.

6 Anthocyanins and betalains

R.L. JACKMAN and J.L. SMITH

6.1 Summary

Anthocyanins occur widely in plants, being responsible for their blue, purple, violet, magenta, red and orange coloration; while betalains, consisting of red-violet betacyanins and yellow-orange betaxanthins, occur exclusively in families of the order Caryophyllales. The occurrence of these two classes of pigments is mutually exclusive. Their stability is markedly influenced by environmental and processing factors such as pH, temperature, oxygen, enzymes and condensation reactions. This and their structural diversity have made the qualitative and quantitative analyses of these pigments difficult. Yet, the speed, resolving power and the ability of high-pressure liquid chromatography to quantitatively separate pigments without preliminary purification has revolutionized pigment analyses. Due to their inherent instability these pigments, and betalains in particular, have not become widely accepted as food colorants. However, structural variants of the anthocyanins, that is, polyacylated and self-associated species, have demonstrated remarkable stability and have great promise as stable natural colorants. Anthocyanin and betalain preparations are currently restricted to those obtained from grapes and beetroots, respectively: they can be used to colour a variety of foods and pharmaceuticals with compatible physicochemical properties, yielding highly coloured and high-quality products. Their application could be enhanced, however, with new sources and stable structural variants, modification of current processes and foods, and technological advances (*e.g.*, industry-scale extractions/purifications, microbial purifications, biotechnology) that would make purer and more stable preparations available. In this chapter, the chemistry and biochemistry of anthocyanins and betalains—their structure and distribution, biosynthesis, factors influencing their stability, methods of extraction and analysis, and current and potential sources and uses—are reviewed.

6.2 Introduction

Anthocyanins are among the best known of the natural pigments, being responsible for the blue, purple, violet, magenta, red and orange colour of a majority of plant species and their products [1, 2]. Anthocyanins are virtually ubiquitous in the plant kingdom. All higher plants may have the potential to produce these pigments; however, colour may be lacking because their stability or synthesis is not favoured by prevailing environmental conditions, or because there is no selective advantage in their expression [3]. Like the more widespread anthocyanins, the betalains, a class of water-soluble pigments consisting of the red-violet betacyanins and the yellow-orange betaxanthins, occur abundantly and uniquely in flowers, fruits and leaves of plants that contain them [4, 5]. Betalains also often accumulate in the stalk, and in the beetroot they are found in high concentration in the root. Interestingly, these biogenetically and structurally distinct pigments, the anthocyanins and betalains, have not been detected together in the same species or even in different species of the same family: their occurrence is mutually exclusive [6].

Although the exact physiological functions of these two classes of pigments have not been clearly established, each is likely to have multiple functions. They play a definitive role as pollination factors, floral and fruit coloration attracting insect, bird and animal vectors, thereby aiding in plant reproduction. However, the function of anthocyanins and betalains in other plant parts (*i.e.* leaves, stems and roots) is obscure. Their accumulation at wound or injury sites in plants that synthesize them normally would indicate that they also function as phytoalexins, whereby they serve to defend the plant against viral or microbial infection, or both. Both anthocyanins and betalains have been shown to partake in biological oxidations, enzyme inhibition, viral resistance and inhibition of microbial growth and respiration [7–10]. Betalains may also serve as nitrogen reservoirs in plants that produce them [11].

The striking colours associated with anthocyanins and betalains often serve as the basis for identification and consumer acceptance of foods containing these pigments. However, despite their obvious aesthetic appeal and familiarity, anthocyanins and betalains have not been widely accepted as colorants in the food industry. This may be due to their inherent instability as a function of pH, temperature and light, low tinctorial strength, low yields, in some cases their association with other properties such as flavour, and the difficulty and expense of extraction and purification from their natural sources. Yet, continued concern over the toxicological safety of synthetic colorants and the banning of several red dyes [12, 13] has encouraged the development of anthocyanins, betalains and other natural colorants for use as food ingredients through technological advances and identification of new sources, as well as modified products. The number of food colorant patents dealing with natural pigments has increased dramatically in the past two decades [12]. Food colorants extracted from 'natural' sources are exempt from the rigorous and costly toxicological testing that synthetic dyes must undergo prior to their clearance as safe food ingredients; natural colorants are generally considered safe due to their presence in edible plant material. The few reports that have provided toxicological data on anthocyanins and betalains [14–17] support the long-held view that these pigments are innocuous in the diet. Moreover, there is some evidence to suggest that these pigments have medicinal properties, for example, anticarcinogenic and anticholesterolemic effects [17, 18], although further investigation of these properties is required. Anthocyanin powder, obtained from the skins of grapes and other by-products of wine manufacture, and beetroot powder, are permitted food colour additives in most countries [19, 20].

A knowledge of the properties of natural colorants and the factors that influence them, including their interaction with other food components, is essential to their successful application in a wide variety of food products. This chapter will review the chemistry and biochemistry of anthocyanins and betalains: their structure and distribution, biosynthesis, factors that influence their stability, methods of extraction and analysis, and current and potential sources and uses. In addition to a book edited by Markakis [21], several comprehensive reviews have previously been published concerning the anthocyanins [22–31]. Betalains and related topics have also been previously reviewed [4, 5, 32–38].

6.3 Anthocyanins

6.3.1 Structure

Anthocyanins are regarded as flavonoid compounds since they possess the characteristic $C_6C_3C_6$ carbon skeleton and the same biosynthetic origin as other natural flavonoids [3, 39]. Yet, they differ from other natural flavonoids by strongly absorbing visible light [40]. The range of colours associated with the anthocyanins results from distinct and varied substitution of the parent $C_6C_3C_6$ nucleus, in addition to various environmental influences.

The anthocyanins are glycosides of sixteen different naturally occurring anthocyanidins, these being polyhydroxy and polymethoxy derivatives of 2-phenylbenzopyrylium (flavylium) salts (Figure 6.1). With the exception of the 3-deoxy forms, which are stable and yellow [41], anthocyanidins are rarely found in their free form in plant tissue [3]. A free 3-hydroxyl group destabilizes the red anthocyanidin chromophores and makes them less soluble than their corresponding glycosides in aqueous media [42]. Therefore, without exception the 3-hydroxyl group is glycosylated, this being assumed to confer stability and solubility to the anthocyanin molecule [43]. Loss of the 3-glycosyl moiety is accompanied by rapid decomposition of the aglycone (*i.e.*, anthocyanidin) with irreversible loss of colour [44]. If a second site is glycosylated it is most often at C-5. Glycosylation of 7-, 3′-, 4′- and/or 5′-hydroxyl groups may also occur; however, steric hindrance generally precludes glycosylation at both the C-3′ and -4′ positions [40]. Based on the position of attachment and number of sugar residues anthocyanins have been classified into eighteen

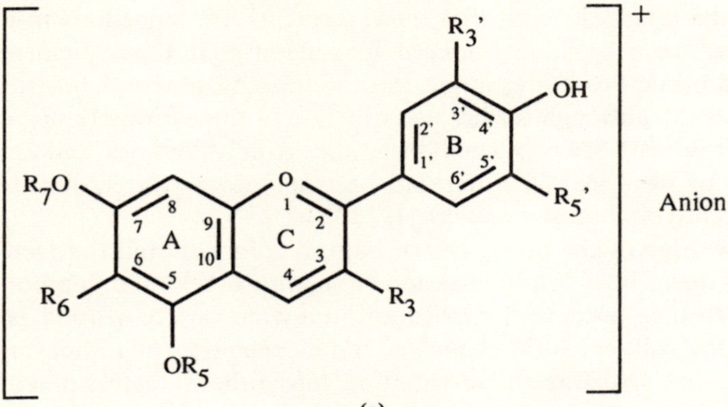

(a)

Anthocyanidin	Substitution Pattern (R)					
	3	5	6	7	3′	5′
Apigeninidin (Ap)	H	OH	H	OH	H	H
Luteolinidin (Lt)	H	OH	H	OH	OH	H
Tricitinidin (Tr)	H	OH	H	OH	OH	OH
Pelargonidin (Pg)	OH	OH	H	OH	H	H
Aurantinidin (Au)	OH	OH	OH	OH	H	H
Cyanidin (Cy)	OH	OH	H	OH	OH	H
Peonidin (Pn)	OH	OH	H	OH	OMe	H
Rosinidin (Rs)	OH	OH	H	OMe	OMe	H
Delphinidin (Dp)	OH	OH	H	OH	OH	OH
Petunidin (Pt)	OH	OH	H	OH	OMe	OH
Pulchellidin (Pl)	OH	OMe	H	OH	OH	H
Europinidin (Eu)	OH	OMe	H	OH	OMe	OH
Malvidin (Mv)	OH	OH	H	OH	OMe	OMe
Hirsutidin (Hs)	OH	OH	H	OMe	OMe	OMe
Capensinidin (Cp)	OH	OMe	H	OH	OMe	OMe

(b)

Figure 6.1 Structures of naturally occurring anthocyanidins. The corresponding anthocyanins are always glycosylated at the C-3 hydroxyl, and may be glycosylated at other free hydroxyl groups as well.

groups [45], the 3-monosides, 3-biosides, 3,5-diglycosides and 3,7-diglycosides being most common [1, 3]. The most common glycosyl moiety is glucose, although several different monosaccharides (*e.g.*, rhamnose, galactose, xylose or arabinose), disaccharides (mainly rutinose, sambubiose or sophorose; lathyrose, gentiobiose or laminariobiose occur less often) or trisaccharides may also glycosylate the anthocyanidins.

As shown in Figure 6.1, methoxylation of anthocyanidins and their glycosides also occurs, most frequently at the C-3′ and -5′ positions but also at positions C-7 and -5. No natural anthocyanin has been reported in which glycosylation or methoxylation occurs at all of the C-3, -5, -7 and -4′ positions. A free hydroxyl group at any of the C-5, -7 or -4′ positions is essential for formation of a quinonoidal (anhydro) base structure [40]. Anthocyanins normally exist as the quinonoidal base in plant tissues where they are localized in cell vacuoles in weakly acidic or neutral (pH 2.5 to 7.5) aqueous solution [46–48].

Acylation of anthocyanins often occurs. The main acylating groups are the phenolic acids ρ-coumeric, caffeic, ferulic or sinapic acids, but may sometimes be ρ-hydroxybenzoic, malonic or acetic acids. Acyl substituents are most often bound to the C-3 sugar [49], esterified to the 6-hydroxyl or, less frequently, to the 4-hydroxyl group of the sugar [50, 51]. Anthocyanins containing two or more acyl groups have been reported [52–57].

6.3.2 Distribution

The structural diversity of anthocyanins has contributed to more than 250 different anthocyanins reported in the literature. A thorough survey of the distribution of these pigments in nature [1] has revealed that they are particularly characteristic of the angiosperms or flowering plants, being more prominent in fruits and berries than in other plant parts [2, 58]. The pigmentation of plants and plant parts is rarely due to a single anthocyanin. This is evident in Table 6.1 in which is listed the individual anthocyanins in several food plants.

The major food sources of anthocyanins belong to the families Vitaceae (grape) and Rosaceae (cherry, plum, raspberry, strawberry, blackberry, apple, peach, *etc.*). Other familes containing anthocyanin pigmented food plants include the Solanaceae (tamarillo, aubergine), Saxifragaceae (red and black currants), Ericaceae (blueberry, cranberry), Oleaceae (black olive) and Cruciferae (red cabbage) (see Table 6.1). Most anthocyanin-containing food plants exhibit colours that are characteristic of their constituent aglycone types; however, it should be emphasized that many of the food plants listed in Table 6.1 also contain pigments other than anthocyanins (*e.g.* chalcones, aurones, carotenoids and chlorophylls), which influence the actual perceived colour (hue, tint, intensity). The most commonly occurring anthocyanins are based on cyanidin, followed by pelargonidin, peonidin and delphinidin, then by petunidin and malvidin [1–3, 58]. In terms of glycoside distribution, 3-glycosides occur approximately two and a half times as often as 3,5-diglycosides, the most ubiquitous anthocyanin being cyanidin-3-glycoside [2].

6.3.3 Biosynthesis

The biosynthesis of anthocyanins follows a pathway common to other flavonoids in plants [39, 116–120]. Anthocyanin accumulation is influenced by several environmental factors, including light, temperature, plant hormones,

Table 6.1 Anthocyanins in selected food plants[a]

Botanical name	Common name (organ)	Anthocyanins present	References
Allium cepa	Red onions	Cy 3-glucoside, 3-galactoside, 3-diglucoside and 3-laminariobioside; Pn 3-glucoside	59, 60
Brassica oleraceae	Red cabbage (leaf)	Cy 3-sophoroside-5-glucoside acylated with malonoyl, ρ-coumaroyl, di-ρ-coumaroyl, feruloyl, diferuloyl, sinapoyl and disinapoyl esters	61–63
Citrus sinensis	Blood orange	Cy and Dp 3-glucosides	64
Cyphomandra betaceae	Tamarillo	Cy, Pg and Dp 3-glucosides and 3-rutinosides	65
Ficus spp.	Fig (skin)	Cy 3-glucoside, 3-rutinoside and 3,5-diglucoside; Pg 3-rutinoside	66, 67
Fragaria spp.	Strawberry	Pg and Cy 3-glucosides	68–71
Glycine maxima	Soybean (seedcoat)	Cy and Dp 3-glucosides	72, 73
Malus pumila	Apple (skin)	Cy 3-glucoside, 3-xyloside, 3-galactoside, 3- and 7-arabinosides; free and acylated	74, 75
Mangifera indica	Mango	Pn 3-galactoside	76
Passiflora edulis	Passion fruit	Pg 3-diglucoside; Dp 3-glucoside	3, 77
Phaseolus vulgaris	Kidney bean (seedcoat)	Pg, Cy and Dp 3-glucosides and 3,5-diglucosides; Pt and Mv 3-glucosides	3, 78
Prunus avium	Sweet cherry	Cy and Pn 3-glucosides and 3-rutinosides	79–81
Prunus cerasus	Sour cherry	Cy 3-glucoside, 3-rutinoside, 3-sophoroside, 3-glucosyl-rutinoside and 3-xylosyl-rutinoside; Pn 3-glucoside and 3-rutinoside	79, 80, 82, 83
Prunus domestica	Plum	Cy and Pn 3-glucosides and 3-rutinosides	71, 80, 84
Punica granatum	Pomegranate	Cy, Dp and Pg 3-glucosides and 3,5-diglucosides	3, 85
Raphanus sativus	Red radish (root)	Pg and Cy 3-sophoroside-5-glucosides acylated with ρ-coumaroyl, feruloyl and caffeoyl esters	49, 86, 87
Rheum rhaponticum	Rhubarb (stem)	Cy 3-glucoside and 3-rutinoside	87–89
Ribes nigrum	Blackcurrant	Cy and Dp 3-glucosides, 3-diglucosides and 3-rutinosides	90, 91

Table 6.1 *Continued*

Botanical name	Common name (organ)	Anthocyanins present	References
Ribes rubrum	Redcurrant	Cy 3-glucoside, 3-rutinoside, 3-sambubioside, 3-sophoroside, 3-(2-glucosylrutinoside) and 3-(2-xylosylrutinoside)	80, 92, 93
Rubus fruticosus	Blackberry	Cy 3-glucoside and 3-rutinoside; free and acylated Mv biosides	80, 94, 95
Rubus ideaus	Red raspberry	Cy and Pg 3-glycosylrutinosides, 3-sophorosides, 3-sambubiosides, 3-rutinosides, 3-glucosides, 3-diglucosides, 3,5-diglucosides and 3-rutinoside-5-glucosides	80, 96–98
Solanum melongena	Aubergine	Dp 3-galactoside, 3-rutinoside, 3,5-diglucoside, 3-rutinoside-5-glucoside and 3-[4-(ρ-coumaroyl)-L-rhamnosyl-(1,6)-glucosido]-5-glucoside	3, 99, 100
Solanum nigrum	Huckleberry	Pt and Mv 3-rutinoside-5-glucoside; free and acylated with ρ-coumaroyl and di-*p*-coumaroyl esters	101, 102
Synsepalum dulcificum	Miracle fruit (skin)	Cy and Dp 3-galactosides, 3-glucosides and 3-arabinosides	103
Vaccinium angustifolium	Lowbush blueberry	Cy, Dp, Pn, Pt, and Mv 3-glucosides, 3-galactosides and 3-arabinosides	104
Vaccinium corymbosum	Highbush blueberry	Cy, Dp, Pn, Pt and Mv 3-glucosides and 3-galactosides; Pn, Dp, Pt and Mv 3-arabinosides	105, 106
Vaccinium macrocarpon	Cranberry	Cy and Pn 3-galactosides, 3-arabinosides and 3-glucosides	107–110
Vitis spp.	Grape	Cy, Pn, Dp, Pt and Mv mono- and diglucosides; free and acylated	111
Zea mays	Purple corn	Cy, Pg and Pn 3-glucosides; Cy 3-galactoside; free and acylated	112–115

[a] Abbreviations: Cy = cyanidin; Pg = pelargonidin; Pn = peonidin; Dp = delphinidin; Pt = petunidin; Mv = malvidin.

plant nutrition, mechanical damage and pathogenic attack, light being regarded as the most important of these [7]. In general, far-red light activates the pigment phytochrome, which induces the enzymes of the flavone-glycoside pathway and, as a result, accumulation of anthocyanins [121, 122]. UV-light is often required in addition to active phytochrome for enzyme induction. Anthocyanins do not accumulate in all cells of plants that synthesize them. Rather, they are typically localized in flower and fruit tissues, and in the epidermal and hypodermal cell layers of leaves and stems [39]. Yet, even within a single tissue type, flavonoid/anthocyanin distribution is non-uniform.

The intracellular localization of anthocyanin biosynthesis is still uncertain. It was originally suggested to occur in highly pigmented spherical organelles within the cell vacuole called anthocyanoplasts [123–125]. These organelles occur widely in both dicotyledons and monocotyledons [39]. However, it is unlikely that anthocyanin biosynthesis takes place in anthocyanoplasts because of their high anthocyanin concentration and low pH: the pH-optima of all enzymes known to be involved in flavonoid biosynthesis are basic [126]. Based on biochemical and immunochemical evidence [125–128] Hrazdina and Wagner [126] proposed that anthocyanin biosynthesis was mediated via a membrane-associated enzyme complex. The first enzyme, phenylalanine ammonia lyase, is located on the lumen face of the endoplasmic reticulum (ER) where it has access to a pool of phenylalanine. Transmembrane cinnamate 4-hydroxylase channels its product to the cytoplasmic face of the ER, where further enzymes of the flavonoid pathway are located. Their products, dihydroflavonoids, are further transformed to anthocyanidins (*i.e.* anhydrobases) by several transmembrane oxido-reductases, and subsequent glycosylation and acylation of anthocyanidins by glycosyl- and acyltransferases occurs on the lumen face of the ER membrane. The resulting anthocyanins are sequestered in specific regions of the ER destined for vesiculation; they are thereby transported to the central or cell vacuole where they accumulate [126].

Anthocyanin biosynthesis is genetically regulated, being governed by a set of dominant genes. A correlation exists between the genotype of plants and enzymes directly involved in anthocyanin biosynthesis, for example, chalcone synthase, chalcone isomerase, flavanone 3-hydroxylase, flavanone 3′-hydroxylase and several glycosyl- and acyltransferases. Anthocyanin biosynthesis proceeds according to the scheme in Figure 6.2. Formation of the anthocyanins occurs via a chalcone synthesized in the plant from condensation of a molecule of activated cinnamic acid (*e.g.* co-enzyme A-ester of mainly 4-coumaric acid and, to a lesser extent, ferulic, caffeic, 5-hydroxyferulic or sinapic acid) and three molecules of malonyl-CoA, catalyzed by the enzyme chalcone synthase. Free activated cinnamic acids are derived from L-phenylalanine by general phenylpropanoid metabolism [117, 119]; malonyl-CoA is formed from the acetyl-CoA carboxylase reaction [129]. Cyclization of the chalcone by chalcone isomerase leads to formation of isomeric flavanones

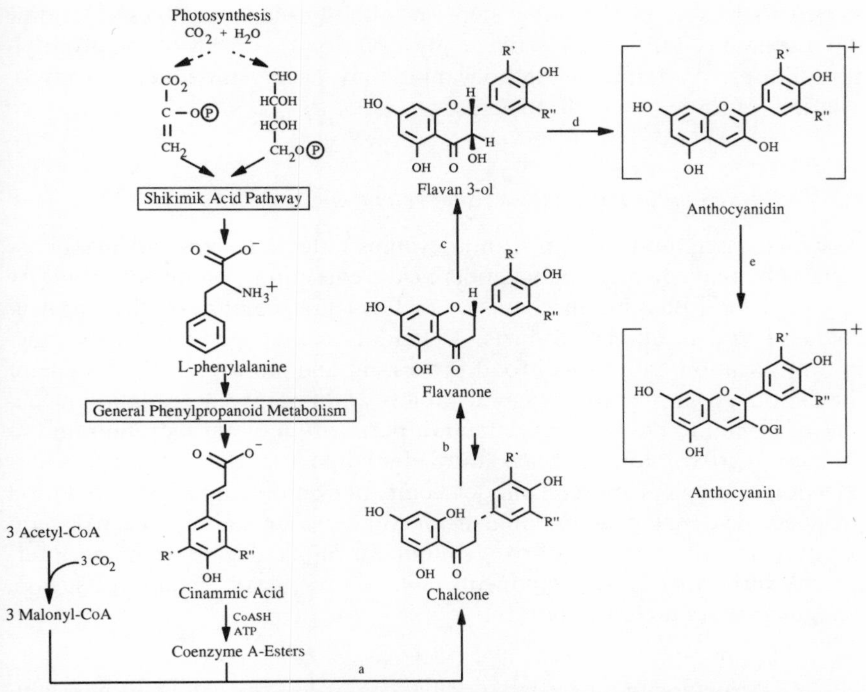

Figure 6.2 Biosynthesis of anthocyanins. Enzymes: chalcone synthase (a); chalcone isomerase (b); flavanone 3-hydroxylase (c); putative oxidoreductase(s) (d); glycosyltransferase (e); Gl = glycosyl group; R′, R″ = H, OH, OCH_3 or OGl.

(*e.g.* naringenin), which themselves are converted to dihydroflavonols (flavan 3-ols) by flavanone 3-hydroxylase, possibly in concert with a 2-hydroxylase and a dehydratase [39, 120]. Dihydroflavonols are presumed to be the immediate precursors of anthocyanidins. Although no enzymes have yet been identified that catalyze the conversion, they are thought to be oxido-reductases with putative intermediates being flavan-3,4-diols (*i.e.* leucoanthocyanidins) [120]. Variation in B-ring substitution may be introduced at the cinnamic acid stage [130], but most often occurs at the stage of a C_{15} intermediate (*e.g.* chalcone/flavanone) later in the pathway [117, 120, 131].

Since the naturally occurring anthocyanidins are unstable in aqueous media [42], under physiological conditions they must be immediately glycosylated at the C-3 hydroxyl group. Alternatively, Brouillard [40] has suggested that anthocyanidin 3-glycosides could derive from dihydroflavonol 3-glycosides. Additional glycosylation, if it occurs, proceeds in a stepwise manner to higher glycosylated forms of the anthocyanins, beginning with attachment of sugars to the 3-glycosyl residue and/or the 5-hydroxyl group. Glycosylation is presumed to be one of the latter steps in anthocyanin biosynthesis. Uridine diphosphate (UDP)-sugars serve as glycosyl donors for all of the glycosyltransferases. Acylation of anthocyanins may occur subsequent to glycosylation, catalyzed by acyltransferases.

6.3.4 Factors influencing anthocyanin colour and stability

As with most natural colorants, anthocyanins suffer from inherent instability. Generally, they are most stable under acidic conditions, but may degrade by any of several possible mechanisms to form first colourless, then brown-coloured and insoluble products. Degradation may occur during extraction/purification and normal food processing and storage. A knowledge of the factors governing anthocyanin stability and putative degradation mechanisms is vital to the efficient extraction/purification of anthocyanins and to their use as food colorants. Such knowledge could also lead to more judicious selection of pigment sources and development of more highly-coloured food products. The major factors influencing anthocyanin stability are pH, temperature and the presence of oxygen, but enzymatic degradation and interactions with other food components (*e.g.* ascorbic acid, metal ions, sugars, copigments) are no less important.

6.3.4.1 Structure and pH. Under acidic conditions the colour of non- and monoacylated anthocyanins is determined largely by substitution in the B-ring of the aglycone (see Figure 6.1). Increased hydroxyl substitution yields a bluer colour, while methoxylation causes the chromophores to become more red [24]. Thus, aqueous extracts containing primarily pelargonidin and/or cyanidin glycosides appear orange-red, those with peonidin glycosides are deep red, and those containing glycosides of delphinidin, petunidin and/or malvidin exhibit bluish-red coloration. Glycosylation and acylation of anthocyanins generally have a blueing effect.

The anthocyanins, because of their structural diversity, are not equally resistant to the deleterious effects of various agents. As the degree of aglycone hydroxylation increases, anthocyanin stability generally decreases; whereas an increase in methoxyl substitution has the opposite effect [132]. Glycosylation of free hydroxyl groups increases anthocyanin stability, similar to the effect of methoxylation, presumably due to blocking of reactive hydroxyl groups. Thus, anthocyanin diglycosides were found to be more stable to decolorization during storage, heat treatment and exposure to light than monoglycosides [133, 134]. The nature of the sugar residues also has an effect, for example, anthocyanins glycosylated with galactose in cranberry juice were more stable than those glycosylated with arabinose during storage [135]. Further work is required to clarify the stabilizing effects of the various glycosyl moieties.

In aqueous media anthocyanins undergo several structural transformations that are pH-dependent. At a given pH, an equilibrium exists between four anthocyanin/aglycone structures: the blue quinonoidal (anhydro) base (A), the red flavylium cation (AH^+) and the colourless carbinol pseudobase (B) and chalcone (C) (Figure 6.3). At neutral or slightly acidic pH the anthocyanins exist predominantly in their non-coloured forms. Yet, stabilization of the coloured species, especially the quinonoidal base, may be conferred by the presence of acyl groups linked to sugar moieties of the pigment molecule, that is, by intramolecular co-pigmentation. Anthocyanins containing two or more acyl groups display excellent colour stability throughout the entire pH range [52–56, 136–139]. The stability conferred by intramolecular copigmentation is attributed to stacking of aromatic residues of the acyl group(s) with the pyrylium ring of the anthocyanin molecule [40, 48, 140]. Deacylated pigments fade immediately after their dissolution in neutral or slightly acidic media, similar to the behaviour of non-acylated anthocyanins [55]. Monoacylated anthocyanins do not display the colour stability of di- or polyacylated anthocyanins, indicating that at least two constituent acyl groups are required for good colour stability/retention in neutral or slightly acidic media [40, 48]. The nature of the acyl moiety, however, has some influence on anthocyanin

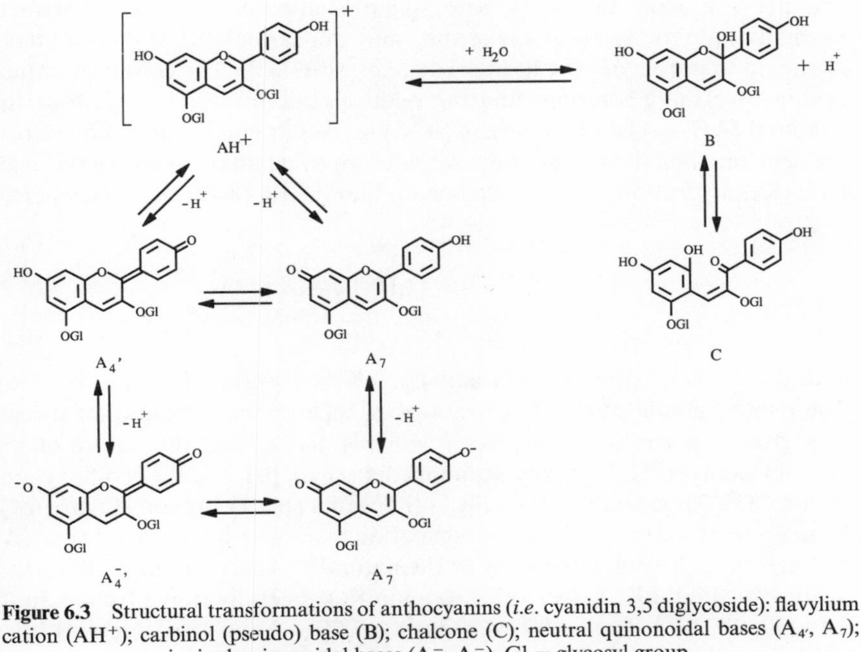

Figure 6.3 Structural transformations of anthocyanins (*i.e.* cyanidin 3,5 diglycoside): flavylium cation (AH^+); carbinol (pseudo) base (B); chalcone (C); neutral quinonoidal bases ($A_{4'}$, A_7); ionized quinonoidal bases ($A_{4'}^-$, A_7^-). Gl = glycosyl group.

stability: at neutral pH anthocyanins acylated with ρ-coumeric acid were found to be less stable than those acylated with caffeic acid [141].

It is not a coincidence that no naturally occurring anthocyanin has yet been reported in which glycosylation and/or methoxylation occurs in all of the C-3, -5, -7 and -4′ positions. For the formation of the quinonoidal species, which is primarily responsible for the pigmentation of flower and fruit tissues, a free hydroxyl group must occur at any of the C-5, -7 or -4′ positions [40, 48]. The quinonoidal base is derived, generally above pH 3.0, from the flavylium cation form of the anthocyanin. The flavylium cation, described as a heterocyclic carboxonium cation, is relatively stable under acid conditions (*e.g.* <pH 3.0) where it exists in as many as six resonance forms: its positive charge is delocalized over the entire heterocyclic structure [40, 142]. Because the highest partial positive charge occurs at the C-2 and -4 positions [143], the flavylium cation is particularly prone to nucleophilic attack at these positions (see section 6.3.4.7). Noteworthy, is that phenyl and, especially, methyl substitution at the C-4 position markedly stabilizes the flavylium chromophore [340, 41, 144–147].

The presence of the oxonium ion adjacent to the C-2 position in anthocyanins is responsible for their characteristic amphoteric nature. Non- and monoacylated anthocyanins thus behave somewhat like pH indicators, existing as either an acid or a base, depending on pH (Figure 6.3). Anthocyanin-containing solutions generally display their most intense red coloration at acid pH (*e.g.* <pH 3.0). With increasing pH, aqueous anthocyanin extracts normally fade to the point where they may appear colourless before finally changing to purple or blue at high pH (*e.g.* >pH 6.0). The colour of anthocyanin-containing solutions and the relative concentrations of each of the coloured (AH^+, A) and colourless (B, C) species at equilibrium (Figure 6.4) are dependent on the values of the equilibrium constants controlling the acid-base (K'_a), hydration (K'_h) and ring-chain tautomeric (K_T) reactions, where:

$$K'_a = ([A]/[A_{H^+}])a_{H^+}, \qquad (6.1)$$

$$K'_h = ([B]/[AH^+])a_{H^+}, \qquad (6.2)$$

$$K_T = [C]/[B], \qquad (6.3)$$

and a_{H^+} is the hydronium ion activity; $pH = -\log a_{H^+}$ [31, 40, 48, 145]. The relative equilibrium concentrations of each of the anthocyanin species at a given temperature varies both with pH/acidity and the nature of the anthocyanin(s) [31]. In highly acidic media (*i.e.* <pH 2.0) the red flavylium cation (AH^+) is essentially the only anthocyanin species present (Figure 6.4). With an increase in pH the flavylium cation yields the quinonoidal base (A) through rapid loss of a proton. For the naturally occurring anthocyanins so far investigated, pK'_a values ($pK'_a = -\log K'_a$) generally range from 3.36 to 4.85 [40, 48, 145]. The hydroxyl groups at C-4′, -7 and -5 (if present) have similar ionization constants; therefore, the quinonoidal species normally

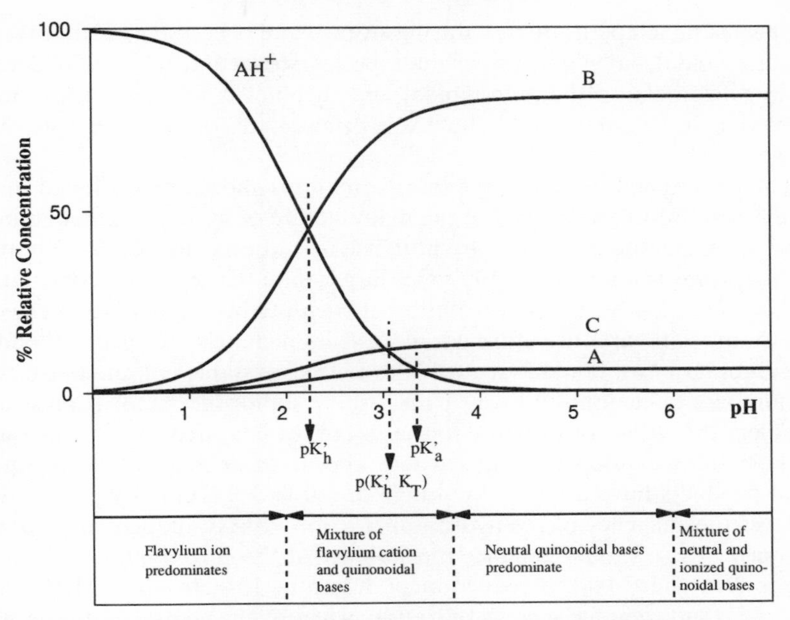

Figure 6.4 pH-distribution of anthocyanin species and predominant chromophores at equilibrium for cyanidin 3,5-diglycoside at 25°C: flavylium cation (AH^+); carbinol base (B); chalcone (c); quinonoidal base(s) (A). $pK'_h = 2.23$; $pK'_a = 3.38$; $p(K'_h K_T) = 3.03$ (adapted from Brouillard [48] and Mazza and Brouillard [145] with permission).

exists as a mixture (*e.g.* A'_4, A_7; see Figure 6.3) [48]. Anthocyanins substituted with more than one free hydroxyl group become further deprotonated at pH 6.0 to 8.0 to yield a mixture of resonance-stabilized quinonoidal anions (*e.g.* A'_4, A_7^-; see Figure 6.3). Unlike flavylium salts unsubstituted at positions C-3 and -5 (*i.e.* synthetic salts), for which hydration reactions occur more readily with quinonoidal base forms than cationic forms [147, 148], equilibrium between the quinonoidal and carbinol bases of naturally occurring anthocyanins occurs exclusively via the flavylium cation [149]. On standing, at pH values of about 3.0 to 6.0, nucleophilic addition of water to the C-2 position of the flavylium cation slowly yields the colourless carbinol pseudobase (B). At an even slower rate, the carbinol equilibrates to the open chalcone species (C), which is also colourless. The hydration-dehydration equilibrium is usually characterized by pK'_h values ($pK'_h = -\log K'_h$) of 2.0 to 3.0 [40, 150].

At room temperature and in slightly acidic aqueous solutions only small amounts of chalcone are formed, since the tautomeric equilibrium constant, K_T, is usually always less than one [151]. However, cation deprotonation by solvent ($AH^+ \rightarrow A \rightarrow A^-$) is exothermic, whereas cation hydration ($AH^+ \rightarrow$ B) and pyrylium ring opening ($B \rightarrow C$) are both endothermic and associated with positive entropy changes [151]. Thus, formation of chalcone is favoured by increasing temperature (*viz*. during storage and processing) at the expense of quinonoidal, flavylium and carbinol species (see section 6.3.4.2). On cooling and/or acidification the quinonoidal and carbinol bases are quickly transformed to the flavylium cation, but transformation of the chalcone is relatively slow.

The pH-dependent structural transformations should be exploited in the analysis of anthocyanins and in the manufacture of anthocyanin-containing food products: too often they are not! The hydration constant K_h' of natural anthocyanins is about 10 to 100 times larger than the acid-base constant K_a' [48]. Consequently, upon dissolution in neutral or slightly acidic media, flavylium salts are transformed almost immediately to neutral and/or quinonoidal bases but slowly evolve to the more stable colourless carbinol pseudobase or its ionized form. Unless some stabilizing factor is present to augment the colour or promote formation of the coloured species, the use of anthocyanins as food colorants would appear to be ineffective since most food products have a pH in the range of 3.0 to 7.0 [26]. Any process that reduces the efficiency of the hydration reaction, that is, induces an apparent decrease of K_h', should enhance the stability of the chromophores. According to Brouillard [48], a reduction of K_h' from 10^{-2} to 10^{-3} M down to 10^{-5} M is sufficient for such stabilization, whereby the neutral and/or ionized quinonoidal bases are both the kinetic and thermodynamic products of equilibrium reactions. *In vivo*, co-pigmentation strongly favours the anthocyanin chromophores (see section 6.3.4.8), by reducing K_h' values, and thereby confers colour to plant products within a pH range in which the anthocyanins generally display poor colour properties (*e.g.* pH 3.0 to 7.0). Anthocyanins may have greater application as food colorants if polyacylated forms or co-pigmentation were exploited by food manufacturers.

6.3.4.2 Temperature. As with most chemical reactions the stability of anthocyanins and the rate of their degradation, in natural and model systems, is markedly influenced by temperature. The thermal stability of anthocyanins varies with their structure, pH, the presence of oxygen and interactions with other components in the system. In general, structural features that lead to increased pH-stability also lead to increased thermal stability, *i.e.* hydroxylation of the aglycone decreases stability, while methoxylation, glycosylation and acylation have the opposite effect [132–134]. In the presence of oxygen, maximum thermal stability of anthocyanidin 3-glycosides has been observed at pH 1.8 to 2.0 [98, 152, 153], while that of anthocyanidin 3,5-diglycosides has been observed at pH 4.0 to 5.0 [132, 154, 155]. Anthocyanin degradation is virtually pH-independent at pH 2.0 to 4.5 in the absence of oxygen [155–157].

With few exceptions [156, 158], anthocyanin degradation, under both aerobic and anaerobic conditions, follows first-order kinetics [152, 153, 159–

162]. The kinetic data accumulated to date suggest that the mechanism of anthocyanin degradation is temperature dependent. At storage temperatures (*e.g.* <40°C) activation energy (E_a) and z values of around 17 kcal/mol and 25°C, respectively, have been reported; while at processing temperatures (*e.g.* >70°C) E_a values of 23 to 27 kcal/mol and z values of around 28°C have been reported [162–165]. (The z value is the temperature change required to change the thermal destruction time by a factor of ten; it reflects the temperature dependence of thermal degradation.) It would appear, then, that anthocyanins have relatively greater stability at higher temperatures, despite that such conditions favour the formation of colourless anthocyanin species, as previously discussed. The protective effect of various constituents in a given system, and of condensation reactions, cannot be overlooked. The concentration of polymeric pigments has been shown to increase with temperature and storage time, and to contribute to the coloration of juices and red wines [166–170]. Several authors have recommended high-temperature short-time processing to achieve maximum pigment retention in anthocyanin-containing food products [159, 165, 170]. The short times involved may be sufficient to prevent significant degradation of anthocyanins and/or transformation to colourless species.

Several pathways for thermal degradation of anthocyanins have been proposed (Figure 6.5), the exact mechanism appearing to not only be temperature-dependent but also anthocyanin-dependent. Hrazdina [171] identified coumarin diglycoside as a common degradation product of anthocyanidin 3,5-diglycosides. He proposed a mechanism whereby the flavylium cation was first transformed to the quinonoidal base, then to several intermediate products before finally breaking down to a coumarin derivative and a compound corresponding to the B-ring of the anthocyanidin (see Figure 6.5(a)). Anthocyanidin 3-glycosides do not form coumarin derivatives [57]. According to Markakis *et al.* [159], the first steps in their thermal degradation involve formation of the colourless carbinol pseudobase and subsequent opening of the pyrylium ring to form the chalcone before hydrolysis of the glycosidic bond occurs (see Figure 6.5(b)). This is consistent with current views of anthocyanin transformations (see Figure 6.3) and the knowledge that heat favours formation of the chalcone. Adams [155, 156, 172] without excluding the possibility of chalcone glycoside formation, offered an alternative mechanism (see Figure 6.5(c)): anthocyanidin 3-glycosides heated at pH 2.0 to 4.0 first undergo hydrolysis of the glycosidic bond (this occurs readily at 100°C) followed by conversion of the aglycone to a chalcone and subsequent formation of an α-diketone. Thermal degradation products of cyanidin 3-glycosides include chalcones and α-diketone [155], protochuic acid, quercetin and phloroglucinaldehyde [173]. It is generally assumed that further degradation of the primary anthocyanin breakdown products (most of which are colourless) leads to formation of brown-coloured products [157, 159, 163, 174, 175].

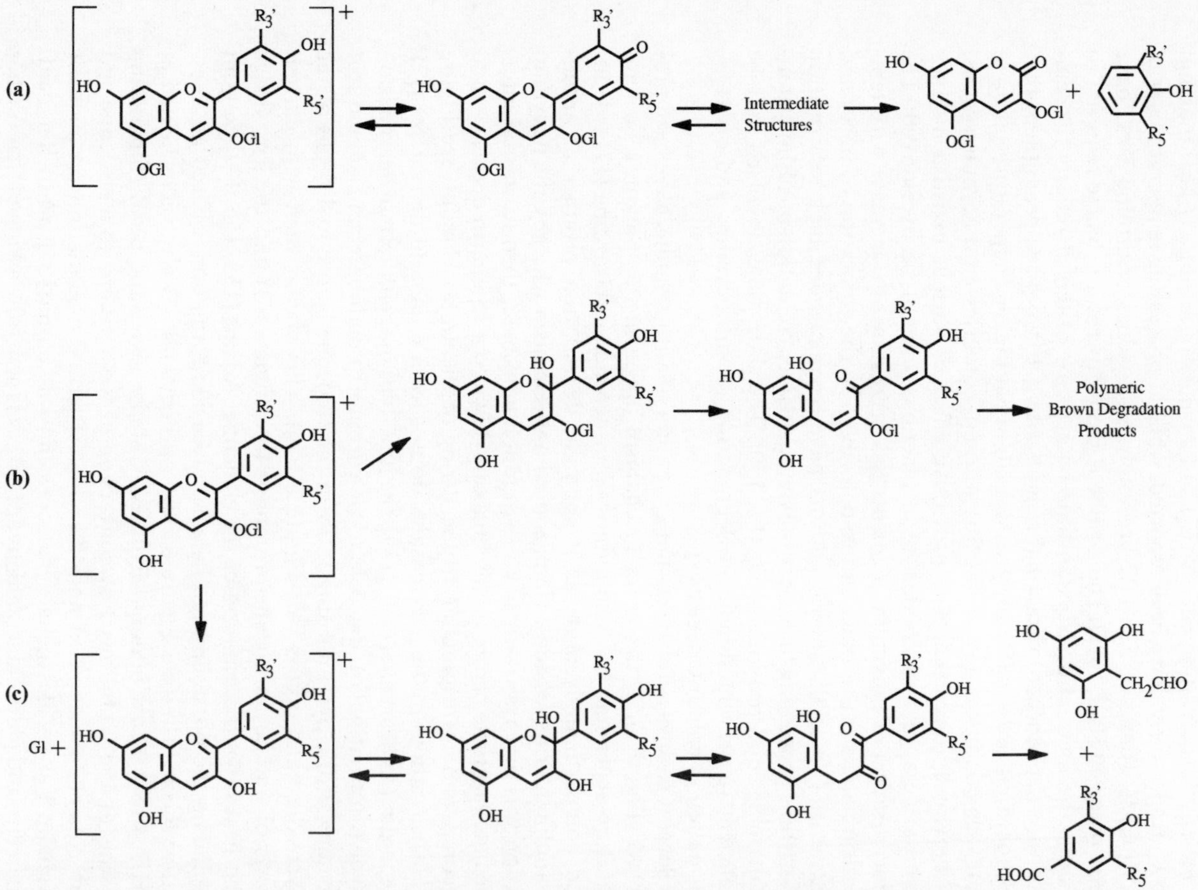

(a)
HO, OGl, OGl, R_3', OH, R_5', +
HO, O, OGl, OGl, R_3', R_5'
Intermediate Structures
HO, O, O, OGl, OGl, +, R_3', OH, R_5'

(b)
HO, OH, OGl, R_3', OH, R_5', +
HO, HO, OH, OGl, R_3', OH, R_5'
HO, HO, OH, O, OGl, R_3', OH, R_5'
Polymeric Brown Degradation Products

(c)
Gl +, HO, OH, OH, R_3', OH, R_5', +
HO, HO, OH, OH, R_3', OH, R_5'
HO, HO, OH, O, O, R_3', OH, R_5'
HO, OH, OH, CH_2CHO
+
HOOC, R_3', OH, R_5'

Figure 6.5 Mechanisms of anthocyanin degradation as proposed by Hrazdina [171] for anthocyanidin 3,5-diglycosides (a), and Markakis *et al.* [159] (b) and Adams [155, 156, 172] (c) for anthocyanidin 3-glycosides. $R_{3'}$, $R_{5'}$ = H, OH, OCH_3 or OGl; Gl = glycosyl group.

6.3.4.3 Oxygen and hydrogen peroxide. Oxygen and temperature have been referred to as the most specific accelerating agents in the destruction of anthocyanins [176]. The deleterious effects of molecular oxygen were alluded to in the previous section; they are well documented in the literature [98, 135, 155–157, 177]. Oxygen may cause degradation of anthocyanins by a direct oxidation mechanism and/or by indirect oxidation whereby oxidized constituents of the medium react with the anthocyanins to yield colourless or brown-coloured products. Precipitate and haze development in fruit juices may result from direct oxidation of the carbinol base [157].

Reports of anthocyanin decolorization from direct attack by ascorbic acid [44, 178] have been discounted [179, 180]. Ascorbic acid-induced destruction of anthocyanins probably results from indirect oxidation by hydrogen peroxide (H_2O_2) formed during the aerobic oxidation of ascorbic acid. Ascorbic acid and oxygen have been shown to act synergistically in anthocyanin degradation [160]. Maximum pigment losses in various anthocyanin-containing fruit juices have occurred under conditions most favourable to ascorbic acid oxidation, that is, at high levels or concentrations of oxygen and ascorbic acid [135, 181]. Significant losses of strawberry-juice colour have also been demonstrated in the presence of added H_2O_2 [182]. Nucleophilic attack at the C-2 position of anthocyanins by H_2O_2 cleaves the pyrylium ring and leads to formation of various colourless esters and coumarin derivatives [51, 183]. The oxidation products may partake in further degradation reactions or polymerization to yield ultimately a brown resinous precipitate [157, 159, 181, 182]. In acid solution, H_2O_2 was shown to oxidize acylated 3,5-diglycosides directly to acylated *O*-benzylphenylacetic acid esters [57].

6.3.4.4 Light. Anthocyanins are generally unstable when exposed to UV or visible light [134, 184, 185] or other sources of ionizing radiation [186, 187]. Their decomposition would appear to be mainly photo-oxidative since *p*-hydroxybenzoic acid has been identified as a minor degradation product [180]. Anthocyanins substituted at the C-5 hydroxyl group, which are known to fluoresce [188], are more susceptible to photochemical decomposition than those unsubstituted at this position [134, 180, 189]. Co-pigmentation can accelerate or retard the decomposition depending on the nature of the co-pigment [190]. The ability of light to yield an anthocyanin excited state via electron transfer would appear to predispose these pigments to photochemical decomposition.

6.3.4.5 Enzymes. Several enzymes that are endogenous in many plant tissues have been implicated in the oxidative decoloration of anthocyanins. These enzymes have been generally termed anthocyanases, but based on their activity two distinct groups of enzymes have been identified:

1. Glycosidases, which hydrolyze the glycosidic bond(s) of anthocyanins to yield free sugar and aglycone, the instability of the latter chromophores resulting in their spontaneous transformation to colourless derivatives [191–194];

2. Polyphenol oxidases (PPO), which act on anthocyanins in the presence of *o*-diphenols via a coupled oxidation mechanism (Figure 6.6) [195–202]. In addition to glycosidase and PPO, Grommeck and Markakis [203] have reported a peroxidase-catalyzed anthocyanin degradation.

PPO is virtually ubiquitous in the plant kingdom, catalyzing the oxidative transformation of catechol and other *o*-dihydroxyphenols to *o*-quinones, which may subsequently either react with each other, with amino acids or proteins (and/or other phenolic compounds including anthocyanins) to yield brown-coloured higher molecular weight polymers; or oxidize compounds of lower oxidation-reduction potential [204]. Anthocyanins are poor substrates for PPO [21, 202]. The quinonoidal base would appear to be more susceptible to oxidative degradation by PPO than the flavylium cation [205]; the rate of decoloration mediated by PPO was suggested to be dependent on the substitution pattern in the B-ring and extent of glycosylation [206].

Various commercial enzyme preparations, usually obtained from fungal sources, have been shown to possess glycosidase [191–193, 207–208] and/or PPO activity [209]. In most cases, secondary activity in commercial enzyme preparations is desirable, and the costs of purification often preclude total removal of such activity. Fungal 'anthocyanase' preparations have been used to remove excess anthocyanin from blackberry jams and jellies that were too dark and unattractive [210]; and similar preparations have been suggested for use in the manufacture of white wines from mature red grapes [191].

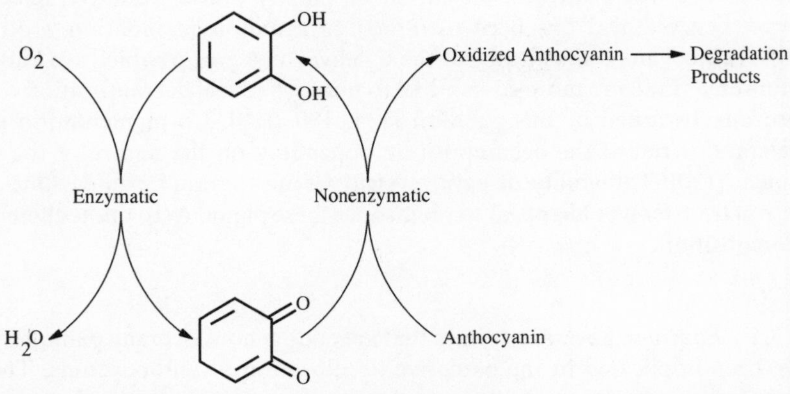

Figure 6.6 Coupled oxidation mechanism involved in polyphenol oxidase decolorization of anthocyanins (from Peng and Markakis [197] with permission).

However, anthocyanase activity, whether endogenous or exogenous, can be problematic if maximum pigment retention is desired. A preliminary steam blanch prior to subsequent processing and/or storage [211] and packing in high concentrations (*e.g.* $>20\%$) of sugar/syrup [212] have proven effective in destroying and inhibiting, respectively, endogenous anthocyanase activity in fruits. Glucose, gluconic acid and glucono-δ-lactone are competitive inhibitors of glycosidases [208]. PPO activity in various anthocyanin-containing fruit extracts has been effectively inhibited by SO_2 [213], bisulphite, dithiothreitol, phenylhydrazine and cysteine [214] and gallotannin [206]. The stabilizing effect of gallotannin is also attributed to co-pigmentation, whereby excess gallotannin complexes with anthocyanin via hydrophobic interactions [206]. Ascorbic acid also has a protective effect on anthocyanin degradation by PPO, by being preferentially oxidized by the *o*-quinone formed from enzymatic oxidation of the mediating phenol. Anthocyanins and their associated colour are not destroyed so long as ascorbic acid is present in the system [205].

6.3.4.6 Nucleophilic agents. Non- and monoacylated anthocyanins are particularly susceptible to nucleophilic attack at the C-2 and/or C-4 positions. The reaction of flavylium salts with nucleophiles occurs readily when the C-5 position is unsubstituted [180]. As previously discussed, nucleophilic attack of the flavylium cation by water occurs at the C-2 position, yielding the colourless carbinol base. For amino acids [215] and carbon nucleophiles such as catechin, phenol, phloroglucinol, 4-hydroxycoumarin and dimedone, attack occurs at the electrophilic C-4 position to yield 4-substituted flav-2-enes [143, 216–219]. These compounds are highly reactive and may undergo further changes depending on the nature of the 4-substituent [217–219]. Loss of anthocyanin colour ultimately results. The decolorizing action of SO_2, an antiseptic agent used extensively in the wine industry, also results from formation of a colourless C-4 adduct [220–223]. This reaction is reversible; however, acidification to about pH 1.0 is required for restoration of red colour [220]. The equilibrium constants and rate of reaction for formation of anthocyanin-bisulphite complexes have been previously published [221–222]. The large values of these constants are indicative of the small amounts of free sulphur dioxide that can decolorize substantial quantities of anthocyanins [163, 222]. Kinetically, anthocyanin-bisulphite complexes are quite stable; the bisulphite moiety presumably deactivates the C-3 glycosidic bond, thereby preventing its hydrolysis and subsequent formation of brown-coloured degradation products [155, 172]. Steric factors also contribute to the stability of anthocyanin-bisulphite complexes.

Synthetic flavylium salts substituted with methyl or phenyl moieties at the C-4 position are virtually resistant to nucleophilic attack, suggesting that C-4 substitution may be a practical means by which to stabilize anthocyanins for use as food colorants [144–146]. The only natural C-4 substituted anthocyanin known is purpurinidin fructo-glucoside extracted from willow bark [224]; a shortage of raw material does not allow commercial use of this orange pigment as a food colorant [58].

At acid pH, anthocyanins possess a positive charge and are, therefore, virtually resistant to electrophilic attack, for example by aldehydes. However, Timberlake and Bridle [225] have shown that pure anthocyanins can react with flavan 3-ols, such as catechin, in the presence of acetaldehyde to yield a product of enhanced colour. The reaction has also been shown to augment and stabilize the colour of crude anthocyanin extracts at pH 4.0 to 6.0 [225–226]. The reaction is purported to involve covalent linkage of the anthocyanin with C-8 of catechin via a CH_3CH linkage. Acetaldehyde alone, and other aldehydes can cause fading of crude anthocyanin extracts by electrophilic attack at the C-6 and/or C-8 positions [225]. Glycosyl substitution at C-5 reduces the nucleophilic character of the C-6 and C-8 positions. Thus, anthocyanidin 3,5-diglycosides are less prone to electrophilic attack at these positions than 3-glycosides [58, 225].

6.3.4.7 Sugars and their degradation products. The use of high sugar concentrations (*i.e.* $>20\%$) or syrups to preserve fruits and fruit products has an overall protective effect on anthocyanin chromophores [212], presumably by lowering water activity (a_w). Reduced a_w is associated with a reduced rate of anthocyanin degradation [174]: the hydration of anthocyanin chromophores to colourless species becomes less favourable as water becomes limiting. Indeed, dried anthocyanin powders ($a_w \leqslant 0.3$) are relatively stable at room temperature for several years when held in hermetically sealed containers [227–229].

Above a threshold level (*e.g.* 100 p.p.m.) [154] sugars and their degradation products have been shown to accelerate the degradation of anthocyanins. Fructose, arabinose, lactose and sorbose demonstrate greater degradative effects on anthocyanins than glucose, sucrose or maltose [152, 153, 230]. The rate of anthocyanin degradation is associated with the rate at which the sugar itself is degraded to furfural-type compounds [98, 152, 153, 159, 231]. These compounds, furfural (derived mainly from aldo-pentoses) and 5-hydroxymethylfurfural (HMF; formed from keto-hexoses), derive from the Maillard reaction [232] or from oxidation of ascorbic acid [233, 234], polyuronic acids [135] or anthocyanins themselves. These degradation products readily condense with anthocyanins by unknown mechanisms, ultimately leading to formation of complex brown-coloured compounds [23]. The advanced sugar degradation products, levulinic and formic acids, which are readily formed from HMF and furfural in the acidic environs of most anthocyanin-containing fruits and their juices [25], do not react nearly as rapidly with anthocyanins as their parent furyl aldehydes [98, 153, 159, 230, 231]. Anthocyanin degradation in the presence of furfural and HMF is directly temperature dependent and more pronounced in natural systems, such as with fruit juice [231]. Oxygen enhances the degradative effects of all sugars and sugar derivatives.

6.3.4.8 Co-pigmentation. Harborne [49, 43] suggested that all anthocyanins may be ionically bound in the cell vacuole to aliphatic organic acids such as malonic, malic or citric acid. Such interactions could provide a mechanism of colour stabilization *in vivo*. Physical absorption of the flavylium cation and/or neutral or anionic quinonoidal bases onto a suitable surface could also stabilize anthocyanin chromophores by taking them out of the bulk solution, thereby preventing colour loss caused by hydration reactions [48]. It is possible that the stability of anthocyanins associated with pectin arises from such interaction [235, 236]. Similarly, the high stability of anthocyanin extracts from the dried flowers of *Clitorea ternatia* used to colour Malaysian rice-cakes was attributed to adsorption of the chromophores onto the starch of the glutinous rice [237].

Anthocyanins are known to form weak complexes with numerous compounds such as proteins, tannins, other flavonoids and polysaccharides through what is referred to as intermolecular co-pigmentation [137, 238]. Most of these compounds themselves are not coloured, but when complexed with anthocyanins they augment the colour and stability of the chromophore. Stability is rendered by the co-pigmentation effect through displacement of the hydration-dehydration equilibrium (see Figure 6.3) towards the chromophores. As an example, the stability constant of the cyanin-quercitin complex was estimated at 2×10^{-3} M^{-1}, and the hydration constant K_h' was reduced from 10^{-2} to 7×10^{-4} M [40]. It is still not known whether the stabilizing effect of intermolecular co-pigmentation is of a kinetic or thermodynamic nature.

Intermolecular co-pigmentation would appear to be less efficient in the stabilization of anthocyanin chromophores than intramolecular co-pigmentation, that is, acylation; yet, at the molecular level, the characteristics of both phenomena are similar [40, 48, 238]. Intermolecular co-pigmentation is more efficient with monoacylated anthocyanins than with non-acylated pigments [239]. Hydrogen bonding has been suggested as the driving force for anthocyanin-co-pigment association [240–242]. However, it is unlikely that the interaction of anthocyanin and co-pigment is due solely to hydrogen bonding in water, since water itself is an excellent hydrogen bond donor and acceptor. Association is more likely to involve a stacking process similar to that which occurs in intramolecular co-pigmentation, hydrophobic forces mediating the process [243]. Strong evidence for vertical stacking of anthocyanin and co-pigments in aqueous media, analogous to that exhibited by nucleosides, has been provided by circular dichroism studies [239, 244–246]. A stacked complex provides the anthocyanin chromophore with protection against nucleophilic attack by water and subsequent loss of colour.

Interaction of co-pigment with anthocyanin can occur with both the flavylium cation and quinonoidal base [137]. Co-pigmentation involving both anthocyanin species results in a bathochromic shift in the visible wavelength of maximum absorption, from red to a stable blue or purple colour; however, increases in maximum absorption or tinctorial strength occur only when complex formation involves the quinonoidal base [137]. Although the nature of the anthocyanin (*i.e.* its substitution pattern) has a marked influence on the co-pigmentation effect [137, 241, 247], neither glycosylation nor hydroxylation is necessary for intermolecular co-pigmentation to occur [48, 190]. All of the common anthocyanins can partake in intermolecular co-pigmentation, complex formation being strongly affected by pigment and co-pigment concentrations [137]. The co-pigmentation effect is more pronounced as anthocyanin concentration and the ratio of co-pigment to anthocyanin increase [47, 137, 248, 249]. However, for a given co-pigment concentration, there is a corresponding molar ratio of pigment to co-pigment below which colour intensity is reduced, that is there is an optimum molar ratio of pigment to co-pigment at which colour intensity and stability are maximized [31, 137, 249].

To date, the effectiveness or efficiency of a co-pigment has been related to the extent of the bathochromic shift and increase of the maximum absorption observed upon addition of co-pigment to an anthocyanin-containing system. These two spectroscopic properties would seem to adequately reflect the stability of a given anthocyanin-co-pigment complex [48]. No measure of a stability constant associated with a given complex has been reported. The most effective co-pigments include the flavonols quercitin and rutin, the aurone aureusidin and *C*-glycosyl flavones such as swertisin (Table 6.2). For

Table 6.2 Co-pigmentation of cyanidin 3,5-diglucoside (2×10^{-3} M) at pH 3.32 (From Asen *et al.* [137] by permission of the authors and Pergamon Press.)

Co-pigments (6×10^{-3} M)	λ_{max} (nm)	$\Delta\lambda_{max}$ (nm)	% Absorbance increase at λ_{max}
None	508	—	—
Aurone			
Aureusidin[a]	540	32	327
Alkaloids			
Caffeine	513	5	18
Brucine	512	4	122
Amino acids			
Alanine	508	0	5
Arginine	508	0	20
Aspartic acid	508	0	3
Glutamic acid	508	0	6
Glycine	508	0	9
Histidine	508	0	19
Proline	508	0	25

Table 6.2 *Continued*

Co-pigments (6×10^{-3} M)	λ_{max} (nm)	$\Delta\lambda_{max}$ (nm)	% Absorbance increase at λ_{max}
Benzoic acids			
Benzoic acid	509	1	18
o-Hydroxybenzoic acid	509	1	9
p-Hydroxybenzoic acid	510	2	19
Protocatechuic acid	510	2	23
Coumarin			
Esculin	514	6	66
Cinnamic acids			
m-Hydroxycinnamic acid	513	5	44
p-Hydroxycinnamic acid	513	5	32
Caffeic acid	515	7	56
Ferulic acid	517	9	60
Sinapic acid	519	11	117
Chlorogenic acid	513	5	75
Dihydrochalcone			
Phloridzin	517	9	101
Flavan 3-ols			
(+)-Catechin	514	6	78
Flavone			
Apigenin 7-glucoside[a]	517	9	68
C-glycosyl Flavone			
8-*C*-Glucosylapigenin (vitexin)	517	9	238
6-*C*-Glucosylapigenin (isovitexin)	537	29	241
6-*C*-Glucosylgenkwanin (swertisin)	541	33	467
Flavonones			
Hesperidin	521	13	119
Naringin	518	10	97
Flavonols			
Kaempferol 3-glucoside	530	22	239
Kaempferol 3-robinobioside-7-rhamnoside (robinin)	524	16	185
Quercetin 3-glucoside (isoquercitrin)	527	19	188
Quercetin 3-rhamnoside (quercitrin)	527	19	217
Quercetin 3-galactoside (hyperin)	531	23	282
Quercetin 3-rutinoside (rutin)	528	20	228
Quercetin 7-glucoside (quercimeritrin)	518	10	173
7-*O*-Methylquercetin-3-rhamnoside (xanthorhamnin)	530	22	215

[a] Formed a slight precipitate.

flavonoids to act as effective co-pigments, a double bond in the 2,3 position of the C-ring would seem to be required. *C*- and *O*-glycosylation of flavones has also proven beneficial for effective co-pigmentation [180]. Sweeny *et al.* [190] noted that flavonosulphonic acids are also effective co-pigments, presumably due to the added attraction of the negative charge of the sulphonic acid groups for the flavylium cation.

When pigment concentrations become relatively high, anthocyanins themselves may act as co-pigments and participate in self-association reactions [137, 250]. Scheffeldt and Hrazdina [251] noted that competition can take place between co-pigmentation and self-association reactions. They demonstrated that with low concentrations of anthocyanin from *Vitis* spp., colour augmentation by co-pigmentation with rutin was intensified, but decreased with corresponding increases in anthocyanin concentration. The net result of competing co-pigmentation and self-association reactions is that colour may increase more than proportionally to pigment concentration [26, 251]. Hoshino *et al.* [244, 245] established that 4′-hydroxyl and 5-glycosyl groups are essential structural elements for self-association. Urea and dimethylsulphoxide disrupt self-associated aggregates, while sodium and magnesium chloride salts promote self-association [244–245].

The co-pigmentation effect, both intra- and intermolecular, has been deemed primarily responsible for the coloration of flower and fruit tissues, fruit juices and red wines since anthocyanins alone are known to be virtually colourless at the pH of these products [31, 40, 48, 53, 137, 235, 238, 241, 249, 252, 253]. That juices obtained from enzyme-treated fruit mashes are more highly coloured than non enzyme-treated press-juices can be attributed to decompartmentalization of various cellular constituents, including flavonoids, alkaloids, amino acids and nucleosides, which may participate in co-pigmentation with anthocyanins to varying degrees. Flavonoids, as co-pigments, are always found in conjunction with anthocyanins [254–256], likely because of their similar biosynthetic pathways [3, 39]. Polymeric flavonoids and anthocyanins have been shown to play an important role in the coloration of grapes and red wines [167–169, 257, 258]. Tannins (condensed flavonoids) have a protective effect on anthocyanins [166, 167]. In addition, during ageing of red wines, monomeric anthocyanins are progressively and irreversibly replaced by polymeric pigments through self-association reactions [167–169]. Such polymeric material is less pH-sensitive and relatively resistant to decoloration by SO_2, ascorbic acid and light [26, 111, 137, 252].

6.3.5 Extraction and purification

The choice of a method to extract anthocyanins from any of their various sources depends largely on the purpose of the extraction, and also on the nature of the constituent anthocyanins themselves. A knowledge of the factors that influence anthocyanin structure and stability is vital. If the extracted pigments are to be subsequently analyzed, qualitatively or quantitatively, it is desirable that a method be chosen that retains the pigments as close to their natural state (*i.e. in vivo*) as possible. Alternatively, if the extracted pigments are to be used as a colorant/food ingredient then maximum pigment yield, tinctorial strength and stability are of greater concern. It is also important that the extraction and clean-up procedures are not too complex, time-consuming or costly [258].

Anthocyanins are not stable in neutral or alkaline solution. Thus, extraction procedures have generally involved the use of acidic solvents, which disrupt plant cell membranes and simultaneously dissolve the water-soluble pigments. The traditional and most common means of anthocyanin extraction involves maceration or soaking of plant material in a low boiling point alcohol containing a small amount of mineral acid (*e.g.* $\leqslant 1\%$ HCl) [59, 260]. Methanol is most often used; however, due to its toxicity, acidified ethanol may be preferred with food sources, although it is a less-effective extractant and more difficult to concentrate due to its higher boiling point [259]. Acidification with hydrochloric acid (HCl) serves to maintain a low pH, thereby providing a favourable medium for the formation of flavylium chloride salts from simple anthocyanins. The use of mineral acids such as HCl may, however, alter the native form of complex pigments by breaking associations with metals, co-pigments, and so on [261–263]. The loss of labile acyl and sugar residues may also occur during subsequent concentration [155]. Thus, to obtain anthocyanins closer to their natural state neutral solvents have been suggested for initial pigment extraction, including 60% methanol, *n*-butanol, ethylene glycol, propylene glycol, cold acetone, acetone/methanol/water mixtures and, simply, boiling water [155, 227, 237, 264]. In addition, weaker organic acids—mainly formic acid but also acetic, citric and tartaric acids—have been used in extraction solvents [228, 261–263, 265].

Extraction of anthocyanins with ethanol containing 200 to 2000 p.p.m. SO_2 has been reported to increase pigment yield and to result in concentrates possessing up to twice the tinctorial strength and flavour intensity as anthocyanin concentrates obtained from extraction with ethanol alone [266]. Extraction of anthocyanins with aqueous SO_2 or bisulphite solutions also led to concentrates of higher purity and pigment stability than those obtained from hot water extraction [185, 267]. However, aqueous SO_2 extraction of anthocyanins would appear to be less selective than extraction with methanol-HCl followed by purification using ion-exchange chromatography [268, 269].

The initial decanted or filtered anthocyanin extracts are usually quite dilute and require concentration prior to subsequent purification or their use as a food colorant. To minimize pigment degradation, concentration is best carried out *in vacuo* at temperatures below 30°C [259]. The acidic aqueous anthocyanin concentrates thus obtained can be used as such, frozen [227], freeze-dried [185, 229, 257, 268, 269] or spray-dried [228].

As emphasized by Markakis [22], none of the solvent systems used for extraction purposes is specific for the anthocyanins. There may be considerable quantities of extraneous material, for example, other polyphenolic substances or pectin, that could influence the stability and/or analysis of these pigments. Anthocyanins are difficult to extract independently of other flavonoids, which react similarly with the common reagents used for phenolic analysis [58]. If appreciable quantities of lipid, chlorophyll or unwanted polyphenols are suspected to be present in the anthocyanin concentrates obtained, these materials may be removed by washing with petroleum ether, ethyl ether, diethyl ether or ethyl acetate [260, 270]. This may improve purification but is not always necessary [260, 270].

Purification of anthocyanins for analytical purposes has been carried out primarily by chromatographic techniques, and traditionally by paper chromatography. Many developing solvents have been used for purification of anthocyanins by paper chromatography, the selection depending on the nature of the anthocyanins in the sample to be purified [3]. Francis [259] has listed several developing solvents that can be used. Typically, at least three chromatographic separations are required, alternating between aqueous- and oily-type developing solvents, to provide individual anthocyanins sufficiently pure for qualitative analyses [259, 271].

Purification of anthocyanins by paper chromatography is time-consuming. Comparable quantities of purified pigment can be obtained in half the time using preparative thin-layer chromatography (PTLC) [272–275]. However, with the development of several column chromatographic techniques that have significantly increased the speed of separation from several days to hours, and allowed for purification of larger quantities of anthocyanins, PTLC has not become widely adopted.

None of the column chromatographic techniques surpass the resolving power of paper chromtography except, perhaps, reverse-phase high-performance liquid chromatography (HPLC). With appropriate selection of eluent, column type and length, flow rate and temperature, HPLC can be used to resolve and quantitate microgram quantities of anthocyanin without the need for preliminary purification of sample concentrates. A major application of HPLC is, therefore, the qualitative and quantitative analysis of anthocyanins. Yet, preparative HPLC has also proven to be a powerful tool for the rapid separation of complex mixtures of anthocyanins from a variety of sources, including grapes (*Vitis* spp.) [168, 276, 277], thornless blackberries [278], Saskatoon berries (*Amelanchier alnifolia* Nutt.) [279], chokeberries (*Aronia melanocarpa*) [280] and berries from several *Vaccinium* spp. [281–283]. The technique is non-destructive and, therefore, separated peaks are readily collected for subsequent analyses.

Among the most popular column chromatographic techniques are those using weak cation-exchange resins, for example, Amberlite CG-50 [82, 284, 285], in addition to insoluble polyvinylpyrrolidone (PVP), for example, Polyclar AT [286–289], polyamide [290–292] and combination polyamide-PVP adsorbents [51, 293, 294]. The successful application of polyamide and PVP for the purification of anthocyanins arises from the ability of these resins to form unusually strong hydrogen bonds with phenolic hydroxyl groups [286, 295], and thereby exhibit a certain degree of selectivity for these pigments.

The quantitative separation/purification of complex mixtures of anthocyanins in relatively high yields has also been at least partially accomplished by simple gel filtration on Sephadex G-25 [7, 258], and column chromatography using polyethylene glycol dimethacrylate gel [296], silicic acid [108, 297] and cellulose resins [90, 97, 194, 298]. Quite recently, droplet count-current chromatography (DCCC) has been applied for anthocyanin purification, using BAW (*n*-butanol-glacial acetic acid-water) as the solvent system [282, 283, 299].

6.3.6 *Qualitative analysis*

Current analytical approaches that allow for unambiguous identification of individual anthocyanins are based on proven *in vitro* techniques. Well-developed chromatographic and spectroscopic methods have been applied for rapid and accurate identification of anthocyanins. However, absolute characterization of anthocyanins cannot be established by chromatographic or spectral methods alone. Complete structural characterization generally involves identification of the aglycone, sugar moieties and acyl groups, if present, and the position(s) of attachment of sugar and acyl groups. Analytical approaches and methods used in the structural characterization of anthocyanins have been previously reviewed [3, 30, 259, 260, 300, 301].

6.3.6.1 Spectral data. Much information regarding the identity of a given anthocyanin and its aglycone can be derived simply by observing its colour in aqueous solution or on paper. The colour generally changes from orange-red to bluish-red as the structure is modified through pelargonidin, cyanidin, peonidin, malvidin, petunidin and delphinidin [259]. In addition, under UV-light, anthocyanins substituted at the C-5 position usually fluoresce [188].

The absorption spectra of anthocyanins and their aglycones in the visible and UV regions have proven very useful in their structural characterization. The spectral characteristics of all the known anthocyanidins and many of the anthocyanins in 0.1% HCl-methanol have been published by Harborne [3, 302]. In acid solution, anthocyanins and their aglycones exhibit two characteristic absorption maxima: one in the visible region between 465 and 550 nm and a smaller one in the UV region at about 275 nm. The wavelength of maximum absorption can be used to tentatively identify the aglycone; however, confirmation by other methods (*e.g.* paper chromatography) is required. A bathochromic shift of 15 to 35 nm of the visible maximum upon addition of 5% alcoholic aluminium chloride ($AlCl_3$) to pigments in 0.1% HCl-methanol indicates the presence of a free *O*-dihydroxyl group in the aglycone, a structural characteristic of cyanidin, delphinidin and petunidin [303]. Glycosylation of anthocyanidins at the C-3 position generally results in a bathochromic shift in the wavelength of maximum absorption, while glycosyl substitution at C-5 produces a shoulder on the absorption curve at 440 nm. The ratio of absorbance at the UV maximum to that at the visible maximum, and the ratio of absorbance at 440 nm to that at the visible maximum, provide valuable information about the extent and position of glycosidic substitution [3, 150, 302]. Acylated anthocyanins exhibit an additional weak absorption maximum in the 310 to 335 nm region, the actual maximum indicating the type of aromatic acylation involved [51, 259].

6.3.6.2 Chromatography. The traditional means by which anthocyanins have been characterized, with or without prior (controlled) acid, alkali, enzyme and/or peroxide hydrolysis, has been through the use of paper chromatography [3, 304]. Thin layer chromatography (TLC) has been widely applied [96, 275, 305]; and analytical HPLC has met with growing popularity [276–283, 306–322], especially when coupled with a photodiode array detector [323, 324]. Pigment characterization by all chromatographic methods requires comparison with authentic anthocyanin, aglycone and/or sugar standards. Although anthocyanins and their aglycones can be purchased from various suppliers, they often require purification prior to their use as reference pigments.

The chromatographic mobility (R_f value) of an anthocyanin has been described as its most important property for identification [3, 304]. The R_f values for many pigments in four solvents using paper chromatography have been published by Harborne [3, 304]. Other solvents with superior resolving power have been used [87, 109]; however, no large body of R_f data is readily available for these solvents. TLC on cellulose strips has been shown to give similar R_f values to those obtained on paper with the same solvent systems [96, 155, 275, 325], thereby permitting direct comparison of results obtained by either method. The general relationship between anthocyanin structure and chromatographic mobility has been reviewed by Harborne [43]. The R_f values obtained using at least three different solvents [259, 271] or HPLC retention times, provide only tentative identification of an anthocyanin and its overall substitution pattern even with corroborative spectral data. Additional controlled hydrolysis tests are usually necessary to confirm anthocyanin structure/identity [271].

Products released upon hydrolysis by enzymes, acid, alkali or oxidation may be identified by comparison with authentic standards using chromatography, for example, paper or HPLC. Generally, acid hydrolysis of anthocyanins yields aglycones and sugars, thereby allowing for identification of these moieties [90]. Sugars are cleaved off the pigment in a more or less random fashion [259]. The course of hydrolysis can be followed, and the glycosylation pattern determined, by the production of intermediate pigments. Blom [326] applied HPLC for measurement of hydrolysis of the glycosidic bond(s) by monitoring the increase in aglycone together with decreasing anthocyanin concentration in heated, acidified anthocyanin solutions.

Alkaline hydrolysis, while it too may be used to determine the nature of the aglycone, is most commonly employed for acyl group determination. Alkaline hydrolysis under nitrogen specifically removes acyl groups, generally leaving the glycoside intact [90]. The removal of oxygen is necessary since anthocyanins possessing *O*-dihydroxy groups are unstable in alkali [259].

Enzymatic hydrolysis procedures generally involve the use of β-glycosidases of fungal origin [191, 208]. The activity of β-glycosidase is generally not influenced by the nature of the aglycone [191]. The glycosyl moieties of simpler glycosides (*e.g.* 3-*O*-β-glycosides) are rapidly hydrolyzed, thereby permitting relatively easy identification and differentiation from more complex anthocyanins: since α-glycosides are hydrolyzed more slowly, if at all, by β-glycosidase, acylated pigments are virtually unaffected during enzymatic hydrolysis tests.

Oxidation of anthocyanins with peroxide induces liberation of the C-3 sugar [327]. Subsequent permanganate treatment releases sugars attached to the readily ruptured heterocyclic ring, and ozonolysis releases sugars attached to enolic and phenolic hydroxyl groups [327]. Peroxidation of anthocyanidin 3,5-diglycosides under neutral conditions yields coumarin derivatives, while peroxidation of anthocyanidin 3-glycosides does not [171, 183].

Other techniques have also been used for the structure determination of complex (*e.g.* polyacylated, co-pigmented) pigments. Goto and co-workers [246, 328–330] were the first to use ^{1}H-NMR to elucidate the structures of several complex polyacylated anthocyanins and to study self-association of anthocyanins. Circular dichroism has also been used to study self-association reactions [241, 243, 244]. Several other reports have appeared in which ^{1}H-NMR has been used to elucidate the structures of anthocyanins [57, 277, 331, 332]. Using fast-atom bombardment mass spectrometry (FAB-MS) several researchers have been able to accurately determine the molecular weights of anthocyanins, these values aiding in pigment identification [57, 277, 332, 333]. Anthocyanins have also been analyzed by gas chromatography/mass spectrometry (GC/MS) after reaction with trimethylchlorosilane and hexamethyldisilazane to quinoline derivatives [281, 334, 335]. Although the infra-red spectra of several anthocyanidins have been reported [336], the information available from these spectra was limited. Resonance Raman spectra of anthocyanins *in vivo* have since been recorded by Statoua *et al.* [337]. Preliminary investigation indicated that these spectra provided limited structural information [48]. However, future development of this and other techniques cited above will surely help to determine the structural elements of anthocyanin chromophores required for their efficient use as food colorants.

6.3.7 *Quantitative analysis*

The quantitative analysis of anthocyanins has been previously reviewed [30, 155, 259, 271, 338]. Quantitative methods are conveniently divided into three groups depending on the needs of the analysis [259]:

1. Determination of total anthocyanin concentration in systems containing little or no interfering substances;
2. Determination of total anthocyanin concentration in systems that contain interfering material; and
3. Quantitation of individual pigments.

In fresh plant extracts/juices there are generally few interfering compounds that absorb energy in the region of maximum absorption of anthocyanins (465 to 550 nm). Total anthocyanin concentration may thus be determined by measuring the absorbance of the sample (diluted with acidified alcohol) at the appropriate wavelength. It is important that Beer's Law be obeyed in the concentration range under study: self-association and co-pigmentation effects cause deviations in Beer's Law and can lead to inaccurate estimates of total anthocyanin concentration [251]. Therefore, considerable sample dilution may be necessary to counteract these effects [251, 270]. Adams [155] suggested that absorbance measurements be carried out at a single pH as low as possible to ensure maximum colour development and low response to changes in pH that may occur (Figure 6.7). Comparative measurements of total anthocyanin may be made at pH 1.0 for samples containing predominantly 3-glycosides [155] and at pH less than 1.0 for samples containing mainly 3,5-diglycosides [221].

For samples containing a mixture of anthocyanins, absorbance measurements at a single low pH value are proportional to their total concentration [339]. Estimates of the concentrations of individual anthocyanins in the mixture can only be obtained with prior knowledge of the proportions of individual pigments therein. Absolute concentrations may then be estimated by the use of weighted average absorptivities and absorbance measurement at weighted average wavelengths [270]. More accurate estimates are obtained with samples in which a single pigment is present or predominates. Molar extinction coefficients (absorptivities) of several anthocyanins have been reported [108, 259, 340]. It is important that the same solvent and pH be used for absorbance measurement as were used for determination of molar absorptivities [40, 108, 270, 304].

The single pH method is subject to interference by and, therefore, cannot be used in the presence of brown-coloured degradation products, for example, from sugar or pigment breakdown [58]. Unpurified pigment extracts may also contain compounds that, at low pH, absorb in the range of maximum absorption of the anthocyanins [155]. Differential or subtractive absorption methods may be used to determine the total anthocyanin concentration in those samples that contain interfering substances.

The subtractive method involves measurement of sample absorbance at the visible maximum, followed by bleaching and remeasurement to give a blank reading. Subtraction gives the absorbance due to anthocyanin, which

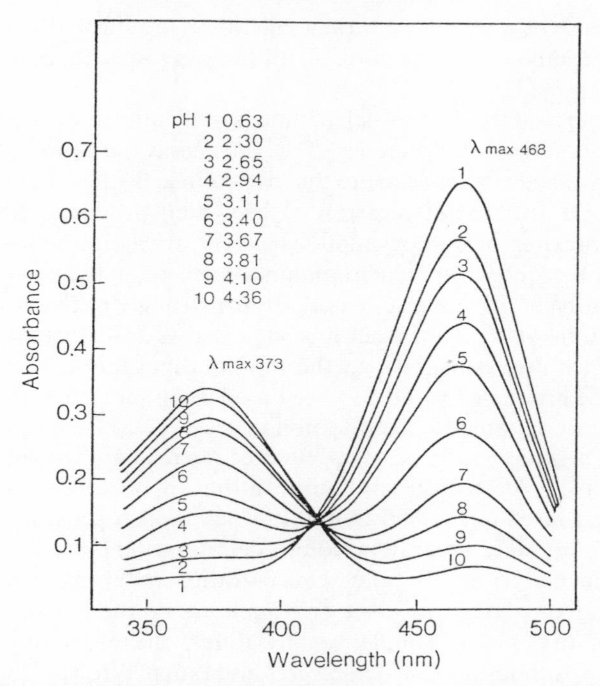

Figure 6.7 Effect of pH on the visible spectrum of 3′-methoxy-4′,7-dihydroxyflavylium chloride (5.2×10^{-3} g/L) in aqueous solution after standing for 2 h (redrawn from Jurd [339a] with permission).

can be converted to anthocyanin concentration with reference to a calibration curve prepared using a standard pigment [155]. The most common bleaching agents have been sodium sulphite [338, 341, 342] and hydrogen peroxide [343]. The subtractive method has come under criticism because the bleaching agents used are presumed to cause a decrease in absorbance of some interfering substances, resulting in erroneously high values of total pigment concentration [259].

The differential method relies on the structural transformations of anthocyanin chromophores as a function of pH (Figures 6.4 and 6.7). It is also based on the observation that the spectral characteristics of interfering brown (degradation) products are generally not altered with changes in pH [167]. The difference in absorbance at two pH values and same wavelength (*e.g.* the anthocyanin visible maximum) is assumed to give a measure of anthocyanin concentration since absorbance due to interfering substances cancels in the subtraction. Several different pH values have been used in the differential method [291, 343, 344] but measurement at pH 1.0 and 4.5 is most common [291]. An equilibrium period is required upon pH adjustment of samples to

4.5 for consistent absorbance measurements. If insufficient time is allowed for equilibration between the different anthocyanin species, erroneous results would be expected.

An advantage of the differential method is that data can be used to calculate an anthocyanin degradation index (DI), defined as the ratio of total anthocyanin determined at a single pH to that determined by the differential method [291]. The DI is a useful measure of browning products that absorb at the visible maxima of anthocyanins. However, it cannot be used with such products as aged red wines, which contain condensed/polymeric anthocyanins that are relatively pH insensitive and absorb strongly at the visible maxima of anthocyanins [167]. They cannot be regarded as anthocyanin degradation products since they contribute to the total anthocyanin concentration and colour of the product. The DI has been used as an indicator of anthocyanin colour and overall quality deterioration in several berry-fruit products [135, 345–348]. The simple ratio of absorbance of red (*e.g.* anthocyanin) pigments measured in the 465 to 550 nm region to that of brown (*e.g.* degradation) products measured in the 400 to 440 nm region also provides an index of anthocyanin or, more accurately, colour deterioration [338, 349, 350].

Measurements of anthocyanin concentration rarely correlate with the actual colour of a given product or sample; since the pH of analysis and endogenous pH of the sample usually differ, the distribution and concentration of anthocyanin chromophores also differ. Wrolstad [338] reviewed methods for determining colour density, polymeric colour, percentage contribution by tannin and monomeric anthocyanin colour, calculated from only a few absorbance readings at the native sample pH. A two-point index—the difference in absorbance at the visible maximum and at 420 nm—has been shown to correlate well with visual appearance [351]. However, for those with available equipment, tristimulus measurements have been shown to correlate much better with human visual responses [259, 350–352].

The quantitative analysis of individual anthocyanins requires their prior separation/purification from a mixture, usually carried out by chromatography. Spectral measurements are usually of limited use since the absorption spectra of most anthocyanins are so similar [302]; yet, as previously mentioned, molar absorptivities have been reported and used for quantitation of several anthocyanins [108, 259, 240]. Quantitation of individual anthocyanins has traditionally involved their separation by paper or thin-layer chromatography, followed by photometric determination of pigments *in situ* by reflectance or transmittance densitometry [82, 110, 135, 353].

The advent of HPLC has revolutionized the qualitative and quantitative analysis of anthocyanins. Manley and Shubiak [294] were the first to apply HPLC to anthocyanins, quantitatively separating the 3-monoglucosides of malvidin, petunidin and peonidin. Twenty different anthocyanins were subsequently separated and quantified from skins of *Vitis vinifera* [276], red grape wines [168, 307] and juices [354]. The separation power of reverse-phase HPLC for the analysis of crude anthocyanin-containing extracts was clearly demonstrated by Casteele *et al.* [315]. Over the past decade, HPLC has usurped traditional approaches for quantification of anthocyanins [277, 278, 280–283, 308, 312–314, 316–321, 323, 324].

6.3.8 Current and potential sources and uses

Several synthetic flavylium salts have been suggested as potential food-colour additives [41, 144, 145, 355]. The production of synthetic anthocyanins identical to their natural counterparts (*i.e.* nature-identical) is also possible [180]. These latter products provide a source of food-colour additives with consistent quality/purity and known properties. All synthetic flavylia, including the nature-identicals, currently require extensive toxicological testing, however, prior to their clearance as safe food-colour additives. Public concern over the safety of food additives has restricted the sources of commercial anthocyanin preparations to food plants naturally containing these pigments; this is not likely to change in the very near future.

The potential of various food plants as commercial sources of anthocyanins is limited by availability of raw material and overall economic considerations. Numerous plants and plant parts have been suggested as potential commercial sources, including cranberry [228, 256, 285, 356], blueberry [268], red cabbage [268, 269], roselle (*Hibiscus sabdariffa*) calyces [227, 264, 357], miracle fruit [103, 358], berries of *Vibernum dentatum* [359], bilberries (*Vaccinium myrtillus*) [360], duhat (*Sysyium cumini* Linn.) [361], black olives [362], flowers of *Clitorea ternatia* [239] and *Ipomea tricolor* [53, 138], and leaves of cherry-plum [363, 364] and *Perilla ocimoides* (*i.e.* shiso) [57]. Of all sources examined to date, grapes certainly provide the greatest quantity of anthocyanins for use as food colorants: production of grapes constitute about a quarter of the annual fruit crop world-wide [58, 185]. Both a spray-dried powder and a concentrated solution containing grape anthocyanins have been marketed for years from Italy under the trade name of Enocolor® (formerly Enocianina) [365].

Economically, the best potential commercial sources of anthocyanins are those from which the pigment is a by-product of manufacture of other value-added products [366]. The extraction of grape anthocyanins into juice or wine is an inefficient process and large quantities of extractable pigment are available in grape pomaces and waste material [58]. Anthocyanin extracts have been produced from these sources [154, 185, 228, 265–267] and used to successfully colour a wide range of commercial products [177]. Anthocyanin extracts have also been prepared from cranberry presscake [228, 256, 285]. Approximately 40% of the anthocyanins from cranberries remains in the presscake following juice extraction [351]. The tropical red berry known as miracle fruit contains a taste modifier, Miraculin, that is currently being investigated as a potential sweetener: anthocyanin extracts have been obtained as a by-product of taste modifier production [103, 358]. The production of expensive food crops for their pigment content alone is not economically feasible. Yet, the availability of highly pigmented inexpensive crops such as red cabbage [268, 269] and bilberries [360] makes the use of these crops as potential sources of commercial anthocyanin preparations viable.

Anthocyanin preparations from natural sources vary considerably in quality [26, 367] since they may contain partially degraded and/or condensed pigments, tannins, co-pigments and various impurities, which are co-extracted with the anthocyanins. Stable, relatively pure anthocyanin preparations have been produced in large quantities by suspension cultures of cells of *Populus* spp. [368], *Daucus carota* [369], *Vitis* spp. [370] and *Euphorbia millii* [371]. However, application of anthocyanins obtained from cultured cells has yet to be reported. The future application of biotechnology, for example tissue culture, to produce anthocyanin preparations for commercial use is dependent on whether the technology can yield colorants that are cheap enough, and which are also acceptable to legislators [372]. Natural anthocyanin preparations used as food colorants suffer the same fate as endogenous pigments: their addition to food systems may encourage reactions with endogenous constituents that can lead to their stabilization or destabilization, and thereby influence product quality. With careful formulation/selection of certain ingredients, choice of appropriate stages during formulation and processing at which the colorant and/or other ingredients are added, and control of processing and storage conditions, a wide range of high-quality products can be attractively coloured with anthocyanins. Counsel *et al.* [373] have reviewed numerous food applications for anthocyanins and other natural pigments, including chewing gums, hard candies (sweets), fruit chews, dry beverage powders, cream fillings, icings and fruit coloration. Anthocyanins may also be successfully used in high-acid foods such as soft drinks, jams and jellies, and in red wines where they contribute to the overall ageing process [28]. For those food products for which endogenous pigments are to serve as the only or major source of coloration, procurement of intensely coloured raw material is necessary to ensure that if some pigment is destroyed during processing/storage, a sufficient portion remains to impart the characteristic product colour [251]. Few practical stabilizing agents have been found for the anthocyanins [366]. Of nineteen different additives, only thiourea, propyl gallate, and quercetin demonstrated a stabilizing effect on colour retention in strawberry juice and buffered pigment solutions [159]. Cysteine, which may act as a reducing agent and/or inhibitor of PPO, was shown to inhibit anthocyanin degradation in Concord grape juice below a temperature of 75°C [201]. Ascorbic acid can render a stabilizing effect on one hand, by being preferentially oxidized by PPO but a destabilizing effect on the other, by indirectly oxidizing the anthocyanins via its oxidation products [180]. Tartaric acid and glutathione have been shown to have a protective effect on anthocyanins by acting as mildly acidic and antioxidant agents, respectively [374]. The addition of phenolic compounds, such as rutin and caffeic acid, also markedly stabilized the colour of blood-orange fruit juice, presumably by means of co-pigmentation [374]. Indeed, in view of the factors that influence pigment stability, the most practical means of stabilizing the colour of anthocyanins within the bounds of maintaining their 'natural' status is via complex formation (*e.g.* surface adsorption), intermolecular co-pigmentation and/or condensation reactions (*e.g.* with proteins, tannins or other polyphenols).

6.4 Betalains

6.4.1 Structure

The structure of the betalain chromophore may be described as a protonated 1,7-diazaheptamethin system (Figure 6.8); betaxanthins and betacyanins are distinguished by substitution on the dihydropyridine moiety by specific R and R′ groups. The yellow betaxanthins are characterized by having R and R′ groups that do not extend the conjugation of the 1,7-diazaheptamethin system. However, if conjugation is extended, whereby R and R′ comprise a substituted aromatic ring (*i.e.* cyclodopa), the chromophore is red and characterized as a betacyanin [38, 40]. All betacyanins are glycosylated [5]. To date, over fifty betacyanins have been identified (Table 6.3), all derived from the aglycones betanidin, isobetanidin (the C-15 epimer of betanidin) and the once reported 2-decarboxybetanidin from *Carpobrotus anci-*

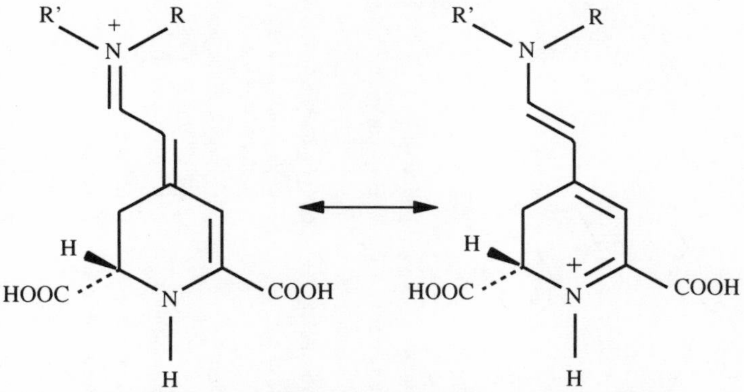

Figure 6.8 Structural transformations of the betalain chromophore.

Table 6.3 Structures of some known betacyanins

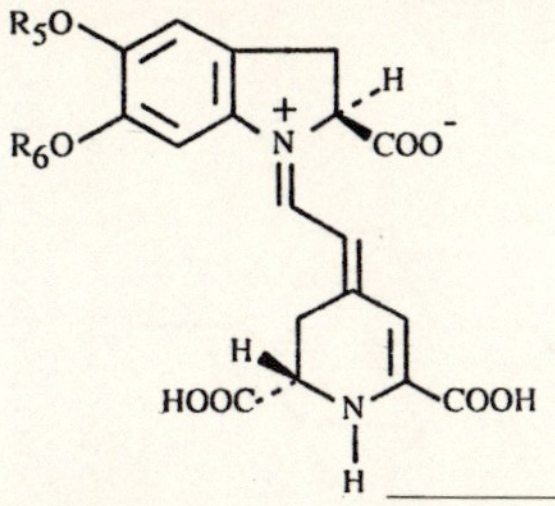

Compound	Substitution pattern: R_5	Substitution pattern: R_6	Botanical source	Reference
Betanin	β-glucose	H	*Beta vulgaris*	375, 376
Amaranthin	2′-*O*-(β-glucoronic acid)-β-glucose	H	*Amaranthus tricolor*	377, 378
Prebetanin	6′-*O*-(SO_3H)-β-glucose	H	*Beta vulgaris*	379
Phyllocactin	6′-*O*-(malonyl)-β-glucose	H	*Phyllocactus hybridus*	380
Celosianin	2′-*O*-[*O*-(*trans*-*p*-feruloyl)-β-glucuronic acid]-*O*-(*trans*-*p*-coumaroyl)-β-glucose	H	*Celosia cristata* L.	380
Irsinin	2′-*O*-(β-glucuronic acid)-6′-*O*-(3-hydroxy-3-methyl-glutaryl)-β-glucose	H	*Iresine herbstii*	380
Lampranthin-I	*O*-(feruloyl-*p*-coumaroyl)-β-glucose	H	*Lampranthus* spp.	381
Lampranthin-II	*O*-(diferuloyl-*p*-coumaroyl)-β-glucose	H	*Lampranthus* spp.	381
Bougainvillein-I	β-sophorose	H	*Bougainvillea glabra* var. *sanderiana*	382
Gomphrenin-I	H	β-glucose	*Gomphrena globosa*	383
Gomphrenin-V	H	6′-*O*-(*trans*-*p*-feruloyl)-β-glucose	*Gomphrena globosa*	383

naciformus [384]. A glycoside of neobetanidin, 14,15-dehydrobetanidin, has also been identified as a natural constituent of some plants [385, 386]. The most common betacyanin exists as the 5-*O*-*β*-glucoside of betanidin and is called betanin. The most common glycosyl moiety is glucose; sophorose and rhamnose occur much less frequently. Acylation of pigments may also occur whereby the acyl group is attached via an ester linkage to the sugar moiety [5]. In total, nine non-acylated betacyanins are known, and the structures of forty acylated derivatives have also been elucidated. The most common acyl groups include sulphuric, malonic, 3-hydroxy-3-methylglutaric, citric, *p*-coumaric, ferulic, caffeic and sinapic acids. The structures of many acylated betacyanins have yet to be fully characterized.

In betaxanthins the cyclodopa unit of betacyanins is displaced by either an amine or amino acid [387]. For example, the R′ group of vulgaxanthin-I from *Beta vulgaris* is a residue derived from glutamic acid; that of indicaxanthin from *Opuntia fiscus-indica* joins with the R group to give a proline moiety (Table 6.4). Other R′ groups of the betaxanthins include glutamine, methionine sulphoxide, tyramine, DOPA [387] and 5-hydroxynorvaline [393]. The R group of betaxanthins is usually a hydrogen. Like the betacyanins, many of the betaxanthins have yet to be structurally characterized [393].

Table 6.4 Structures of some known betaxanthins

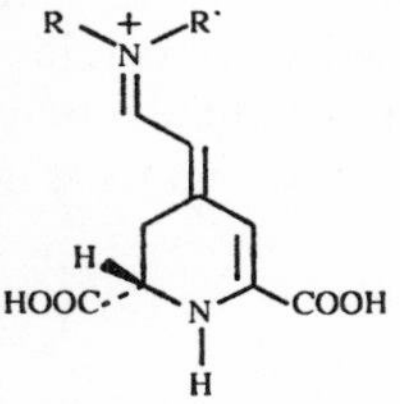

Compound	Substitution pattern R	R′	Botanical source	Reference
Indicaxanthin	R and R′ together form proline		*Opuntia fiscus-indica*	388
Portulaxanthin	R and R′ together form hydroxyproline		*Portulaca grandiflora*	389
Vulgaxanthin-I	H	glutamine	*Beta vulgaris*	390
Vulgaxanthin-II	H	glutamic acid	*Beta vulgaris*	390
Miraxanthin-I	H	methionine sulphoxide	*Mirabilis jalapa*	391
Miraxanthin-II	H	aspartic acid	*Mirabilis jalapa*	391
Miraxanthin-III	H	tyramine	*Mirabilis jalapa*	391
Miraxanthin-IV	H	dopamine	*Mirabilis jalapa*	391
Dopaxanthin	H	L-DOPA	*Glottiphyllum longum*	392
Humilixanthin	H	hydroxynorvaline	*Rivina humilis*	393
Muscaaurin-I	H	ibotenic acid	*Amanita muscaria*	394
Muscaaurin-II	H	stizolobic acid	*Amanita muscaria*	394
Muscaaurin-VII	H	histidine	*Amanita muscaria*	394

6.4.2 *Distribution*

Unlike the more widely distributed anthocyanins, the structurally unrelated betalains have been detected only in red-violet, orange and yellow-pigmented botanical species belonging to a few closely related families of the order Caryophyllales [5, 395]. Although remaining a point of phylotaxonomic contention, ten betalain-producing families have been identified: Aizoaceae, Amaranthaceae, Basellaceae, Cactaceae, Chenopoiaceae, Didiereaceae, Holophytaceae, Nyctaginaceae, Phytolaccaceae (including Stegnospermaceae) and Portulacaceae [5]. Betalains of fungal origin have also been found. A violet betacyanin, muscapurpurin, and seven yellow betaxanthins, muscaaurins-I to -VII, have been isolated from the poisonous mushroom *Amanita muscaria* (fly agaric; order Agaricinales) [394]. The unusual amino acid stizolobic acid and a cyclized form of this amino acid are found in muscaaurin-II and muscapurpurin, respectively. Some of the muscaaurins have identical structures to those of known betaxanthins produced by members of the Caryophyllales, for example, indicaxanthin, vulgaxanthin-I and -II, miraxanthin-III [387, 394].

Betalains, as with the anthocyanins, accumulate in cell vacuoles of the flowers, fruits and leaves of the plants that synthesize them. Moreover, betalains often accumulate in plant stalks and are found in high concentrations in underground parts of the beetroot [4]. Of the numerous natural sources of the betalains (*e.g.* beetroot, cacti, cockscomb, pokeberry), the beetroot and prickly pear are the only food products containing this class of pigments [37, 396]. The major betacyanin/betalain in beets, betanin, accounts for 75 to 95% of the total betacyanin content; the remainder is comprised of isobetanin, prebetanin and isoprebetanin. These latter two pigments are sulphate monoesters of betanin and isobetanin, respectively. The major yellow pigments in the beetroot are vulgaxanthin-I and -II [396].

6.4.3 *Biosynthesis*

The dihydropyridine moiety present in all betalains is synthesized *in vivo* from two molecules of L-5,6-dihydroxyphenylalanine (L-DOPA), one of which must first undergo 4,5-extra-diol oxidative cleavage and recyclization [5, 397] (Figure 6.9). The cleaved and recyclized intermediate has been identified as betalamic acid [398]. The enzyme catalyzed conversion of L-DOPA to betalamic acid and its subsequent condensation with cyclodopa or a derivative of cyclodopa leads to betacyanidins; condensation with other amines or amino acids results in betaxanthins. Glycosylation of betacyanidins occurs late in the biosynthetic pathway [399], mediated by glycosyl transferases. Acylation, if it occurs, takes place subsequent to glycosylation, catalyzed by acyl transferases.

Betalain biosynthesis is controlled/regulated by numerous factors (*e.g.*

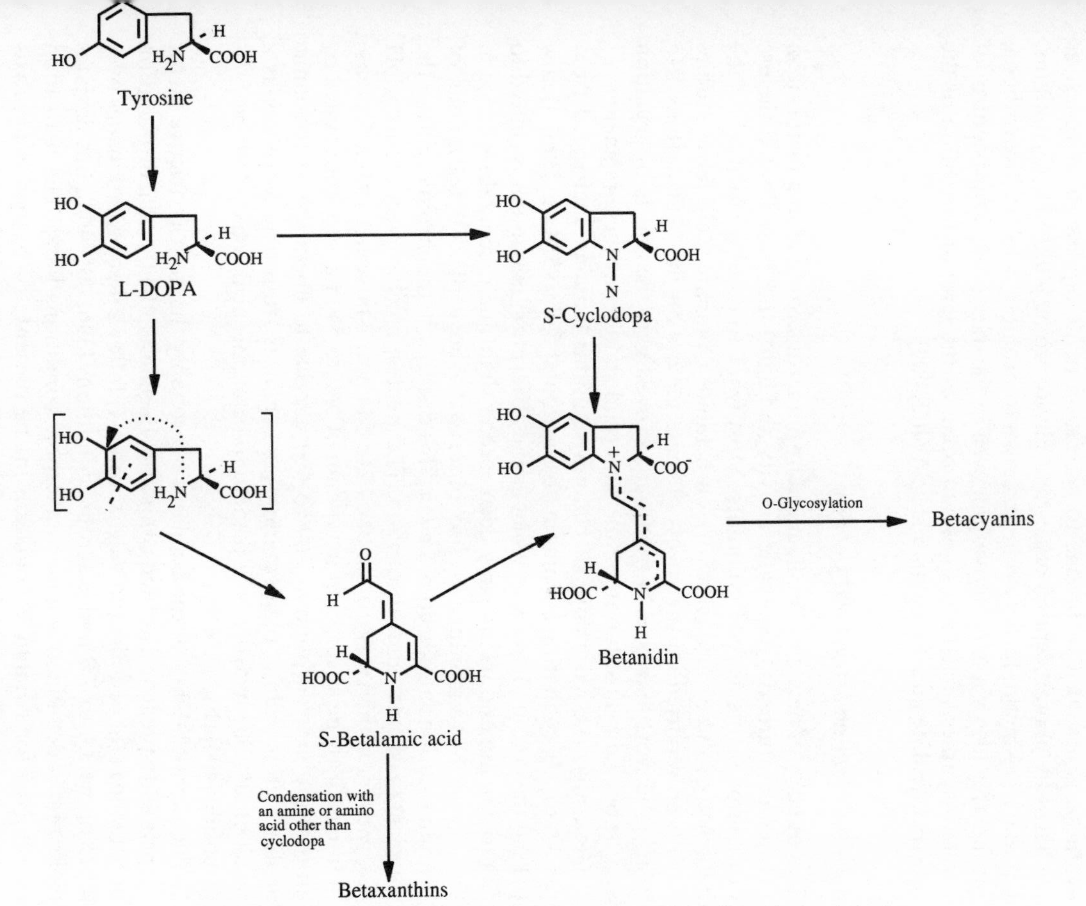

Figure 6.9 Biosynthesis of betalains (redrawn from Mabry [5] with permission).

light, temperature, precursor availability, cytokinins, abscisic acid and phenolic compounds), the most important of these being light [387]. Light is an absolute requirement for pigment synthesis in some plants (*e.g. Amaranthus tricolor*), but not others (*e.g. Beta vulgaris*) [400, 401]. Light-induced betalain synthesis is assumed to be due to activation or de-repression of genes, as mediated by photoreceptors other than phytochrome [402, 403]. Cytokinins, particularly kinetin, have also been shown to promote betalain biosynthesis, presumably also via gene activation, even in the dark [400]. Photocontrol of betalain synthesis has been shown to occur at the level of formation of the dihydropyridine moiety, betalamic acid [402, 403].

6.4.4 Colour and structural stability

The stability of betalains is influenced by numerous confounding factors (*e.g.* pH, temperature, oxygen, water activity and light) that have limited the use of these pigments as food colorants. Most work to date has focused on the influence of these various factors on betanin. Comparatively fewer studies have been reported on the effects of these factors on the betaxanthins. No systematic work has been carried out to investigate the effects of acylation and/or betalain substitution on the structural stability of these pigments.

Generally, the red colour of betanin solutions remains unchanged from pH 3.0 to 7.0, exhibiting maximum absorption at 537 to 538 nm [396]. Below pH 3.0 the colour changes to violet as the absorption maximum is shifted to 534 to 536 nm and its intensity decreased; a slight increase in absorbance in the 570 to 640 nm range is also observed. Above pH 7.0 the colour of betanin solutions also becomes bluer, this due to a bathochromic shift in the wavelength of maximum absorption. The greatest blueing effect occurs at pH 9.0, with maximum absorption at 543 to 544 nm. Above pH 10.0 a decrease in intensity at the absorption maximum of 540 to 550 nm is accompanied by an increase in absorption at 400 to 460 nm due to liberation of betalamic acid, which is yellow: pigmentation thus changes from blue to yellow as a result of alkaline hydrolysis of betanin to betalamic acid and cyclodopa-5-*O*-glycoside [404] (Figure 6.10).

The thermolability of the betalains is probably the most restrictive factor in their widespread use as food colorants. In general, the thermal degradation of betalains (*i.e.* betanin and vulgaxanthin-I) follows first-order kinetics over a pH range 3.0 to 7.0 under aerobic conditions [396, 405–408], but deviates from first-order kinetics under anaerobic conditions [409]. The activation energy (E_a) for betanin degradation in the presence of oxygen is approximately 20 kcal/mol at pH 3.0 to 7.0 [407, 408, 410, 411]. Under similar conditions, vulgaxanthin-I was shown to be more sensitive than betanin to thermal degradation with an E_a of around 16 kcal/mol [407]. Based on half-lives, the thermal stability of the betalain pigments is greatest between pH 5.0 and 6.0 in the presence of oxygen [405, 407, 412] and pH 4.0 and 5.0 in the absence of oxygen [405, 408]. The rate of degradation of betanin and vulgaxanthin-I is more rapid in model systems compared with that in beet juice, suggesting a protective effect conferred by other constituents in the natural system (*e.g.* by polyphenols, antioxidants, *etc.*) and/or the presence of betalamic acid and cyclodopa-5-*O*-glycoside, which may undergo Schiff's base condensation to regenerate betanin (Figure 6.10) [405, 412–414]. Betanin may also yield isobetanin and/or decarboxylated betanin, their formation being favoured upon heating at pH 3.0–4.0. These latter compounds do not differ

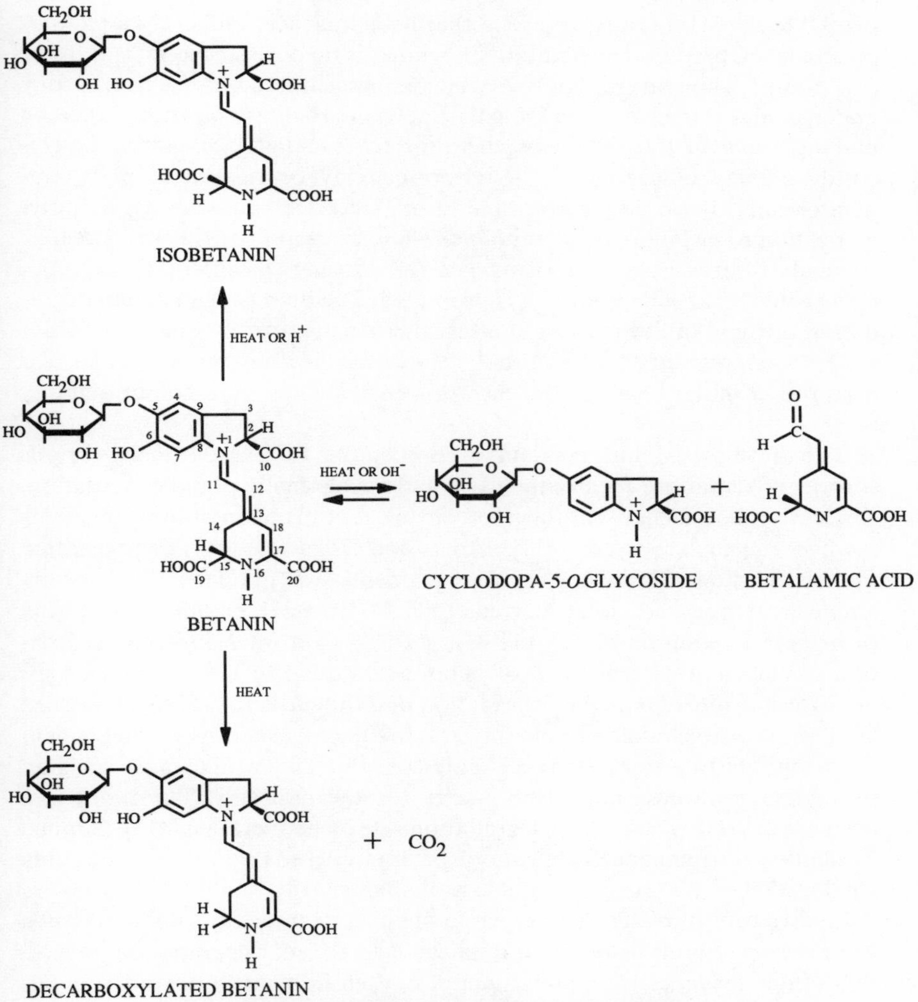

ISOBETANIN

HEAT OR H^+

BETANIN

HEAT OR OH^-

CYCLODOPA-5-*O*-GLYCOSIDE + BETALAMIC ACID

HEAT

DECARBOXYLATED BETANIN + CO_2

Figure 6.10 Mechanisms of betanin degradation.

significantly from betanin in their absorption/colour properties and they yield similar thermal and pH degradation products (*i.e.* betalamic acid and cyclodopa-5-*O*-glycoside) [404].

The primary steps in betanin degradation, as mediated by temperature and/or pH, are assumed to involve nucleophilic attack (*e.g.* by water) at the C-11 position yielding cyclodopa-5-*O*-glycoside and betalamic acid (Figure 6.10). As already alluded to, this reaction is reversible, the regeneration of betanin being greatest in the pH range in which the pigment is most stable [408, 413, 415]. While maximum stability of betalamic acid occurs at pH 6.0 to 8.0 [416], that of cyclodopa-5-*O*-glycoside occurs at acid pH (*e.g.* pH 3.0) [408]. The E_a for regeneration of pigment ranges from 0.6 to 3.5 kcal/mol at pH 3.0 to 5.0 [414]. Thus, provided that betalamic acid and cyclodopa-5-*O*-glycoside are present, regeneration of betanin is favoured, especially at lower (*e.g.* room) temperature. However, betalamic acid is heat sensitive: it may undergo aldol condensation or participate in Maillard reactions, thereby making it unavailable for the regeneration reaction [408]. Similarly, the glycosidic moiety of cyclodopa-5-*O*-glycoside may be cleaved at high temperatures. It is also very susceptible to oxidation reactions, initiating polymerization to melanin-type compounds [408]. Thus as temperature increases, particularly in the presence of oxygen (*viz.* during storage or processing), irreversible betanin degradation is promoted. The quantity of betanin degradation during, and regeneration after, thermal treatment is dependent not only on temperature and pH, but also initial betanin concentration, for example as initial betanin concentration increases so does colour stability [414].

Like the anthocyanins, betalains are susceptible to degradation by various sources of radiation, for example UV and visible light, gamma irradiation. There is some evidence that the photochemical destruction of these pigments involves molecular oxygen [411]. Sapers and Hornstein [417] demonstrated that the photo-oxidation of beetroot pigments was pH-dependent, for example greater degradation occurred at pH 3.0 than pH 5.0. Photo-oxidation of betanin in solution (40°C, pH 4.0, 400 ft-c light intensity) follows first-order kinetics [411]. Exposure of betanin solutions to UV or visible light increases the rate of pigment degradation by as much as 15% [396]. Absorbed light in the visible range presumably acts to excite π electrons of the pigment chromophore to a more energetic state (*i.e.* π^*). This would cause a higher reactivity or lowered activation energy for the molecule. The result is a decrease in the dependence of degradation rate on temperature [411]. Gamma irradiation also enhances the rate of betanin degradation: doses exceeding 100 krad (1000 Gy) result in total loss of colour [418].

Involvement of molecular oxygen in pigment degradation is not restricted to photo-oxidation; it has been implicated in thermal degradation as well. Substantial evidence for involvement of oxygen in pigment degradation has been provided by reduction of pigment loss through nitrogen purging of purified betanin solutions [405] and through application of glucose oxidase (an oxygen scavenger) to beetroot concentrates [419]. Metal ions are common food constituents, or food contaminants from processing equipment, and may function as pro-oxidants. Metal cations (*e.g.* iron, copper, tin and aluminium) at 100 p.p.m. accelerated the rate of degradation of buffered (pH 5.0) betanin solutions saturated with oxygen, with copper having the greatest effect [420]. As electron donors or acceptors, metal ions, depending on their oxidation state, may function to destabilize the electrophilic centre of betalain (*i.e.* betanin) resulting in rearrangement of associated bonds, destruction of the chromophore and loss of colour [420]. Since betanin, and betalains in general, is sensitive to oxidation, antioxidants have been studied for their potential ability to improve pigment stability. No protective effect by addition of 100 p.p.m. ascorbic acid or α-tocopherol was observed; extremely elevated concentrations of ascorbic acid (*i.e.* 1000 p.p.m.) decreased pigment shelf-life, as ascorbic acid acted as a pro-oxidant. High concentrations (10 000 p.p.m.) of citric acid and EDTA have been found to improve betanin stability by 1.5 times [420]. These results suggest that degradation of betanin does not occur via a free radical mechanism but rather an oxidation mechanism. The mechanism of betanin oxidation may involve nucleophilic attack at the quaternary amino nitrogen or electrophilic attack at an adjacent bond; the protective effect of citric acid and EDTA may be the result of partial neutralization of the electrophilic centre (*i.e.* they may sequester betanin) [420] and/or the result of sequestering metal ions that may otherwise act as pro-oxidants.

The effect of water activity (a_w) on betanin stability has also been documented [406]. First-order rate constants for betanin degradation varied exponentially with respect to a_w, a four-fold increase in pigment stability being observed as a_w decreased from 1.0 to 0.37. Reduced a_w may increase betalain stability by reducing reactant mobility and/or decreasing oxygen solubility.

As with other natural pigments, decolorizing enzymes also influence betalain stability. Products of enzymatic betacyanin degradation are the same as those from thermal destruction or alkaline hydrolysis [421]. Decolorizing enzyme activity has been demonstrated in subcellular tissue extracted from beetroot [422–424]. Whether there are distinct betacyanin and betaxanthin decolorizing enzymes remains obscure. Available evidence suggests that betacyanin decoloration is mediated by a peroxidase [425, 426]. Conditions for optimal catalytic pigment degradation are pH 3.4 and 40°C [421, 424]. Decolorizing activity is generally concentrated in those portions of the beetroot where most pigment accumulates, that is, in tissue separated from the epidermal layers. Furthermore, the enzyme(s) is reported to be covalently attached to the cell wall, although enzyme accounting for up to 25% of total activity may also be either soluble (*i.e.* free in intracellular, intact tissue) or ionically bound to cell-wall fragments following tissue homogenization [427]. Systematic investigation into the identity and physicochemical characterization of betalain decolorizing enzyme(s) has yet to be carried out. However, their presence is of consequence during extraction/purification and subsequent use of betalain pigments as food colorants; inactivation, for example through mild thermal treatment, may be required to prevent or reduce enzyme catalysed fading of pigment extracts.

Betalains may also participate in condensation reactions. Smith and Croker [428] demonstrated that the polyamines spermine, putrescine and spermidine complexed with betacyanins from beetroot resulting in loss of colour. The authors suggested that the decomposition of pigments was via interaction of the polyamine with their carboxyl groups, leading to subsequent cleavage to cyclodopa-5-*O*-glycoside and betamalic acid.

6.4.5 *Extraction and purification*

The availability of commercial preparations of betalains is currently restricted by legislation in most countries [19, 20] to beet juice concentrates produced by concentrating juice under vacuum to 60 to 65% total solids, or beet powders produced by spray- or freeze-drying the concentrate. Beet juices have traditionally been prepared by hydraulic press operations in which raw beets are blanched, chopped, pressed and the expressed juice filtered. Less than 50% recovery of betalains can be expected from most conventional press operations unless macerating enzymes are used to facilitate pressing. However, up to 90% recovery of betalains from beetroots has been reported with use of a continuous diffusion apparatus [429, 430]. Since commercial preparations are not pure they vary in colour, depending on the proportions of betacyanin and betaxanthin extracted, and often possess beet-like odour and taste [38]. Purification of the betalains is possible on both small and large scale, although the methods used generally give poor yields.

Separation and purification of betalains from crude plant extracts is usually necessary if qualitative and/or quantitative analyses are to be performed. Several different approaches have been reviewed by Mabry [5]. Preliminary separation of betalains in a variety of crude aqueous plant extracts has been accomplished by non-ionic absorption on strongly acid resins (*e.g.* Dowex 50W) followed by polyamide column chromatography, using increasing concentrations of methanol in aqueous citric acid as the developing solvent [37, 388, 431, 432]. Citric acid is removed from resolved betalains by resin treatment. Sequential chromatography on CM-Sephadex C-25, Sephadex G-10 and DEAE-Sephadex A-25 columns has also provided good separation of betalains [394]. Adams and von Elbe [433] reported that gel-filtration column chromatography on Sephadex or polyacrylamide (Bio-Gel P-6) gels could be used to rapidly and efficiently separate the betalains from beet juice. They suggested that the primary mechanism for retention of pigments on gel supports was adsorption rather than size exclusion.

Numerous other chromatographic [379, 434–438] and electophoretic procedures [439–444] have been used to separate/purify betalains from a variety of sources. However, most of these procedures are time-consuming. Since its initial application for analysis of the betalains [445], HPLC methods have become widely adopted for betalain separation and analysis [385, 393, 438, 446–450].

6.4.6 *Qualitative analysis*

The structural characterization and identification of betalains has been reviewed by Piattelli [37, 395], whose methods used are outlined below. They generally involve direct comparison of spectroscopic, chromatographic and electrophoretic properties with authentic standards, before and after controlled hydrolysis. Since the betalains differ from anthocyanins in being more sensitive to acid hydrolysis, in their colours with changes in pH, and in their chromatographic and electrophoretic properties, they may be easily differentiated by simple colour tests [5, 260]. Since the occurrence of these two pigment types is mutually exclusive, colour tests can usually be performed using crude plant extracts [260].

6.4.6.1 Chromatographic and spectral characterization. Betacyanins and betaxanthins possess similar chromatographic and electrophoretic properties; thus, similar techniques are used for their isolation/purification. However, polyamide chromatography is only moderately successful for the separation of betaxanthins. Preparative paper electrophoresis and chromatography using alternate supports are commonly used for separation of individual betaxanthins from co-occurring betacyanins and other plant constituents [37]. HPLC is also a valuable means by which to separate and analyse the betalains: the number of betaxanthin structures was found to be considerably greater using HPLC than was known from more traditional separation techniques [393]. Undoubtedly, more betalains will be identified as HPLC techniques become further developed.

Tentative indentification of betalains can be deduced from their chromatographic or electrophoretic behaviour [37]. For example, on polyamide resins:

- the retention of betacyanins decreases with increased glycosyl substitution;
- the retention of 6-glycosides exceeds that of the 5-glycosides;
- iso-derivatives are retained slightly longer than their corresponding C-15 epimers; and
- acyl groups, aromatic more than aliphatic, increase the retention of pigments.

Each pigment possesses characteristic electrophoretic mobility between 0.3

and 1.78, relative to betanin, usually determined at both pH 4.5 and 2.4 [34, 432].

Corroborative data may be provided by analysis of absorption spectra. All betacyanins exhibit visible maxima between 534 and 552 nm [432], while all betaxanthins have absorption maxima in the range 474 to 486 nm [34, 389]. Acylated betalains generally exhibit a second absorption maximum in the UV region of 260 to 320 nm, where the absorption of non-acylated pigments is weak [37, 380]. As with the anthocyanins, the number of acyl residues in betalains can be estimated from the ratio of absorption at the visible maximum to that at the UV maximum [37]. Further valuable structural information may be obtained from ^{1}H-NMR spectra [380, 386, 388, 391, 393, 451]. Supporting molecular weight data have been obtained by FAB-MS and/or GC/MS [386, 393]. The IR spectra of betalains have been found to provide little useful information [380].

6.4.6.2 Controlled hydrolysis procedures. Betacyanins occur naturally as C-15 epimers, the betanidin derivatives generally being present in greater quantities than their corresponding stereoisomers. Strong acid hydrolysis of betanin gives, along with glucose, a mixture of both aglycones [37, 452], while milder controlled acid hydrolysis [453], or β-glucosidase hydrolysis [388], yields only betanidins. It is thus possible to establish whether two betacyanins of unknown structure are C-15 epimers; if this is the case, subsequent analyses can be conveniently carried out using a mixture of both pigments [37]. In addition to acid and enzyme treatment, glycosyl moieties of pigments may also be obtained by oxidative destruction of the aglycone with alkaline hydrogen peroxide [378]. The cleaved sugar group(s) may be identified by conventional paper chromatography or HPLC; the aglycone and its degradation products may be identified by comparison with authentic standards using paper chromatography or electrophoresis, or HPLC. From acid hydrolysis, the amine or amino acid attached to the dihydropyridine moiety of betaxanthins can be isolated and subsequently identified [5]. The position(s) of glycosylation of betacyanins may be determined by treatment with excess diazomethane, which converts the aglycone into an *O*-methyl neobetacyanidin trimethylester glycoside in which the originally free phenolic hydroxyl group of the aglycone has been methylated and thus labelled for identification [37]. In acidic media, the neobetacyanidin exhibits a bathochromic shift of about 100 nm relative to the parent glycoside [454, 455]. Diazomethane methylation similarly converts betaxanthins into neopigments, which exhibit absorption maxima at 340 to 360 nm and show a bathochromic shift of 80 to 100 nm in acid solution [391]. Alkaline fusion of the neo-pigments always yields 4-methylpyridine-2,6-dicarboxylic acid [5, 375].

The identity and position of acyl groups linked to glycosyl moieties may be determined by ^{1}H-NMR. More traditionally, acyl residues have been removed from the more structurally complex betalains by cold alkaline hydrolysis under nitrogen (*i.e.* oxygen must be absent), isolated and identified by chromatographic techniques [380]. Their position of attachment has been determined by:

1. Mild periodate oxidation, subsequent borohydride reduction, mild acid hydrolysis, a second borohydride reaction, then chromatographic identification of the resulting polyols; or
2. Pigment methylation and subsequent identification of the products released after hydrolysis of the permethylated compound [37, 380].

6.4.7 *Quantitative analysis*

In the absence of interfering substances (*e.g.* fresh extracts/juices or purified samples), the quantitation of betalains can be accomplished using spectrophotometry, whereby absorbance at the visible maximum is expressed in terms of concentration by use of appropriate absorptivities [376, 390, 398, 432, 452, 456]. Nilsson [436] developed a spectrophotometric method that directly determines betacyanin and betaxanthin pigments in beets without their initial separation. The method is based on the observation that while vulgaxanthin-I absorbs only at 476 to 478 nm, betanin exhibits a maximum at 535 to 540 nm but also absorbs at the visible maximum of vulgaxanthin-I. Calculation of the ratio of absorbance at 538 nm to that at 477 nm as a function of concentration (*i.e.* dilution) leads to determination of the measurable concentration of vulgaxanthin-I. Absorbance of the sample at 538 nm is due only to betanin, permitting its direct determination. Results are expressed in terms of total betacyanin and total betaxanthin concentrations since minor constituent betalains also contribute to measured absorbance.

If interfering substances are present in the sample, for example due to betalain degradation, the betalain pigments must first be purified. The procedures generally used, for example paper or column chromatography [433] or electrophoresis [444], not only purify but also separate pigments, thereby allowing for quantitation of total and individual betalains simultaneously. Due to its speed and high resolving power, and the fact that preliminary purification is not always necessary, HPLC has become the method of choice for quantitative analysis of individual and total betalains [404, 411, 420, 446, 449, 450]. This trend is likely to continue.

6.4.8 *Current and potential sources and uses*

The use of betalain pigments as food colorants dates back to at least the turn of the century when juice from pokeberries, which contain betanin, was added to wine to impart a more desirable red colour [457]. With passing of legislation against the adulteration of foods, this practice was eventually prohibited. Current legislation restricts betalain colorants to concentrates or powders obtained from aqueous extracts of beetroots (*Beta vulgaris*). Because of the relatively high pigment concentrations therein, beetroots are the only plant food source from which extraction of betalains for commercial use as a colorant is economically feasible. Commercial beet colorants typically contain 0.4 to 1.0% pigment (expressed as betanin), 80% sugar, 8% ash and 10% protein [20]. Their colour can vary considerably, depending on their content of yellow betaxanthins; thus their colour varies with beet variety, beetroot quality and age at harvest, and method of pigment extraction [38].

Improvements in the pigment content of commercial beet colorants could be achieved if legislative restrictions on the use of purified pigments were lifted. Purification of beet extracts could be carried out via industrial-scale chromatography [38]. Although yields have not generally been good to date, purification of crude extracts by ion-exchange and gel-filtration chromatography has been proven successful on a laboratory scale [38, 434]. Further developments should make chromatographic systems very useful for large-scale production of purified betalain colorants. Purification of red beet extracts has been carried out by fermentation using *Aspergillus niger* [458, 459], *Candida utilis* [460, 461] and *Saccharomyces oviformis* [450], resulting in products with improved stability and no 'off' flavours. The practical application of microbial purification for large-scale preparation of commercial betalain colorants is uncertain. Similarly, the accumulation of betalains in beetroot tissue culture has been reported [462, 263], but the practical application of this or other biotechnologies to the production of highly purified betalain preparations for use as food colorants is questionable [372].

Commercial and purified betalain preparations have been shown to be effective colorants of foods with compatible chemical and/or physical properties [38]. Microbiologically purified beetroot preparations have demonstrated suitable stability for use in various pharmaceutical forms [464–466]. Beetroot colorants may be used alone to produce hues resembling raspberry or cherry, or in combination with other colorants such as annatto to obtain strawberry shades [20, 373]. Considering the factors that influence pigment stability, beetroot preparations may be successfully used to colour products with short shelf-lives that are marketed in a dry state, that are packaged to reduce exposure to light, oxygen and humidity, and/or that do not receive high or prolonged heat treatment. If thermal processing is required, pigment degradation can be minimized by adding the colorant after the heat treatment or as near the end of the heating cycle as possible [20]. As reviewed by Counsell *et al.* [393] and von Elbe [38], beetroot colorant can be effectively used to colour hard candies and fruit chews, dairy products such as yoghurt and ice cream [467], salad dressing, frostings, cake mixes, gelatin desserts, meat substitutes [396], poultry meat sausages [468, 469], gravy mixes, soft drinks and powdered drink mixes [470].

6.5 Conclusions

The successful use of anthocyanins and betalains (especially betacyanins) to colour a variety of foods and pharmaceuticals has been clearly demonstrated. Although these pigments may not replace the more stable synthetic red dyes, they do provide suitable natural alternatives for the production of highly coloured and high-quality products. Their successful application is obviously limited to products with compatible physicochemical properties. However, if current legislation restricting the use of more highly purified extracts and sources other than grapes (and grape by-products) and beetroots were to be lifted, wider application of anthocyanins and betalains as food colorants could be realized.

At present, the anthocyanins would appear to be more stable than the betacyanins. However, systematic research to investigate the stabilizing influence of acylation, various condensation reactions and/or surface adsorption has not been carried out on the betacyanins/betalains. Polyacylated anthocyanins, and those that partake in condensation reactions (*e.g.* self-association and co-pigmentation) and/or are adsorbed onto surfaces, demonstrate marked stability and offer great promise as stable colorants. With further technological advances, for example in the development of industry-scale chromtographic purifications/separations and application of biotechnology (*e.g.* tissue culture) for large-scale pigment production, discovery of new pigment sources and structural variants with enhanced stability, and modification of foods to allow for stabilization of anthocyanins and betalains, the potential of these natural pigments for wider application as colorants would seem to be unlimited. A greater knowledge of the structural features necessary for enhanced stability and colour properties is required, however. Further investigations into the physicochemistry of anthocyanins and betalains is necessary if these natural colorants are to be more competitive and viable alternatives to the synthetic colorants currently used.

Acknowledgements

The authors gratefully appreciate the help of Cathy Young and Allen Mao in the preparation of this chapter, and the Department of Food Science, University of Guelph, for financial assistance.

References

1. Timberlake, C.F. and Bridle, P. in *The Flavonoids* (J.B. Harborne, T.J. Mabry and H. Mabry, Eds), Chapman & Hall, London (1975), Chapter 5.
2. Timberlake, C.F. and Bridle, P. in *Anthocyanins as Food Colors* (P. Markakis, Ed.), Academic Press, New York (1982) Chapter 5.

3. Harborne, J.B. *The Comparative Biochemistry of the Flavonoids*, Academic Press, New York (1967) Chapter 1.
4. Mabry, T.J. in *Chemistry of the Alkaloids* (S.W. Pelletier, Ed.), Von Nostrand Reinhold, New York (1970) Chapter 13.
5. Mabry, T.J. in *Secondary Plant Products* (E.A. Bell and B.V. Charlwood, Eds), Springer-Verlag, Berlin (1980) Chapter 11.
6. Kimler, L., Mears, J., Mabry, T.J. and Rosler, H. *Taxon.* **19** (1970) 875–878.
7. McClure, J.W. in *The Flavonoids* (J.B. Harborne, T.J. Mabry and H. Mabry, Eds), Chapman & Hall, London (1975) Chapter 18.
8. Sosnova, V. *Biol. Plant.* **12** (1970) 424–427.
9. Kimler, L.M. *Abstr. Bot. Soc. Am.* 70th Annual Meet. (1975) 36.
10. Stenlid, G. *Phytochem.* **15** (1976) 661–663.
11. Mabry, T.J. in *Comparative Phytochemistry* (T. Swain, Ed.), Academic Press, London (1966) Chapter 14.
12. Francis, F.J. in *Developments in Food Colors—2* (J. Walford, Ed.), Applied Science Publishers, London (1984) Chapter 7.
13. Meggos, H.N. *Food Technol.* **38** (1984) 70–74.
14. Singleton, V.L. and Esau, P.L. *Phenolic Substances in Grapes and Wines and Their Significance*, Academic Press, New York (1969).
15. Haveland-Smith, R.B. *Mut. Res.* **91** (1981) 285–290.
16. Von Elbe, J.H. and Schwartz, S.J. *Arch. Toxicol.* **49** (1981) 93–98.
17. Pourrat, A., Lejeune, B., Grand, A., Bastide, P. and Bastide, J. *Med. Nut.* **23** (1987) 166–172.
18. Igarashi, K., Abe, S. and Satoh, J. *Agricol. Biol. Chem.* **54** (1990) 171–175.
19. Huschke, G. in *Natural Colours for Food and Other Uses* (J.N. Counsell, Ed.), Applied Science Publishers, London (1981) Chapter 8.
20. Marmion, D.M. *Handbook of U.S. Colorants for Foods, Drugs and Cosmetics*. 2nd edn. John Wiley & Sons, New York (1984) Chapter 8.
21. Markakis, P. (Ed.) *Anthocyanins as Food Colors*. Academic Press, New York (1982).
22. Markakis, P. *CRC Crit. Rev. Food Technol.* **4** (1974) 437–456.
23. Shrikhande, A.J. *CRC Crit. Rev. Food Sci. Nutr.* **7** (1976) 193–218.
24. Francis, F.J. in *Current Aspects of Food Colorants* (T.E. Furia, Ed.), CRC Press, Cleveland (1977) Chapter 2.
25. Eskin, N.A.M. *Plant Pigments, Flavors and Textures: The Chemistry and Biochemistry of Selected Compounds*, Academic Press, New York (1979) Chapter 2.
26. Timberlake, C.F. *Food Chem.* **5** (1980) 69–80.
27. Hrazdina, G. *Lebensm. -Wiss. u. Technol.* **7** (1974) 103–108.
28. Hrazdina, G. *Lebensm. -Wiss. u. Technol.* **14** (1981) 283–286.
29. Jackman, R.L., Yada, R.Y., Tung, M.A. and Speers, R.A. *J. Food Biochem.* **11** (1987) 201–247.
30. Jackman, R.L., Yada, R.Y. and Tung, M.A. *J. Food Biochem.* **11** (1987) 297–308.
31. Mazza, G. and Brouillard, R. *Food Chem.* **25** (1987) 207–225.
32. Mabry, T.J. in *Taxonomic Biochemistry and Serology* (C.A. Leone, Ed.), Ronald Press, New York (1964) Chapter 12.
33. Mabry, T.J. *Comparative Phytochemistry*, Academic Press, New York (1966) Chapter 14.
34. Mabry, T.J. and Dreiding, A.S. *Recent Adv. Phytochem.* **1** (1968) 145–160.
35. Mabry, T.J., Kimler, L. and Chang, C. *Recent Adv. Phytochem.* **5** (1972) 105–134.
36. Reznik, H. *Ber. Deutsch. Bot. Ges.* **88** (1975) 179–190.
37. Piattelli, M. in *Chemistry and Biochemistry of Plant Pigments*, 2nd edn. Vol. 1 (T.W. Goodwin, Ed.), Academic Press, New York (1976) Chapter 11.
38. von Elbe, J.H. in *Current Aspects of Food Colorants* (T.E. Furia, Ed.), CRC Press, Cleveland (1977) Chapter 3.
39. Grisebach, H. in *Anthocyanins as Food Colors* (P. Markakis, Ed.), Academic Press, New York (1982) Chapter 3.
40. Brouillard, R. in *Anthocyanins as Food Colors* (P. Markakis, Ed.), Academic Press, New York (1982) Chapter 1.
41. Sweeny, J.G. and Iacobucci, G.A. *J. Agricol. Food Chem.* **31** (1983) 531–533.
42. Timberlake, C.F. and Bridle, P. *Nature* **212** (1966) 158–159.

43. Harborne, J.B. *Recent Adv. Phytochem.* **12** (1979) 457–474.
44. Jurd, L. *Adv. Food Res.* Suppl. **3** (1972) 123–142.
45. Harborne, J.B. in *Chemical Plant Taxonomy* (T. Swain, Ed.), Academic Press, New York (1963) Chapter 13.
46. Stewart, R.N., Norris, K.H. and Asen, S. *Phytochem.* **14** (1975) 937–942.
47. Asen, S. *Acta Hortic.* **63** (1976) 217–223.
48. Brouillard, R. *Phytochem.* **22** (1983) 1311–1323.
49. Harborne, J.B. *Phytochem.* **3** (1964) 151–160.
50. Gueffroy, D.E., Kepner, R.E. and Webb, A.D. *Phytochem.* **10** (1971) 813–819.
51. Hrazdina, G. and Franzese, A.J. *Phytochem.* **13** (1974) 225–229.
52. Saito, N., Osawa, Y. and Hayashi, K. *Phytochem.* **10** (1971) 445–447.
53. Asen, S., Stewart, R.N. and Norris, K.H. *Phytochem.* **16** (1977) 1118–1119.
54. Yoshitama, K. and Abe, K. *Phytochem.* **16** (1977) 591–593.
55. Yoshitama, K. *Bot. Mag. Tokyo* **91** (1978) 207–212.
56. Yoshitama, K. *Phytochem.* **20** (1981) 186–187.
57. Yoshida, K., Kondo, T., Kameda, K. and Goto, T. *Agricol. Biol. Chem.* **54** (1990) 1745–1751.
58. Timberlake, C.F. and Bridle, P. in *Developments in Food Colors—1* (J. Walford, Ed.), Applied Science Publishers, London (1980) Chapter 5.
59. Fuleki, T. *J. Food Sci.* **36** (1971) 101–104.
60. Du, C.T., Wang, P.L. and Francis, F.J. *J. Food Sci.* **39** (1974) 1265–1266.
61. Tanchev, S.S. and Timberlake, C.F. *Phytochem.* **8** (1969) 1825–1827.
62. Lanzarini, G. and Morselli, L. *Ind. Conserve* **49** (1974) 16–20.
63. Hrazdina, G., Iredale, H. and Mattick, L.R. *Phytochem.* **16** (1977) 297–299.
64. Chandler, B.V. *Nature* **182** (1958) 933.
65. Wrolstad, R.E. and Heatherbell, D.A. *J. Sci. Food Agricol.* **35** (1974) 1221–1228.
66. Ishikura, N. and Sugahara, K. *Bot. Mag. Tokyo* **92** (1979) 57–61.
67. Puech, A.A., Rebeiz, C.A., Catlin, P.B. and Crane, J.C. *J. Food Sci.* **40** (1975) 775–779.
68. Sondheimer, E. and Karash, C.B. *Nature* **178** (1956) 648–649.
69. Co, H. and Markakis, P. *J. Food Sci.* **33** (1968) 281–283.
70. Wrolstad, R.E., Hildrum, K.I. and Amos, J.F. *J. Chromatogr.* **50** (1970) 311–318.
71. Ishikura, N. *Bot. Mag. Tokyo* **88** (1975) 41–45.
72. Kuroda, C. and Wada, M. *Proc. Imp. Acad.* (*Tokyo*) **11** (1935) 189–191.
73. Yoshikura, K. and Hamaguchi, Y. *J. Jap. Soc. Food Nutr.* **22** (1969) 367–370.
74. Sun, B.H. and Francis, F.J. *J. Food Sci.* **32** (1967) 647–649.
75. Timberlake, C.F. and Bridle, P. *J. Sci. Food Agricol.* **22** (1971) 509–513.
76. Proctor, J.T. and Creasy, L.L. *Phytochem.* **8** (1969) 2108.
77. Pruthi, J.S., Susheela, R. and Lal, G. *J. Food Sci.* **26** (1961) 385–388.
78. Yoshikura, K. and Hamaguchi, Y. *J. Jap. Soc. Food Nutr.* **24** (1971) 275–278.
79. Li, K. and Wagenknecht, A.C. *Nature* **182** (1958) 657.
80. Harborne, J.B. and Hall, E. *Phytochem.* **3** (1964) 453–463.
81. Lynn, D.Y. and Luh, B.S. *J. Food Sci.* **29** (1964) 735–743.
82. Dekazos, E.D. *J. Food Sci.* **35** (1970) 237–241.
83. Shrikhande, A.J. and Francis, F.J. *J. Food Sci.* **38** (1973) 649–651.
84. Tanchev, S.S. and Vasilev, V.N. *Gradinar. Lozar. Nauka* **10** (1973) 23–28.
85. Du, C.T., Wang, P.L. and Francis, F.J. *J. Food Sci.* **40** (1975) 417–418.
86. Ishikura, N. and Hayashi, K. *Bot. Mag. Tokyo* **78** (1965) 91–96.
87. Fuleki, T. *J. Food Sci.* **34** (1969) 365–369.
88. Gallop, R.A. *Variety, Composition and Colour in Canned Fruits, Particularly Rhubarb.* Fruit and Veg. Canning Res. Assoc., Chipping Campden, Glos. Science Bulletin No. 5 (1965).
89. Wrolstad, R.E. and Heatherbell, D.A. *J. Food Sci.* **33** (1968) 592–594.
90. Chandler, B.V. and Harper, K.A. *Aust. J. Chem.* **15** (1962) 114–120.
91. Casoli, U., Cultrera, R. and Gherardi, S. *Ind. Conserve* **42** (1967) 255.
92. Nybom, N. *Frukt. Baer.* (1970) 106–118.
93. Oeydin, J. *Hortic. Res.* **14** (1974) 1–7.
94. Barritt, B.H. and Torre, L.C. *J. Chromatogr.* **75** (1973) 151–155.
95. Jennings, D.L. and Carmichael, E. *New Phytol.* **84** (1980) 505–513.
96. Nybom, N. *J. Chromatogr.* **38** (1968) 382–387.

97. Daravingas, G. and Cain, R.F. *J. Food Sci.* **31** (1966) 927–936.
98. Daravingas, G. and Cain, R.F. *J. Food Sci.* **33** (1968) 138–142.
99. Watanabe, S., Sakamura, S. and Obata, Y. *Agricol. Biol. Chem.* **30** (1966) 420–422.
100. Tanchev, S.S., Ruskov, P.J. and Timberlake, C.F. *Phytochem.* **9** (1970) 1681–1682.
101. Saito, N., Hotta, R., Imai, K. and Hayashi, K. *Proc. Jap. Acad.* **41** (1965) 593–598.
102. Francis, F.J. and Harborne, J.B. *J. Food Sci.* **31** (1966) 524–528.
103. Buckmire, R.E. and Francis, F.J. *J. Food Sci.* **41** (1976) 1363–1365.
104. Francis, F.J., Harborne, J.B. and Barker, W.G. *J. Food Sci.* **31** (1966) 583–587.
105. Ballinger, W.E., Maness, E.P. and Kushman, L.J. *J. Am. Soc. Hort. Sci.* **95** (1970) 283–285.
106. Ballinger, W.E., Maness, E.P., Galletta, G.J. and Kushman, L.J. *J. Am. Soc. Hort. Sci.* **97** (1972) 381–384.
107. Sakamura, S. and Francis, F.J. *J. Food Sci.* **26** (1961) 318–321.
108. Zapsalis, C. and Francis, F.J. *J. Food Sci.* **30** (1965) 396–399.
109. Fuleki, T. and Francis, F.J. *Phytochem.* **6** (1967) 1705–1708.
110. Fuleki, T. and Francis, F.J. *J. Food Sci.* **33** (1968) 471–478.
111. Ribereau-Gayon, P. in *Anthocyanins as Food Colors* (P. Markakis, Ed.), Academic Press, New York (1982) Chapter 8.
112. Straus, J. *Plant Physiol.* **34** (1959) 536–541.
113. Harborne, J.B. and Gavazzi, G. *Phytochem.* **8** (1969) 999–1001.
114. Nakatani, N., Fukuda, H. and Fuwa, H. *Agricol. Biol. Chem.* **43** (1979) 389–391.
115. Lawanson, A.O. and Osude, B.A. *Z. Pflanzenphysiol.* **67** (1972) 460–463.
116. Alston, R.E. in *Biochemistry of Phenolic Compounds* (J.B. Harborne, Ed.), Academic Press, New York (1964) Chapter 5.
117. Hahlbrock, K. and Grisebach, H. in *The Flavonoids* (J.B. Harborne, T.J. Mabry and H. Mabry, Eds.), Chapman & Hall, London (1975) Chapter 16.
118. Wong, E. in *Chemistry and Biochemistry of Plant Pigments* 2nd edn. Vol 1. (T.W. Goodwin, Ed.), Academic Press, London (1976) Chapter 9.
119. Hahlbrock, K. and Grisebach, H. *Ann. Rev. Plant Physiol.* **30** (1979) 105–130.
120. Grisebach, H. in *The Biochemistry of Plant Phenolics* (C.F. Van Sumere and P.J. Lea, Eds), Clarendon Press, Oxford (1985) Chapter 10.
121. Mohr, H. *Lectures on Photomorphogenesis.* Springer-Verlag, Berlin (1972).
122. Schopfer, P. *Ann. Rev. Plant Physiol.* **28** (1977) 223–252.
123. Pecket, R.C. and Small, C.J. *Phytochem.* **19** (1980) 2571–2576.
124. Hemleben, V. *Z. Naturforsch. Teil C.* **36** (1981) 925–927.
125. Small, C.J. and Pecket, R.C. *Planta* **154** (1982) 97–99.
126. Hrazdina, G. and Wagner, G.J. in *The Biochemistry of Plant Phenolics* (C.F. Van Sumere and P.J. Lea, Eds), Clarendon Press, Oxford (1985) Chapter 7.
127. Hrazdina, G., Wagner, G.J. and Siegelman, H.W. *Phytochem.* **17** (1978) 53–56.
128. Hrazdina, G. and Wagner, G.J. *Arch. Biochem. Biophys.* **237** (1985) 88–100.
129. Egin-Buhler, B., Loyal, R. and Ebel, J. *Arch. Biochem. Biophys.* **203** (1980) 90–100.
130. Kamsteeg, J. Doctoral thesis. Utrecht (1980).
131. Ebel, J. and Hahlbrock, K. in *The Flavonoids: Advances in Research* (J.B. Harborne and T.J. Mabry, Eds), Chapman & Hall, London (1982) Chapter 11.
132. Hrazdina, G., Borzell, A.J. and Robinson, W.B. *Am. J. Enol. Vitic.* **21** (1970) 201–204.
133. Robinson, W.D., Weiers, L.D., Bertino, J.J. and Mattick, L.R. *Am. J. Enol. Vitic.* **17** (1966) 178–183.
134. Van Buren, J.P., Bertino, J.J. and Robinson, W.B. *Am. J. Enol. Vitic.* **19** (1968) 147–154.
135. Starr, M.S. and Francis, F.J. *Food Technol.* **22** (1968) 1293–1295.
136. Saito, N.N., Osawa, Y. and Hayashi, K. *Bot. Mag. Tokyo* **85** (1972) 105–110.
137. Asen, S., Stewart, R.N. and Norris, K.H. *Phytochem.* **11** (1972) 1139–1144.
138. Asen, S., Stewart, R.N. and Norris, K.H. *US Patent No.* 4172902 (1979).
139. Du, C.T. and Francis, F.J. *J. Food Sci.* **40** (1975) 1101–1102.
140. Brouillard, R. *Phytochem.* **20** (1981) 143–145.
141. Yoshitama, K. and Hayashi, K. *Bot. Mag. Tokyo* **87** (1974) 33–40.
142. Ingold, C.K. *Structure and Mechanism in Organic Chemistry.* Cornell University Press, Ithica (1969).
143. Bendz, G., Martensson, O. and Nilsson, E. *Ark. Kemi* **27** (1967) 65–77.

144. Timberlake, C.F. and Bridle, P. *Chem. Ind.* (1968) (Oct.) 1489.
145. Mazza, G. and Brouillard, R. *J. Agricol. Food Chem.* **35** (1987) 422–426.
146. Baranac, J., Amic, D. and Vukadinovic, V. *J. Agricol. Food Chem.* **38** (1990) 932–936.
147. McClelland, R.A. and McGall, G.H. *J. Org. Chem.* **47** (1982) 3730–3736.
148. Amic, D., Baranac, J. and Vukadinovic, V. *J. Agricol. Food. Chem.* **38** (1990) 936–940.
149. Brouillard, R. and Dubois, J.E. *J. Am. Chem. Soc.* **99** (1977) 1359–1364.
150. Timberlake, C.F. and Bridle, P. *J. Sci. Food Agricol.* **18** (1967) 473–478.
151. Brouillard, R. and Delaporte, B. *J. Am. Chem. Soc.* **99** (1977) 8461–8468.
152. Meschter, E.E. *J. Agricol. Food Chem.* **1** (1953) 574–579.
153. Tinsley, I.J. and Bockian, A.H. *Food Res.* **25** (1960) 161–173.
154. Calvi, J.P. and Francis, F.J. *J. Food Sci.* **43** (1978) 1448–1456.
155. Adams, J.B. *Changes in the Polyphenols of Red Fruits During Processing—The Kinetics and Mechanism of Anthocyanin Degradation.* Campden Food Press Res. Assoc., Chipping Campden, Glos. Tech. Bull. P. 22 (1972).
156. Adams, J.B. *J. Sci. Food Agricol.* **24** (1973) 747–762.
157. Lukton, A., Chichester, C.O. and MacKinney, G. *Food Technol.* **10** (1956) 427–432.
158. M. Simard, R.E., Bourzeix, M. and Heredia, N. *Lebensm. -Wiss. u. Technol.* **15** (1982) 177–180.
159. Markakis, P., Livingston, G.E. and Fellers, C.R. *Food Res.* **22** (1957) 117–129.
160. Keith, E.S. and Powers, J.J. *J. Agricol. Food Chem.* **13** (1965) 577–579.
161. Segal, B. and Negutz, G. *Nahrung* **13** (1969) 531–535.
162. Tanchev, S. *Proc. 6th Int. Congr. Food Sci. Technol.* (1983) 96.
163. Markakis, P. in *Anthocyanins as Food Colors* (P. Markakis, Ed.), Academic Press, New York (1982) Chapter 6.
164. Mishkin, M. and Saguy, I. *Z. Lebensm. u. Forsch.* **175** (1982) 410–412.
165. Havlikova, L. and Mikova, K. *Z. Lebensm. u. Forsch.* **181** (1985) 427–432.
166. Sastry, L.V.L. and Tischer, R.G. *Food Technol.* **6** (1952) 264–268.
167. Somers, T.C. *Phytochem.* **10** (1971) 2175–2186.
168. McCloskey, L.P. and Yengoyan, L.S. *Am. J. Enol. Vitic.* **32** (1981) 257–261.
169. Bakker, J. and Timberlake, C.F. *J. Sci. Food Agricol.* **37** (1986) 288–292.
170. Adams, J.B. and Ongley, M.H. *J. Food Technol.* **8** (1973) 139–145.
171. Hrazdina, G. *Phytochem.* **10** (1971) 1125–1130.
172. Adams, J.B. *Food Manufact.* (1973) (Feb.) 19–20, 41.
173. Tanchev, S.S. and Ioncheva, N. *Nahrung* **20** (1976) 889–893.
174. Erlandson, J.A. and Wrolstad, R.E. *J. Food Sci.* **37** (1972) 592–595.
175. Abers, J.E. and Wrolstad, R.E. *J. Food Sci.* **44** (1979) 75–78, 81.
176. Nebesky, E.A., Esselsen Jr., W.B., McConnell, J.E.W. and Fellers, C.R. *Food Res.* **14** (1949) 261–274.
177. Clydesdale, F.M., Main, J.H., Francis, F.J. and Damon Jr., R.A. *J. Food Sci.* **43** (1978) 1687–1692, 1697.
178. Poei-Langston, M.S. and Wrolstad, R.E. *J. Food Sci.* **46** (1981) 1218–1236.
179. King, G.A., Sweeny, J.G., Radford, T., Iacoubucci, G.A. *Bull. Liaison—Groupe Polyphenols* **9** (1980) 121–128.
180. Iacobucci, G.A. and Sweeny, J.G. *Tetrahedron* **39** (1983) 3005–3038.
181. Sondheimer, E. and Kertesz, A.I. *Food Res.* **18** (1953) 475–479.
182. Sondheimer, E. and Kertesz, A.I. *Food Res.* **18** (1952) 288–297.
183. Hrazdina, G. *Phytochem.* **9** (1970) 1647–1652.
184. Tressler, D.K. and Pederson, C.S. *Food Res.* **1** (1936) 87–97.
185. Palamidis, N. and Markakis, P. *J. Food Sci.* **40** (1975) 1047–1049.
186. Hannan, K.S. *Research on the Science and Technology of Food Preservation by Ionizing Radiations.* Chemical Publications Co., New York (1956).
187. Markakis, P., Livingston, G.E. and Fagerson, J.S. *Food Res.* **24** (1959) 520–528.
188. Pellegrino, F., Sekular, P. and Alfano, R.R. *Photobiochem. Photobiophys.* **2** (1981) 15–20.
189. Santhanam, M., Hautala, R.R., Sweeny, J.G. and Iacobucci, G.A. *Photochem. Photobiol.* **38** (1984) 477–480.
190. Sweeny, J.G., Wilkinson, M.M. and Iacobucci, G.A. *J. Agricol. Food Chem.* **29** (1981) 563–567.
191. Huang, H.T. *J. Agricol. Food Chem.* **3** (1955) 141–146.

192. Huang, H.T. *Nature* **177** (1956) 39.
193. Huang, H.T. *J. Am. Chem. Soc.* **78** (1956) 2390–2393.
194. Forsyth, W.G.C. and Quesnel, V.C. *Biochem. J.* **65** (1957) 177–179.
195. Van Buren, J.P., Scheiner, D.M. and Wagenknecht, A.C. *Nature* **185** (1960) 165–166.
196. Wagenknecht, A.C., Scheiner, D.M. and van Buren, J.P. *Food Technol.* **14** (1960) 47–49.
197. Peng, C.Y. and Markakis, P. *Nature* **199** (1963) 597–598.
198. Sakamura, S. and Obata, Y. *Agricol. Biol. Chem.* **27** (1963) 121–127.
199. Sakamura, S., Watanabe, S. and Obata, Y. *Agricol. Biol. Chem.* **29** (1965) 181–190.
200. Segal, B. and Segal, R.M. *Rev. Ferment. Ind. Aliment.* **24** (1969) 22–24.
201. Skalski, C. and Sistrunk, W.A. *J. Food Sci.* **38** (1973) 1060–1062.
202. Wesche-Ebeling, P. and Montgomery, M.W. *J. Food Sci.* **55** (1990) 731–734, 745.
203. Grommeck, R. and Markakis, P. *J. Food Sci.* **29** (1964) 53–57.
204. Matthew, A.G. and Parpia, H.A.B. *Adv. Food Res.* **19** (1971) 75–145.
205. Pifferi, P.G. and Cultrera, R. *J. Food Sci.* **39** (1974) 786–791.
206. Sakabura, H. and Ichinose, H. *Agricol. Chem. Soc. Jap.* **56** (1982) 517–524.
207. Blom, H. *Food Chem.* **12** (1983) 197–204.
208. Blom, H. and Thomassen, M.S. *Food Chem.* **17** (1985) 157–168.
209. Tanchev, S., Vladimirov, G. and Ioncheva, N. *Nauchni Trudove, Vissh Inst. Khranit. Vkusova Promyshl.* **16** (1969) 77–82.
210. Yang, H.Y. and Steele, W.F. *Food Technol.* **12** (1958) 517–519.
211. Siegel, A., Markakis, P. and Bedford, C.L. *J. Food Sci.* **36** (1971) 962–963.
212. Wrolstad, R.E., Skrede, G., Lea, P. and Enersen, G. *J. Food Sci.* **55** (1990) 1064–1065, 1072.
213. Goodman, L.P. and Markakis, P. *J. Food Sci.* **30** (1965) 135–137.
214. Cash, J.N., Sistrunk, W.A. and Stotte, C.A. *J. Food Sci.* **41** (1976) 1398–1402.
215. Shriner, R.L. and Sutton, R. *J. Am. Chem. Soc.* **85** (1963) 3989–3991.
216. Jurd, L. *Tetrahedron* **21** (1965) 3707–3714.
217. Jurd, L. *Tetrahedron* **23** (1967) 1057–1064.
218. Jurd, L. *J. Heterocycl. Chem.* **18** (1981) 429–430.
219. Jurd, L. and Waiss Jr., A.C. *Tetrahedron* **21** (1965) 1471–1483.
220. Jurd, L. *J. Food Sci.* **29** (1964) 16–19.
221. Timberlake, C.F. and Bridle, P. *J. Sci. Food Agricol.* **18** (1967) 479–485.
222. Brouillard, R. and El Hage Chahine, J.M. *Bull. Liaison-Groupe Polyphenols* **9** (1980) 77–78.
223. Brouillard, R. and El Hage Chahine, J.M. *J. Am. Chem. Soc.* **102** (1980) 5375–5378.
224. Bridle, P., Scott, K.G. and Timberlake, C.F. *Phytochem.* **12** (1973) 1103–1106.
225. Timberlake, C.F. and Bridle, P. *J. Sci. Food Agricol.* **28** (1977) 539–544.
226. Green, R. and Mazza, G. *29th Annu. Conf. Can. Inst. Food Sci. Technol.*, Calgary, June 1986.
227. Esselen, W.B. and Sammy, G.M. *Food Prod. Dev.* **9** (1975) 37–38, 40.
228. Main, J.H., Clydesdale, F.M. and Francis, F.J. *J. Food Sci.* **43** (1978) 1693–1694, 1697.
229. Bronnum-Hansen, K. and Flink, J.M. *J. Food Technol.* **20** (1985) 725–733.
230. MacKinney, G., Lukton, A. and Chichester, C.O. *Food Technol.* **9** (1955) 324–326.
231. Debicki-Pospisil, J., Lovric, T., Trinajstic, N. and Sabljic, A. *J. Food Sci.* **48** (1983) 411–416.
232. Hodge, J.E. *J. Agricol. Food Chem.* **1** (1953) 928–943.
233. Kurata, T. and Sakurai, Y. *Agricol. Biol. Chem.* **31** (1967) 170–176.
234. Sloan, J.L., Bills, D.D. and Libbey, L.M. *J. Agricol. Food Chem.* **17** (1969) 1370–1372.
235. Asen, S., Stewart, R.N., Norris, K.H. and Massie, D.R. *Phytochem.* **9** (1970) 619–627.
236. Bayer, F., Fink, A., Nether, K. and Wegmann, K. *Angew Chem., Inter. Ed. Engl.* **5** (1966) 791–798.
237. Lowry, J.B. and Chew, L. *Econ. Bot.* **28** (1974) 61–62.
238. Osawa, Y. in *Anthocyanins as Food Colors* (P. Markakis, Ed.), Academic Press, New York (1982) Chapter 2.
239. Hoshino, T., Matsumoto, U. and Goto, T. *Phytochem.* **19** (1980) 663–667.
240. Chen, L. and Hrazdina, G. *Phytochem.* **20** (1981) 297–303.
241. Williams, M. and Hrazdina, G. *J. Food Sci.* **44** (1979) 66–68.
242. Somers, T.C. and Evans, M.E. *J. Sci. Food Agricol.* **30** (1979) 623–633.

243. Goto, T., Hoshino, T. and Takase, S. *Tetrahedron Lett.* **20** (1979) 2905–2908.
244. Hoshino, T., Matsumoto, U. and Goto, T. *Phytochem.* **20** (1981) 1971–1976.
245. Hoshino, T., Matsumoto, U., Harada, N. and Goto, T. *Tetrahedron Lett.* **22** (1981) 3621–3624.
246. Hoshino, T., Matsumoto, U., Goto, T. and Harada, N. *Tetrahedron Lett.* **23** (1982) 433–436.
247. Yazaki, Y. *Bot. Mag. Tokyo* **89** (1976) 45–57.
248. Asen, S. *Acta Hortic.* **41** (1974) 57–68.
249. Scheffeldt, P. and Hrazdina, G. *J. Food Sci.* **43** (1978) 517–520.
250. Nakayama, T.O.M. and Powers, J.J. *Adv. Food Res.* **Suppl. 3** (1972) 193–199.
251. Timberlake, C.F. and Bridle, P. in *Sensory Quality in Foods and Beverages: Definition, Measurement and Control* (A.A. Williams and R.R. Atkins, Eds), Ellis Horwood, Chichester (1982) Chapter 3.2.
252. Jurd, L. and Asen, S. *Phytochem.* **5** (1966) 1263–1271.
253. Asen, S., Stewart, R.N. and Norris, K.H. *Phytochem.* **14** (1975) 2677–2682.
254. Hooper, F.C. and Ayres, A.D. *J. Sci. Food Agricol.* **1** (1950) 5–8.
255. Harborne, J.B. *J. Chromatogr.* **2** (1959) 581–604.
256. Chiriboga, C. and Francis, F.J. *J. Food Sci.* **38** (1973) 464–467.
257. Somers, T.C. *Nature* **209** (1966) 368–370.
258. Somers, T.C. *J. Sci. Food Agricol.* **18** (1967) 193–196.
259. Francis, F.J. in *Anthocyanins as Food Colors* (P. Markakis, Ed.), Academic Press, New York (1982) Chapter 7.
260. Harborne, J.B. *Phytochemical Methods. A Guide to Modern Techniques of Plant Analysis* 2nd edn, Chapman & Hall, London (1984).
261. Anderson, D.W., Julian, E.A., Kepner, R.E. and Webb, A.D. *Phytochem.* **9** (1970) 1569–1578.
262. Moore, A.B., Francis, F.J. and Clydesdale, F.M. *J. Food Protect.* **45** (1982) 738–743.
263. Moore, A.B., Francis, F.J. and Jason, M.E. *J. Food Protect.* **45** (1982) 590–593.
264. Esselen, W.B. and Sammy, G.M. *Food Prod. Dev.* **7** (1973) 80, 82, 86.
265. Philip, T. *J. Food Sci.* **39** (1974) 859.
266. Peterson, R.J. and Jaffe, E.B. *US Patent No.* 3484254 (1969).
267. Langston, M.S.K. *US Patent No.* 4500556 (1985).
268. Shewfelt, R.L. and Ahmed, E.M. *Food Prod. Dev.* **11** (1977) 52, 57–59, 62.
269. Shewfelt, R.L. and Ahmed, E.M. *J. Food Sci.* **43** (1978) 435–438.
270. Fuleki, T. and Francis, F.J. *J. Food Sci.* **33** (1968) 72–77.
271. Francis, F.J. *Hort. Sci.* **2** (1967) 170–171.
272. Asen, S. *J. Chromatogr.* **18** (1965) 602–603.
273. Quarmby, C. *J. Chromatogr.* **34** (1968) 52–58.
274. Barritt, B.H. and Torre, L.C. *J. Chromatogr.* **75** (1973) 151–155.
275. Andersen, O.M. and Francis, G.W. *J. Chromatogr.* **318** (1985) 450–455.
276. Wulf, L.W. and Nagel, C.W. *Am. J. Enol. Vitic.* **29** (1978) 42–49.
277. Bakker, J. and Timberlake, C.F. *J. Sci. Food Agricol.* **36** (1985) 1315–1324.
278. Sapers, G.M., Hicks, K.B., Burgher, A.M., Hargrave, D.L., Sondey, S.M. and Bilyk, A. *J. Am. Soc. Hort. Sci.* **111** (1986) 945–950.
279. Mazza, G. *J. Food Sci.* **51** (1986) 1260–1264.
280. Oszmianski, J. and Sapis, J.C. *J. Food Sci.* **53** (1988) 1241–1242.
281. Baj, A., Bombardelli, E., Gabetta, B. and Martinelli, E.M. *J. Chromatogr.* **279** (1983) 365–372.
282. Andersen, O.M. *J. Food Sci.* **50** (1985) 1230–1232.
283. Andersen, O.M. *J. Food Sci.* **52** (1987) 665–666, 680.
284. Fuleki, T. and Francis, F.J. *J. Food Sci.* **33** (1968) 266–274.
285. Chiriboga, C. and Francis, F.J. *J. Amer. Soc. Hort. Sci.* **95** (1970) 233–236.
286. Wrolstad, R.E. and Putnam, T.B. *J. Food Sci.* **34** (1969) 154–155.
287. Hrazdina, G. *J. Agricol. Food Chem.* **18** (1970) 243–245.
288. Wrolstad, R.E. and Struthers, B.J. *J. Chromatogr.* **55** (1971) 405–408.
289. van Teeling, C.G., Cansfield, P.E. and Gallop, R.A. *J. Chromatogr. Sci.* **9** (1971) 505–509.
290. Chandler, B.V. and Swain, T. *Nature* **183** (1959) 989.
291. Fuleki, T. and Francis, F.J. *J. Food Sci.* **33** (1968) 78–83.

292. Strack, D. and Mansell, R.L. *J. Chromatogr.* **109** (1975) 325–331.
293. Hrazdina, G. *Lebensm. -Wiss. u. Technol.* **8** (1975) 111–113.
294. Manley, C.H. and Shubiak, P. *Can. Inst. Food Sci. Technol. J.* **8** (1975) 35–39.
295. Anderson, R.A. and Sowers, J.A. *Phytochem.* **7** (1968) 293–301.
296. Goldstein, G. *J. Chromatogr.* **129** (1976) 466–468.
297. Li, K.C. and Wagenknecht, A.C. *J. Am. Chem. Soc.* **78** (1956) 979–980.
298. Daravingas, G. and Cain, R.F. *J. Food Sci.* **30** (1965) 400–405.
299. Francis, G.W. and Andersen, O.M. *J. Chromatogr.* **283** (1984) 445–448.
300. Harborne, J.B. *Phytochemical Methods*, Chapman & Hall, London (1973).
301. Ribereau-Gayon, J. *Plant Phenolics*, Oliver & Boyd, London (1972).
302. Harborne, J.B. *Biochem. J.* **70** (1958) 22–28.
303. Geissman, T.A., Jorgenson, E.C. and Harborne, J.B. *Chem. Ind.* (1953) (Dec.) 1389.
304. Harborne, J.B. *J. Chromatogr.* **1** (1958) 473–488.
305. Morton, A.D. *J. Chromatogr.* **28** (1967) 480–481.
306. Wilkinson, M., Sweeny, J.G. and Iacobucci, G.A. *J. Chromatogr.* **132** (1977) 349–351.
307. Nagel, C.W. and Wulf, L.W. *Am. J. Enol. Vitic.* **30** (1978) 111–116.
308. Camire, A.L. and Clydesdale, F.M. *J. Food Sci.* **44** (1979) 926–927.
309. Ballinger, W.E., Galetta, G.J. and Maness, E.P. *J. Am. Soc. Hort. Sci.* **104** (1979) 554–557.
310. Ballinger, W.E., Maness, E.P. and Ballington, J.R. *Can. J. Plant. Sci.* **62** (1982) 683–687.
311. Bronnum-Hansen, K. and Hansen, S.J. *J. Chromatogr.* **262** (1983) 385–392.
312. Karppa, J. *Lebensm. -Wiss. u. Technol.* **17** (1984) 175–176.
313. Bakker, J. and Timberlake, C.F. *J. Sci. Food Agricol.* **36** (1985) 1325–1333.
314. Bakker, J., Preston, N.W. and Timberlake, C.F. *Am. J. Enol. Vitic.* **37** (1986) 121–126.
315. Casteele, K.V., Geiger, H., Deloose, R. and van Sumere, C.F. *J. Chromatogr.* **259** (1983) 291–300.
316. Bakker, J. *Vitis* **25** (1986) 203–214.
317. Ballington, J.R., Ballinger, W.E. and Maness, E.P. *J. Am. Soc. Hort. Sci.* **112** (1987) 859–864.
318. Ballington, J.R., Ballinger, W.E., Maness, E.P. and Luby, J.J. *Can. J. Plant Sci.* **68** (1988) 241–246.
319. Ballington, J.R., Ballinger, W.E. and Maness, E.P. *Can. J. Plant Sci.* **68** (1988) 247–253.
320. Andersen, O.M. *J. Food Sci.* **54** (1989) 383–384, 387.
321. Lamikrana, O. *Food Chem.* **33** (1989) 225–237.
322. Werner, D.J., Maness, E.P. and Ballinger, W.E. *Hort. Sci.* **24** (1989) 488–489.
323. Hebrero, E., Santos-Buelga, C. and Rivas-Gonzalo, J.C. *Am. J. Enol. Vitic.* **39** (1988) 227–233.
324. Hebrero, E., Garcia-Rodriguez, C., Santos-Buelga, C. and Rivas-Gonzalo, J.C. *Am. J. Enol. Vitic.* **40** (1989) 283–291.
325. Lees, D.H. and Francis, F.J. *J. Food Sci.* **36** (1971) 1056–1060.
326. Blom, H. *Rev. Food Sci. Technol.* **7** (1983) 587–589.
327. Chandler, B.V. and Harper, K.A. *Aust. J. Chem.* **14** (1961) 586–595.
328. Goto, T., Takase, S. and Kondo, T. *Tetrahedron Lett.* **19** (1978) 2413–2416.
329. Goto, T., Kondo, T., Imagawa, H. and Miura, I. *Tetrahedron Lett.* **22** (1981) 3213–3216.
330. Goto, T., Kondo, T., Tamura, H., Imagawa, H., Iino, A. and Takeda, T. *Tetrahedron Lett.* **23** (1982) 3695–3698.
331. Cornus, G., Wyler, H. and Lauterwein, J. *Phytochem.* **20** (1981) 1461–1462.
332. Tamura, H., Kumaoka, Y. and Sugisawa, H. *Agricol. Biol. Chem.* **53** (1989) 1969–1970.
333. Saito, N., Timberlake, C.F., Tucknott, O.G. and Lewis, I.A.S. *Phytochem.* **22** (1983) 1007–1009.
334. Bombardelli, E., Bonati, A., Gabetti, B., Martinelli, E.M., Mustich, G. and Danieli, B. *J. Chromatogr.* **120** (1976) 115–122.
335. Bombardelli, E., Bonati, A., Gabetti, B., Martinelli, E.M. and Mustich, G. *J. Chromatogr.* **139** (1977) 111–120.
336. Ribereau-Gayon, P.R. and Josien, M.L. *Bull. Soc. Chim. Fr.* (1960) 934–937.
337. Statoua, A., Merlin, J.C., Del Haye, M. and Brouillard, R. *Raman Spectrosc., Proc. 8th Int. Conf.* (1982) 629–630.

338. Wrolstad, R.E. *Color and Pigment Analyses in Fruit Products. Bull. 624*. Oregon Agricol. Exp. Sta., Corvallis (1976).
339. Ponting, J.D., Sanshuck, W.D. and Brekke, J.E. *Food Res.* **25** (1960) 471–478.
339a. Jurd, L. *J. Org. Chem.* **28** (1963) 987–991.
340. Niketic-Aleksic, G.K. and Hrazdina, G. *Lebensm. -Wiss. u. Technol.* **5** (1972) 163–165.
341. Dickinson, D. and Gawler, J.H. *J. Sci. Food Agricol.* **7** (1956) 699–705.
342. Somers, T.C. and Evans, M.E. *J. Sci. Food Agricol.* **25** (1974) 1369–1379.
343. Swain. T. and Hillis, W.E. *J. Sci. Food Agricol.* **10** (1959) 63–68.
344. Sondheimer, E. and Kertesz, Z.I. *Anal. Chem.* **20** (1948) 245–248.
345. Tanchev, S. *Z. Lebensm. u. Forsch.* **150** (1972) 28–30.
346. Tanchev, S. *Nahrung* **18** (1974) 303–308.
347. Lees, D.H. and Francis, F.J. *Hort. Sci.* **7** (1972) 83–84.
348. Godek, S. *Przemysl Fermentacyjny i Owocowo-Warzywny* **25** (1981) 27–30.
349. Speers, R.A., Tung, M.A. and Jackman, R.L. *Can. Inst. Food Sci. Technol. J.* **20** (1987) 15–18.
350. Spayd, S.E. and Morris, J.R. *J. Food Sci.* **46** (1981) 414–418.
351. Staples, L.C. and Francis, F.J. *Food Technol.* **22** (1968) 611–614.
352. van Buren, J.P., Hrazdina, G. and Robinson, W.B. *J. Food Sci.* **39** (1974) 325–328.
353. Torre, L.C. and Barritt, B.H. *J. Food Sci.* **42** (1977) 488–490.
354. Williams, M., Hrazdina, G., Wilkinson, M.M., Sweeny, J.G. and Iacobucci, G.A. *J. Chromatogr.* **155** (1978) 389–398.
355. Jurd, L. *Food Technol.* **18** (1964) 559–561.
356. Clydesdale, F.M., Main, J.H. and Francis, F.J. *J. Food Protect.* **42** (1979) 196–201.
357. Clydesdale, F.M., Main, J.H. and Francis, F.J. *J. Food Protect.* **42** (1979) 204–207.
358. Buckmire, R.E. and Francis, F.J. *J. Food Sci.* **43** (1978) 908–911.
359. Francis, F.J. *Food Technol.* **29** (1975) 52, 54.
360. Anonymous. *Food Chem.* **5** (1975) 69–80.
361. Martinez, S.B. and Del Valle, M.J. *UP Home Econ. J.* **9** (1981) 7–10.
362. Coudonis, M., Katsaboxakis, K. and Papanicolaou, D. *Rev. Food Sci. Technol.* **7** (1983) 567–572.
363. Baker, C.H., Johnston, M.R. and Barber, W.D. *Food Prod. Dev.* **8** (1974) 83–84, 86–87.
364. Baker, C.H., Johnston, M.R. and Barber, W.D. *Food Prod. Dev.* **8** (1974) 65–70.
365. Anonymous. *Enocolor. The Importance of the Natural Colour.* Reggiana Antociani, Industria Coloranti Naturali, Italy (1981).
366. Taylor, A.J. in *Developments in Food Colours—2* (J. Walford, Ed.), Applied Science Pub., London (1984) Chapter 5.
367. Lancrenon, X. *Process Biochem.* **16** (1978) 16.
368. Matsumoto, T., Nishida, K., Noguchi, M. and Tamaki, E. *Agricol. Biol. Chem.* **37** (1973) 561–567.
369. Dougal, D.K. and Weyrauch, K.W. *Biotech. Bioeng.* **22** (1980) 337–352.
370. Yamakawa, T., Kato, S., Ishida, K., Kodama, T. and Minoda, Y. *Agricol. Biol. Chem.* **47** (1983) 2185–2192.
371. Yamamoto, Y., Kinoshita, Y., Watanabe, S. and Yamada, Y. *Agricol. Biol. Chem.* **53** (1989) 417–423.
372. Bell, E.R.J. and White, E.B. *Inter. Ind. Biotech.* **9** (1989) 20–26.
373. Counsell, J.N., Jeffries, G.S. and Knewstubb, C.J. *Natural Colors for Food and Other Uses* (J.N. Counsell, Ed.), Applied Science Publishers, New York (1981) Chapter 7.
374. Maccarone, E., Maccarone, A. and Rapisarda, P. *J. Food Sci.* **50** (1985) 901–904.
375. Piattelli, M., Minale, L. and Prota, G. *Ann. Chim. (Rome)* **54** (1964) 955–962.
376. Wilcox, M.E., Wyler, H., Mabry, T.J. and Dreiding, A.S. *Helv. Chim. Acta* **48** (1965) 252–258.
377. Piattelli, M., Minale, L. and Prota, G. *Ann. Chim. (Rome)* **54** (1964) 963–968.
378. Piattelli, M. and Minale, L. *Ann. Chim. (Rome)* **56** (1966) 1060–1064.
379. Wyler, H., Rosler, H., Mercier, M. and Dreiding, S. *Helv. Chim. Acta* **50** (1967) 545–561.
380. Minale, L., Piattelli, M., de Stefano, S. and Nicolaus, R.A. *Phytochem.* **5** (1966) 1037–1052.
381. Piattelli, M. and Impellizzeri, G. *Phytochem.* **8** (1969) 1595–1596.
382. Minale, L., Piattelli, M. and de Stefano, S. *Phytochem.* **6** (1967) 703–709.

383. Piattelli, M. and Imperato, F. *Phytochem.* **9** (1970) 455–458.
384. Piattelli, M. and Impellizzeri, G. *Phytochem.* **9** (1970) 2553–2556.
385. Strack, D., Engel, U. and Wray, V. *Phytochem.* **26** (1987) 2399–2400.
386. Alard, D., Wray, V., Grotjahn, L., Reznik, H. and Strack, D. *Phytochem.* **24** (1985) 2383–2385.
387. Liebisch, H.W. in *Biochemistry of Alkaloids* (K. Mothes, H.R. Schutte and M. Luckner, Eds.), VEB Deutscher Verlag der Wissenshaften, Berlin (1985) Chapter 14.
388. Piattelli, M., Minale, L. and Prota, G. *Tetrahedron* **20** (1964) 2325–2329.
389. Piattelli, M., Minale, L. and Nicholaus, R.A. *Rend. Accad. Sci. Fis. Mat., Naples* **32** (1965) 55–56.
390. Piattelli, M., Minale, L. and Prota, G. *Phytochem.* **4** (1965) 121–125.
391. Piattelli, M., Minale, L. and Nicolaus, R.A. *Phytochem.* **4** (1965) 817–823.
392. Impellizzeri, G., Piattelli, M. and Sciuto, S. *Phytochem.* **12** (1973) 2293–2294.
393. Strack, D., Schmitt, D., Reznik, H., Boland, W., Grotjahn, L. and Wray, V. *Phytochem.* **26** (1987) 2285–2287.
394. Musso, H. *Tetrahedron* **35** (1979) 2843–2853.
395. Piattelli, M. in *The Biochemistry of Plants. Vol. 7. Secondary Plant Products* (P.K. Stumpf and E.E. Conn, Eds.), Academic Press, New York (1981) Chapter 18.
396. von Elbe, J.H. and Maing, I.-V. *Cereal Sci. Today* **18** (1973) 263–264.
397. Impellizzeri, G. and Piattelli, M. *Phytochem.* **11** (1972) 2499–2502.
398. Chang, C., Kimler, L. and Mabry, T.J. *Phytochem.* **13** (1974) 2771–2775.
399. Sciuto, S., Oriente, G. and Piattelli, M. *Phytochem.* **11** (1972) 2259–2262.
400. Piattelli, M., Giudici de Nicola, M. and Castrogiovanni, V. *Phytochem.* **10** (1971) 289–293.
401. Wohlpart, A. and Mabry, T.J. *Plant Physiol.* **43** (1968) 457–459.
402. Giudici de Nicola, M., Piattelli, M. and Amico, V. *Phytochem.* **12** (1973) 2163–2166.
403. Giudici de Nicola, M., Amico, V., Sciuto, S. and Piattelli, M. *Phytochem.* **14** (1975) 479–481.
404. Schwartz, S.J. and von Elbe, J.H. *Z. Lebensm. u. Forsch.* **176** (1983) 448–453.
405. von Elbe, J.H., Maing, I.-V. and Amundson, C.H. *J. Food Sci.* **39** (1974) 334–337.
406. Pasch, J.H. and von Elbe, J.H. *J. Food Sci.* **40** (1975) 1145–1146.
407. Saguy, I.J. *J. Food Sci.* **44** (1979) 1554–1555.
408. Huang, A.S. and von Elbe, J.H. *J. Food Sci.* **52** (1987) 1689–1693.
409. Attoe, E.L. and von Elbe, J.H. *J. Agricol. Food Chem.* **30** (1982) 708–712.
410. Saguy, I., Kopelman, I.J. and Mizrahi, S. *J. Agricol. Food Chem.* **26** (1978) 360–362.
411. Attoe, E.L. and von Elbe, J.H. *J. Food Sci.* **46** (1981) 1934–1937.
412. Savolainen, K. and Kuusi, T. *Z. Lebensm. u. Forsch.* **166** (1978) 19–22.
413. von Elbe, J.H., Schwartz, S.J. and Hildenbrand, B.E. *J. Food Sci.* **46** (1981) 1713–1715.
414. Huang, A.S. and von Elbe, J.H. *J. Food Sci.* **50** (1985) 1115–1120, 1129.
415. Bilyk, A. and Howard, M. *J. Agricol. Food Chem.* **30** (1982) 906–908.
416. Attoe, E.L. and von Elbe, J.H. *Z. Lebensm. u. Forsch.* **179** (1984) 232–236.
417. Sapers, G.M. and Hornstein, J.S. *J. Food Sci.* **44** (1979) 1245–1248.
418. Aurstad, K. and Dahle, H.K. *Z. Lebensm. u. Forsch.* **151** (1972) 171–174.
419. Mikova, K. and Kyzlink, V. *Sbornik Vysokeskoly Chemicko-Technologicke v. Praze, E.* **58** (1985) 9–15.
420. Pasch, J.H. and von Elbe, J.H. *J. Food Sci.* **44** (1979) 72–74.
421. Elliott, D.C., Schultz, C.G. and Cassar, R.A. *Phytochem.* **22** (1983) 383–387.
422. Soboleva, G.A., Ul'yanova, M.S., Zakharova, N.S. and Bokuchava, M.A. *Biokhimiya* **41** (1976) 968–973.
423. Lashley, D. and Wiley, R.C. *J. Food Sci.* **44** (1979) 1568–1569.
424. Shih, C.C. and Wiley, R.C. *J. Food Sci.* **47** (1981) 164–166, 172.
425. Wasserman, B.P. and Guilfroy, M.P. *Phytochem.* **22** (1983) 2657–2660.
426. Wasserman, B.P., Eiberger, L.L. and Guilfroy, M.P. *J. Food Sci.* **49** (1984) 536–538, 557.
427. Wasserman, B.P. and Guilfroy, M.P. *J. Food Sci.* **49** (1984) 1075–1077, 1084.
428. Smith, T.A. and Croker, S.J. *Phytochem.* **24** (1985) 2436–2437.
429. Wiley, R.C. and Lee, Y.-N. *J. Food Sci.* **43** (1978) 1056–1058.
430. Wiley, R.C., Lee, Y.-N., Saladini, J.J., Wyss, R.C. and Topalain, H.H. *J. Food Sci.* **44** (1979) 208–212.

431. Piattelli, M. and Minale, L. *Phytochem.* **3** (1964) 307–311.
432. Piattelli, M. and Minale, L. *Phytochem.* **3** (1964) 547–557.
433. Adams, J.P. and von Elbe, J.H. *J. Food Sci.* **42** (1977) 410–414.
434. Wilkins, C.K. *Internat. J. Food Sci. Technol.* **22** (1987) 571–573.
435. Bilyk, A. *J. Food Sci.* **46** (1981) 298–299.
436. Nilsson, T. *Lantbrukshogskolans Annaler* **36** (1970) 179–219.
437. Dopp, H. and Musso, H. *Z. Naturforsch.* **29C** (1974) 640–642.
438. Singer, J.W. and von Elbe, J.H. *J. Food Sci.* **45** (1980) 489–491.
439. Powrie, W.D. and Fennema, O.J. *J. Food Sci.* **28** (1963) 214–220.
440. Lindstedt, G. *Acta Chem. Scand.* **10** (1956) 689–699.
441. Wyler, H. *Chem. Unserer Zeit* **3** (1969) 111–115.
442. Wyler, H. *Chem. Unserer Zeit* **3** (1969) 146–151.
443. Kremer, B.P. *Biol. Unserer Zeit* **5** (1975) 155–157.
444. von Elbe, J.H., Sy, S.H., Maing, I.-Y. and Gabelman, W.H. *J. Food Sci.* **37** (1972) 932–934.
445. Vincent, K.R. and Scholz, R.G. *J. Agricol. Food Chem.* **26** (1978) 812–816.
446. Schwartz, S.J. and von Elbe, J.H. *J. Agricol. Food Chem.* **28** (1980) 540–543.
447. Strack, D. and Reznik, H. *Z. Pflanzenphysiol.* **94** (1979) 163–167.
448. Strack, D., Engel, D. and Reznik, H. *Z. Pflanzenphysiol.* **101** (1981) 215–222.
449. Pourrat, A., Lejeune, B., Grand, A. and Pourrat, H. *J. Food Sci.* **53** (1988) 294–295.
450. Drdak, M., Vallova, M., Daucik, P. and Greif, G. *Z. Lebensm. u. Forsch.* **188** (1989) 547–550.
451. Wilcox, M.E., Wyler, H. and Dreiding, A.S. *Helv. Chim. Acta* **48** (1965) 1134–1147.
452. Wyler, H. and Dreiding, A.S. *Helv. Chim. Acta* **42** (1959) 1699–1702.
453. Schmidt, O.Th., Becher, P. and Hubnerr, M. *Chem. Ber.* **93** (1960) 1296–1304.
454. Mabry, T.J., Wyler, H., Sassy, G., Mercier, M., Parikh, I. and Dreiding, A.S. *Helv. Chim. Acta* **45** (1962) 640–647.
455. Mabry, T.J., Wyler, H., Parikh, I. and Dreiding, A.S. *Tetrahedron* **23** (1967) 3111–3127.
456. Piattelli, M., Giudici de Nicola, M. and Castrogiovanni, V. *Phytochem.* **8** (1969) 731–736.
457. Wyler, H. and Dreiding, A.S. *Helv. Chim. Acta* **44** (1961) 249–257.
458. Pourrat, A., Lejeune, B., Jean, D. and Pourrat, H. *Ann. Pharm. Fr.* **38** (1980) 261–266.
459. Pourrat, H., Lejeune, B., Regerat, F. and Pourrat, A. *Biotech. Lett.* **5** (1983) 381–384.
460. Adams, J.P., von Elbe, J.H. and Amundson, C.H. *J. Food Sci.* **41** (1976) 78–81.
461. Avila, A., Garcia-Hernandez, F. and Santos, E. *Dev. Ind. Microbiol.* **24** (1983) 553–561.
462. Constabel, F. and Nassif-Makki, H. *Ber. Dtsch. Bot. Ges.* **84** (1971) 629–636.
463. Weller, T.A. and Lasure, L.L. *J. Food Sci.* **47** (1981) 162–163.
464. Lejeune, B., Pouget, M.P. and Pourrat, A. *Labo-Pharma* **334** (1983) 638–643.
465. Lejeune, B., Pourrat, A. and Pouget, M.P. *Ann. Pharm. Fr.* **44** (1986) 461.
466. Lejeune, B., Grand, A. and Pourrat, A. *S.T.P. Pharma* **3** (1987) 400.
467. Pasch, J.H., von Elbe, J.H. and Sell, R.J. *J. Milk Food Technol.* **38** (1975) 25–28.
468. von Elbe, J.H., Klement, J.T., Amundson, C.H., Cassens, R.G. and Lindsay, R.C. *J. Food Sci.* **39** (1974) 128–132.
469. Dhillon, A.S. and Mauer, A.J. *Poultry Sci.* **54** (1975) 1272–1277.
470. Kopelman, I.J. and Saguy, I. *J. Food Proc. Pres.* **1** (1977) 217–224.

7 Miscellaneous colorants

F.J. FRANCIS

7.1 Acylated β-ring substituted anthocyanins

For many years anthocyanins were believed to be sugar-substituted only in the A-ring. This concept was surprising since β-ring substitution was well-known in the closely related yellow flavonoid group. However, the first β-ring sugar-substituted anthocyanin was isolated by Yoshitama [1] from *Lobelia* (Figure 7.1). It was delphinidin 3-rutinoside-5,3′,5′-triglucoside acylated with caffeic and *p*-coumaric acids. Since then, similar pigments have been found in *Tradescantia*, *Bromeliaceae*, *Commelinaceae*, *Compositae*, *Gentianaceae*, *Leguminoseae*, *Liliaceae*, and *Lobeliaceae* [2]. This group of pigments usually has a complex structure with several sugar and acyl groups. For example, ternatin (see Fig. 7.1) from *Clitoria ternatea* is delphinidin 3,3′,5′-triglucoside with four more sugar units, four molecules of *p*-coumaric acid and one molecule of malonic acid [3]. With a molecular weight of approximately 2310, it has replaced the anthocyanin from Heavenly Blue morning glories (*Ipomoea tricolor*) [4] as the largest monomeric anthocyanin. Figure 7.1 presents the structure of some selected highly acylated anthocyanins; however, there are some disagreements in structure in the literature. For example, Stirton and Harborne [5] reported that zebrinin had three caffeic acid units whereas Idaka *et al.* [6] reported the presence of four units. Idaka *et al.* also reported that setcreasin from *Setcreasea purpurea* contained cyanidin-3,7,3^1-triglucoside with four ferulic acid units. Shi *et al.* [7] reported that *Tradescantia pallida* anthocyanin (TPA) from *T. pallida* (a synonym of *S. purpurea*) contained the same basic structure except with three units of ferulic acid and one extra glucose molecule. The number of complex acylated anthocyanins is likely to increase in view of more sophisticated methods of structure determination. The methods of structure determination that work so well on the simpler anthocyanins [8] are not adequate to determine the structure of the highly acylated anthocyanins. The methods of choice today are high-performance liquid chromatography (HPLC), for clean separation, with concurrent three-dimensional photodiode array detection in order to yield good spectra. The number, identity and points of attachment of the acyl molecules can be determined by mass spectroscopy of the intact molecule and selected products of hydrolysis. The newer physical methods of analysis such as fast atom-bombardment mass spectroscopy (FAB-MS), COSY (H-H correlated spec-

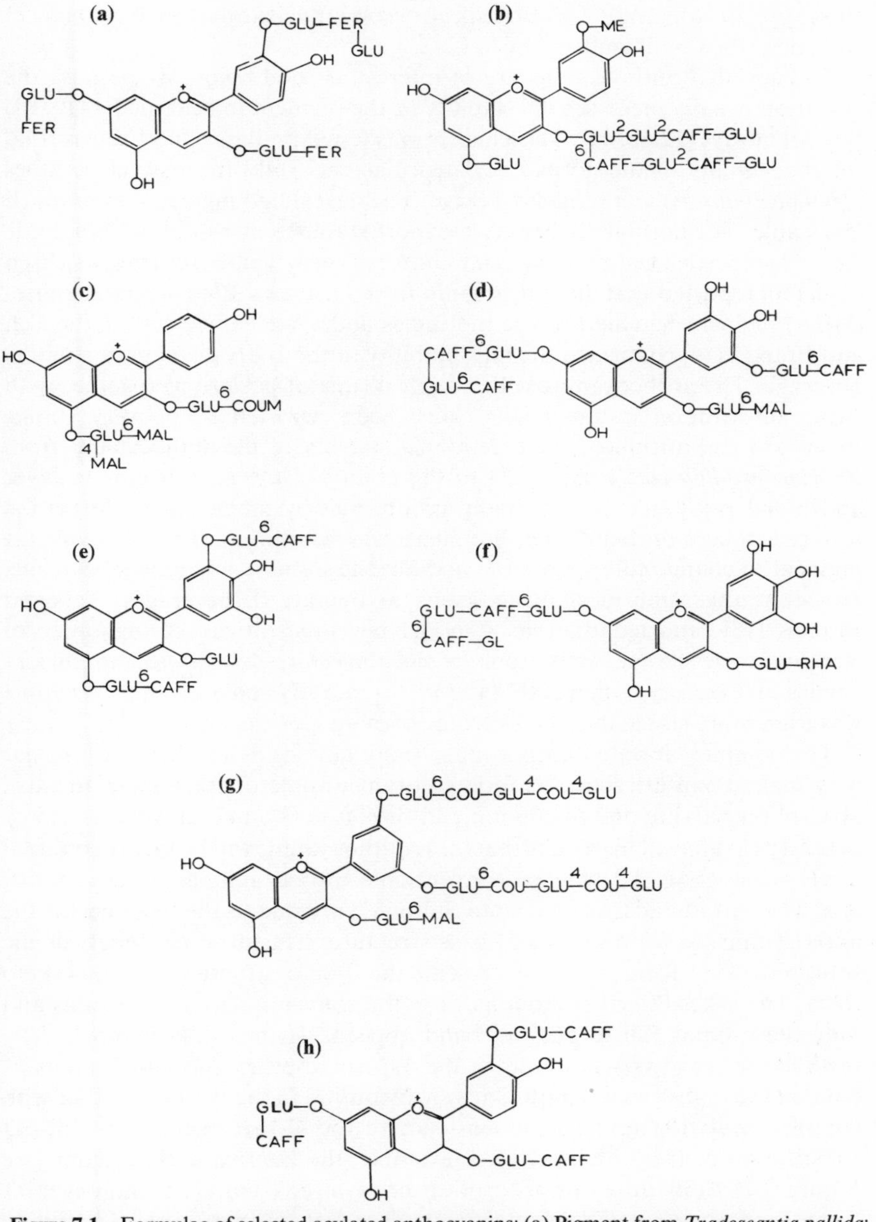

(a) (b) (c) (d) (e) (f) (g) (h)

Figure 7.1 Formulae of selected acylated anthocyanins: (a) Pigment from *Tradescantia pallida*; (b) HBA from morning glories; (c) Monardein; (d) Cinerarin; (e) Gentiodelphin; (f) Platyconin; (g) Ternatin D; and (h) Zebrinin. Abbreviations Glu and Rha refer to the sugars glucose and rhamnose. The abbreviations Caff, Coum, Fer and Mal refer to the acyl acids caffeic, coumaric, ferulic and malonic acid. The superscript refers to the carbon linkage of the sugar acid (*e.g.* 6). HBA = heavenly blue anthocyanin.

troscopy), proton and ^{13}C NMR spectroscopy (nuclear magnetic resonance) were described by Brouillard [9].

The acylated anthocyanins are of interest as food colorants because the acylation usually increases the stability of the pigment in acid media [10, 11] but not always [12a,b, 13]. This concept was tested by Bassa and Francis [14a] for the stability of the normal acylated pigments [14b] from sweet potatoes (*Ipomoea batatas*) in a model beverage. The deacylated pigments were much less stable. The normal pigments were more stable than cyanidin-3-glucoside from blackberries and a commercial sample of oenocyanin from grapes. Asen *et al.* [10] reported that an anthocyanin from Heavenly Blue morning glories (HBA) pigment had six glucose molecules and three caffeic acid units. Teh and Francis [15] confirmed the high stability of the HBA pigment in a model beverage. The anthocyanins with a high degree of acylation, together with sugar substitution in the β-ring, have been reported to be very stable. Brouillard [16] attributed the exceptional stability of the anthocyanins from *Zebrina pendula* (see Figure 7.1) to the ability of the acyl groups to layer above and below the anthocyanidin group, thereby inhibiting the formation of a pseudobase or a chalcone. Brouillard was referring to the stability of the pigment to change colour with pH and not the stability in a model beverage. However, the same mechanism seems to operate in beverages. Teh and Francis [15] reported that, in a model beverage, in decreasing order of stability, were the pigments from *Ipomoea tricolor*, *Zebrina pendula*, blackberries and oenocyanin. Terahara *et al.* [3] recently reported that ternatin I was even more stable than HBA from morning glories.

The complex acylated anthocyanins may also have another feature that may make them attractive as food colorants. Some of them have an extra absorption band at 560 to 600 nm and/or 600 to 640 nm, at weakly acid or neutral pH values. This means that these anthocyanins will be highly coloured at pH values above 4.0, whereas conventional anthocyanins are nearly colourless. The extra bands, according to Saito [17] are due to the tendency of the anthocyanins to form quinoidal base structures at neutral or weakly acidic solutions. Two forms may be present, the 7-keto structure or the 4′-keto form. The 7-keto form predominates in the conventional anthocyanins and only the normal 520 to 560 nm band appears. If the 7-OH is blocked by methylation or glycosylation, then the 4′-form appears; this shows an extra band at the higher wavelength. Sugar substitution in the B-ring together with complex acylation favours the existence of the 4′-keto form even without substitution of the 7 hydroxyl. For example, the *Zebrina* anthocyanins (see Figure 7.1) show an extra absorption band at 587 nm. The anthocyanins from *Tradescantia pallida* show an extra band at 583 nm at pH 5.5 (Figure 7.2). Platyconin and cinerarin show the extra band. Gentiodelphin [18] shows only a shoulder at pH 5.5 and 560 nm, because the 7 position is free and only one acyl sugar group is in the 3′ position. The HBA anthocyanin does not show the extra bands because all the acyl sugars are on the 3 position.

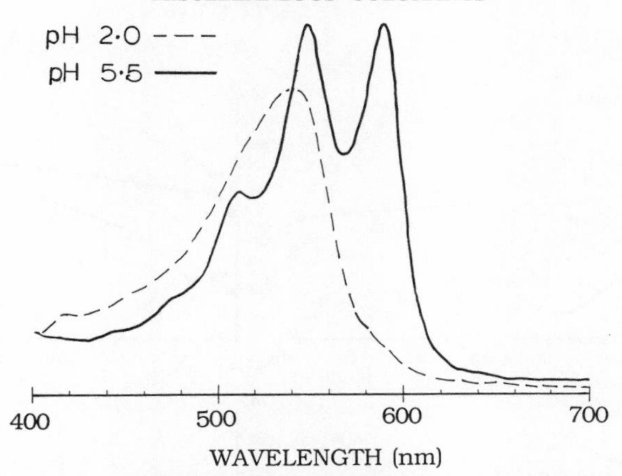

Figure 7.2 Absorption spectra of the major anthocyanin in *Tradescantia pallida* at pH values 2.0 and 5.5.

Figure 7.3 shows typical stability data for the anthocyanins from *T. pallida* at pH 3.5 and 5.5 [19]. Colour data for a beverage is appropriately expressed using Hunter tristimulus data. The data for Hunter L *versus* storage time for *T. pallida* are relatively stable at both pH values. The values for Theta[1] are redder than those for cyanidin-3-glucoside (Cn-3-G) or oenocyanin. The pigment retention data show that the *T. pallida* pigments are more stable particularly at higher pH values. This is not surprising since conventional anthocyanins are known to be unstable and nearly colourless at pH values above 4.0.

The presence of an absorption band at 583 nm and pH 5.5 and its absence at pH 2.0 suggests that the difference in absorption at the two pH values would provide a simple analytical method. Min *et al.* [20] proposed that a single absorption method at pH 5.5 and 583 nm be used for fresh tissue and the difference in absorption at pH 5.5 and 2.0 be used for stored samples. The $A^{1\%}_{1\,cm}$ 583 nm values were 111 and 97 respectively. The anthocyanin content of *T. pallida*, using this method was reported to be 120 mg/100 g [20].

Several sources of acylated pigments have been patented as potential food colorants. These include morning glories [21, 22], red cabbage [23–33], *Gibasis geniculata* [34], *Zebrina purpusii* [35] and sweet potatoes [36]. The patent literature for all food colorants from 1969 to 1984 was surveyed by Francis [37]. Murai and Wilkins [38] and LaBell [39] described the recent introduction of a stable red colorant (trade name San Red RC) from red cabbages.

[1] Theta is the angle that a line joining a point in Hunter space makes with the origin. In this work, a lower Theta value indicates a redder sample.

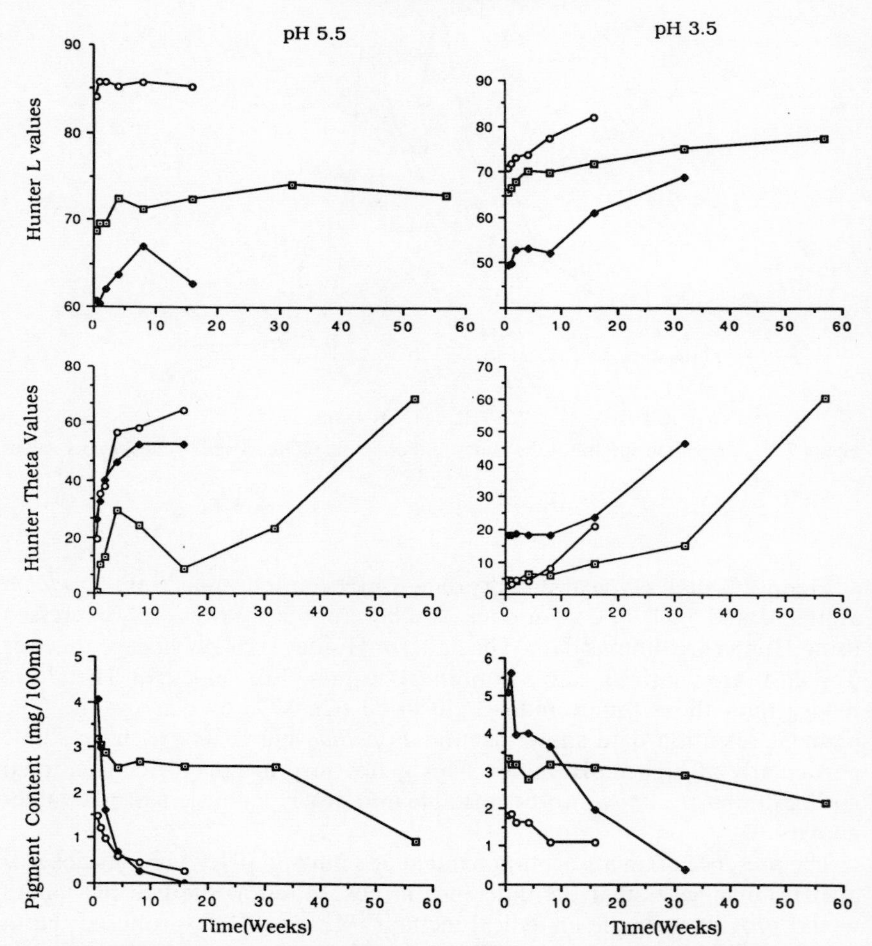

Figure 7.3 Typical stability data for the anthocyanins from *T. pallida* (□), blackberries (Cn-3-G) (◆), and oenocyanin (○) from grapes. The Hunter L value is a measure of lightness with white = 100 and black = 0. Theta is a function of hue. In these examples, which are all in the +a, +b Hunter quadrant, a theta = 0 represents a bluish-red colour whereas 90 indicates a yellow colour.

7.2 Annatto

Annatto is a yellow-orange carotenoid preparation obtained from the seeds of the plant *Bixa orellana*, which are grown primarily in Central and South America. The pigments in annatto are a mixture of bixin, the mono-methyl ester of a dicarboxylic carotenoid compound, and norbixin, the dicarboxylic derivative of the same carotenoid as in bixin (Figure 7.4). Both bixin and norbixin normally occur in the *cis* form and a small percentage of both is

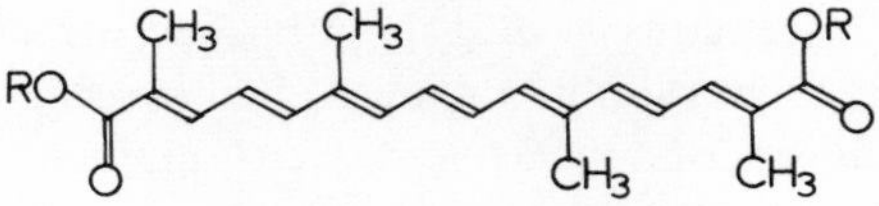

CROCIN R = gentiobiose
CROCETIN R = H

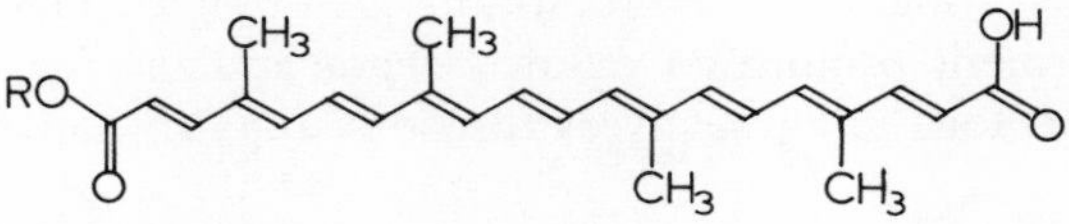

trans-BIXIN

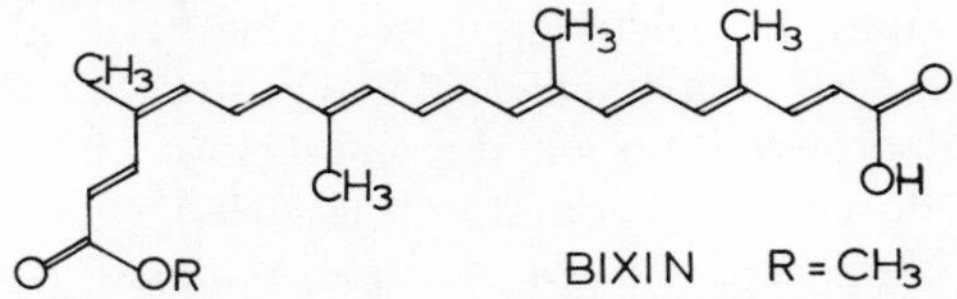

BIXIN

BIXIN R = CH_3
NORBIXIN R=H

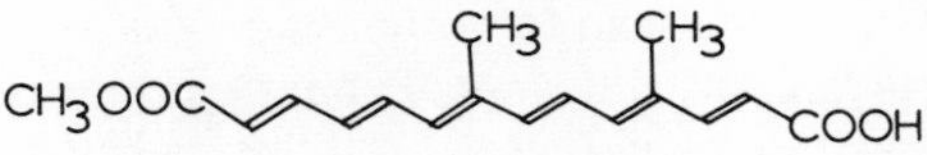

CH_3OOC ... COOH

C_{17} Yellow Pigment

Figure 7.4 Carotenoid pigments of annatto (bixin, norbixin, trans-bixin and C_{17} yellow pigment) and saffron (crocin and crocetin).

changed to the more stable *trans* form upon heating. A yellow degradation product, termed C17 yellow pigment, with the formula shown in Figure 7.4. is also produced on heating. The *cis* forms are redder than the *trans* forms or the yellow C17 compound, thus a series of yellow to red colours are available.

A colorant can be extracted from the seeds by treatment with water. Rozinkov *et al.* [40] patented a continuous countercurrent extraction with water while submitting the seeds to intensive mechanical friction. Bixin, on treatment with alkalis, is hydrolysed to the water-soluble dibasic acid form norbixin. Normal commercial procedure is to soften the seeds with steam and extract the pigments with propylene glycol containing potassium hydroxide. An oil-soluble form can be produced by treating steam-softened seeds with a solvent such as ethanol, a chlorinated hydrocarbon or a vegetable oil. The pigments are primarily in the outer coats of the seed and can also be removed mechanically. Guimares *et al.* [41] published a method to isolate pigment from seeds using agitation in a spouted bed holding 50 kg of seeds. The dust from the agitation contained over 20% of pigment. A dry powder is useful for colouring dry and instant foodstuffs.

The pigments in annatto can be separated and analysed using HPLC methods. Rouseff [42] published a method using a Zorbax ODX column with a 58/42 mixture of water/tetrahydrofuran. Earlier methods used paper chromatography or TLC plates [43] but usually required separate analyses for annatto and turmeric. The two colorants are often found together because addition of turmeric produces a yellower shade and also lends stability [44]. The HPLC methods allow analyses for both annatto and turmeric in the same sample.

The pigments of annatto are unstable to oxidation, as are all carotenoid compounds, but they are relatively more stable than the other carotenoid colorants used in food. Najar *et al.* [45] reported the effects of light, air, antioxidants and pro-oxidants on the stability of annatto extracts. Light proved to be the most destructive and 10% ascorbyl palmitate was an effective antioxidant. Heltiarachchy and Muffett [46] patented the use of inorganic polyvalent cations together with hydrocolloids containing carboxyl groups to produce stabilized colorant complexes with annatto. Ford and Mellor [47] patented a process to use demineralized water or demineralized glucose syrup together with 0.5% ascorbic acid to prevent fading of annatto in a beverage. Berset and Marty [48] reported the use of annatto in extrusion cooking. They were able to demonstrate good thermal stability at between 175 and 185°C. Ford and Draisey [49] patented the use of annatto together with cloudifier preparation containing beeswax and a natural gum to minimize fading of annatto in fruit squashes. Ikawa and Kagemoto [50] reported that a stable colorant could be made from bixin by addition of ethanol, sucrose acetate hexaisobutyrate, coconut oil and gum arabic. Mori *et al.* [51] reported stabilization of bixin with vanillin eugenol or vitamin E. Harrison [52] patented the production of a yellow and blue twin curd cheese by adding annatto to half the batch. Both batches were remelted, coagulated, mixed and ripened. The cheese had an attractive mottled appearance.

Annatto has been used in foods, especially dairy products, for many years. Technically it is a good colorant and the existence of both water-soluble and oil-soluble forms has provided flexibility. The worldwide movement towards natural colorants in the past two decades has led to a great increase in the use of annatto.

7.3 Saffron

Pigments similar to those in annatto are found in saffron. They are crocetin, a dicarboxylic carotenoid (Figure 7.4) together with its digentiobioside ester, crocin. Gentiobiose is a diglucoside with a beta 1-6 linkage. The sugar portion makes the molecule water soluble. It is one of the few carotenoids found in nature that is freely water soluble and this is one reason for its application in the food and pharmaceutical industries. Saffron is obtained from the stigmas of the flowers of *Crocus sativus*. The same pigments also occur in *C. albifloris*, *C. luteus*, *Cedrela toona*, *Nyctanthes arbor-tristis*, *Verbascum phlomoides*, and the Cape Jasmine (*Gardenia jasminoides*). Until recently, *Crocus sativus* was the only commercial source of crocin and crocetin and, since it requires about 64 000 stigmas to produce one pound of colorant, the price is obviously very high (>$500/lb). Also production is limited due to the cost of labour. Saffron, of course, provides both colour and flavour and is considered to be a gourmet item.

The principal substance responsible for the bitter taste of saffron is picrocrocin [53]. Picrocrocin hydrolyses to safranal (2,6,6-trimethyl-3-cyclohexadiene-carboxaldehyde) and glucose. Safranal is formed in saffron extracts during processing and is the main component responsible for the odour. Alonso *et al.* [53] studied the auto-oxidation of saffron at 40°C and 75% relative humidity using a spectrophotometric analysis. Picrocrocin, safranal and crocin absorb at 257, 297 and 327 nm respectively [54]. The loss of colour followed first-order kinetics and the loss of bitterness followed second-order kinetics. Solinas and Cichelle [55] were able to analyse for the colour (crocin and four similar compounds) and flavour (picrocrocin and safranal) using an HPLC approach. This method was effective in judging the quality and authenticity of saffron extracts.

Klaui *et al.* [56] patented the synthesis of di-n-dodecyl, di-n-decyl, and di-n-tetradecyl esters of crocetin. These compounds, together with alpha-tocopherol and ascorbyl palmitate provided a stable yellow colorant for ice-cream and beverages. The same concept was extended to the synthesis of several other esters [57] but the preferred compounds are listed above.

Saffron extracts contain other carotenoids such as β-carotene and zeaxanthin, as well as crocin and crocetin. They also contain, as might be expected, a range of flavonoid compounds. Garrido *et al.* [58] reported derivatives of myricetin, quercetin, kempferol, delphinidin and petunidin.

Saffron is a very old food colorant and spice. Its pure-yellow colour and odour has made it an attractive ingredient in specialty foods such as risotto, curry, bakery products, sausages, beverages, and so on. Its current price and availability will limit it to specialty gourmet foods.

7.4 Gardenia pigments

The technological success of the pigments in annatto and saffron combined with their relatively high price led to searches for other plant sources with the same pigments. It soon became obvious that the same pigments (but not

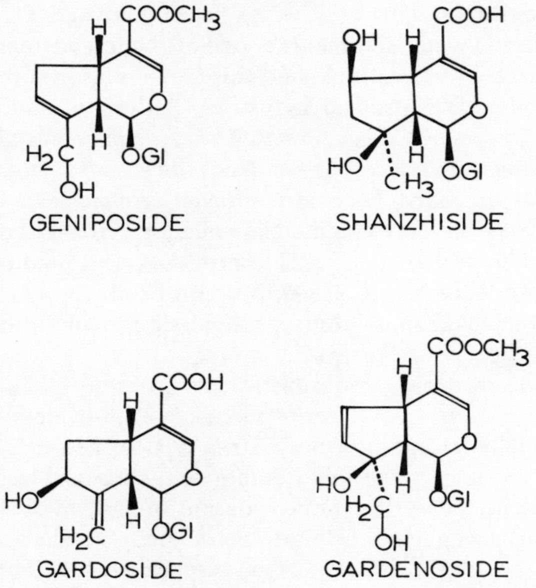

GENIPOSIDE SHANZHISIDE

GARDOSIDE GARDENOSIDE

Gl = glucose

Figure 7.5 Iridoid pigments of gardenia.

the flavour) could be obtained in much greater quantities and at a lower price from the fruits of the Cape Jasmine.

The fruits of gardenia contain three major groups of pigments, crocins, iridoid pigments and flavonoids. The carotenoid crocin and related compounds develop in the fruit of *G. jasminoides* [59] during the eighth to twenty-third weeks of growth but the iridoid pigments develop 1 to 6 weeks after flowering. The structure of the iridoid pigments gardenoside, geniposide, shanzhiside, gardoside, methyldeacetylasperuloside, genipingentiobioside, beniposidic acid, acetylgeniposide and scandoside methylester are shown in Figures 7.5 and 7.6 [60]. The third group comprises a series of flavonoid compounds as illustrated by the five flavonoids isolated by Gunatilaka *et al.* [61, 62] (Figure 7.7) from *G. fosbergii*. This is a different species from *G. jasminoides* but flavonoid pigments tend to be similar in closely related species. The flavonoid pigments are yellow but their contribution to the overall colour of gardenia preparations is unclear. Gardenia fruits and other portions of the plant have been a part of the folklore of Oriental culture for many years

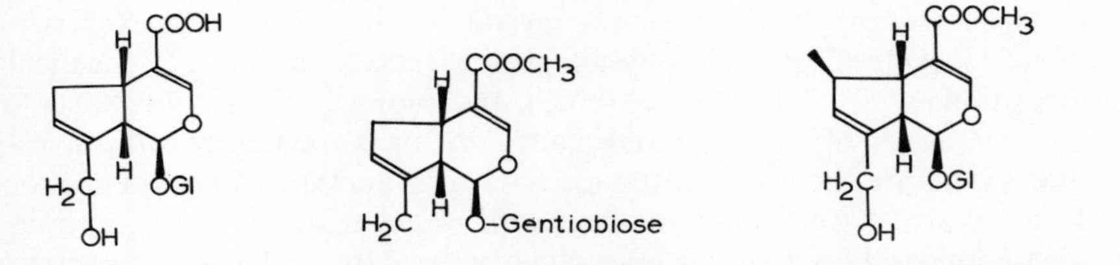

GENIPOSIDIC ACID

GENIPINGENTIOBIOSIDE

SCANDOSIDE METHYLESTER

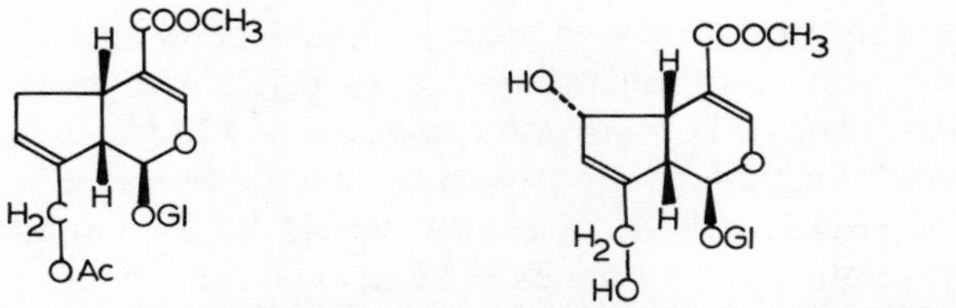

ACETYLGENIPOSIDE

METHYLDEACETYLASPERULOSIDE

Figure 7.6 Additional iridoid pigments of gardenia.

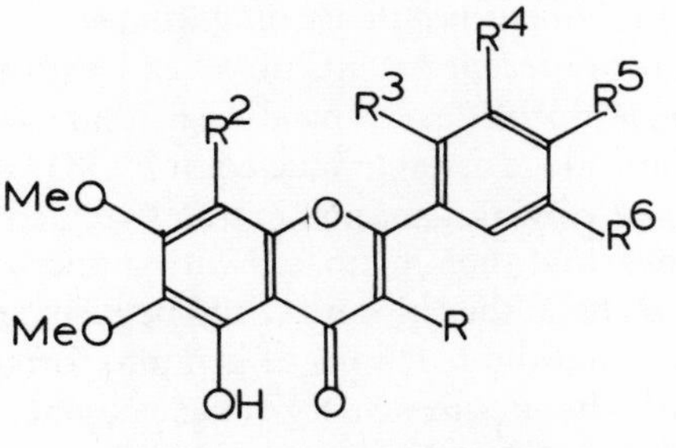

Flavonoid compounds

CPD.	R^1	R^2	R^3	R^4	R^5	R^6
1	H	H	H	OMe	OMe	OMe
2	OMe	H	H	OH	OMe	OH
3	H	H	OMe	OMe	H	OH
4	OMe	H	H	OMe	OMe	OH
5	OMe	OMe	H	H	OH	H

Figure 7.7 Flavonoid compounds of gardenia.

[63]. There appears to have been a revival of interest in recent years in terms of biosynthesis [64–69], physiological effects [60, 63, 70–72], chemical composition [59, 73–77], analysis [78, 79], and antimicrobial activity [80].

The production of food colorants from gardenia is apparently being investigated at the present time but the mechanism is unclear. Most patent procedures involve extraction of the fruit with water, treatment with enzymes having β-glucosidic activity or proteolytic activity (bromelain) and reaction with primary amines from either amino acids or proteins such as those in soy. Manipulation of the reaction conditions such as time, pH, temperature, oxygen content, degree of polymerization and conjugation of the primary amino reaction products, and so on, enables a series of colorants to be produced that vary from yellow to green, red, violet and blue. Several of the patents involve culturing preparations of gardenia with micro-organisms such as *Bacillus subtilis* [81], *Aspergillus japonicus* or a *Rhizopus* [82]. Depending on the conditions, preparations with the colour described above can be produced. The overall colour, of course, would be a combination of the yellow-orange contributed by the carotenoids and the colours produced by the iridoid pigments and their derivatives. Apparently, seven preparations involving four greens, two blues and one red have been commercialized in Japan. Flow sheets for their production together with their physical properties such as solubility in various solvents, heat and light stability, pH effects, hygroscopicity, reaction with metals, antioxidants, and so on, were published by Yoshizumi *et al.* [60]. Nineteen patents on gardenia colorants were published up to 1984 [37]. More recent patents involved hydrolysis of the iridoid glycoside geniposide by β-glucosidase to produce genipin. The genipin is then reacted with taurine to produce a stable blue colorant [83, 84]. Tomara *et al.* [85] patented a process involving gardenia extracts treated with acylase for hydrolysis of geniposides, and then incubated with a micro-organism. After enzyme inactivation with heat, the colorant could be recovered with ethanol. Fujikawa *et al.* [86] reported the reactions of genipin, formed by hydrolysis of the geniposides, with the amino acids glycine, alanine, leucine, phenylalanine and tyrosine. The resultant 'brilliant sky-blue' pigment was stable for 2 weeks at 40°C in 40% ethanol. Fujikawa *et al.* [87] also reported the continuous production of a blue pigment using *B. subtilis* cells immobilized on a calcium alginate gel. This micro-organism had β-glucosidase activity and assimilated genipin as a carbon source. The colorants from gardenia have been described in several reviews (*e.g.* Toyama and Moritome, [88]).

Colorant preparations from gardenia were suggested by Yoshizumi (1980a) for use with candies, sweets, ices, noodles, imitation crab (Kamabago), fish eggs, glazed chestnuts, boiled beans, dried fish substitutes (furikaki), hot cakes, liquor, etc. Mitsuta *et al.* [26] patented a green colour for beverages using colorants from gardenia. In view of the wide range of available colours and their apparent excellent stability, colorants from gardenia appear to have good potential.

7.5 Cochineal and related pigments

Cochineal has been suggested as 'the best of all natural pigments' [89]. It has been an article of commerce for many years in the form of extracts from the bodies of the coccid insects from the families *Coccoidea* and *Aphidoidea*. Several preparations are known under various names [80]. Armenian Red is obtained from the insect *Porphyrophera hameli*, which grows on the roots and lower stems of several grasses in Azerbaijan and Armenia. Kermes is obtained from the Old World insects *Kermes ilicis* or *Kermococcus vermilis*, which grows above ground on several species of oak, particularly *Quercus coccifera*—the 'Kermes Oak'. Polish cochineal is obtained from *Margarodes polonicus* or *Porphyrophera pomonica* found on the roots of *Scleranthus perennis*, a grass found in Central and Eastern Europe. Lac is obtained from *Laccifera lacca*, which is found on the trees *Schleichera oleosa*, *Ziziphus mauritiana* and *Butea monosperma* found in India and Malaysia. The lac insects are better known as sources of shellac. American cochineal is obtained from insects of the family *Dactylopiidae* primarily *Dactylopius coccus* Costa found as parasites on the aerial parts of cactus, *Opuntia* and *Nopalea* spp.—especially *N. cochenillifera*. There are 80 000 to 100 000 insects in each kilogram of raw, dried cochineal as exported from Central and South America. World production today is a small fraction of that available in the sixteenth to nineteenth centuries. For example, the Canary Islands alone produced 3000 tons in 1875, but the availability of the synthetic food colorants in the twentieth century and the high cost of cochineal reduced their usage.

Cochineal is extracted from the bodies of female insects just prior to egg-laying time, and they may contain as much as 22% of their dry weight as pigment. Historically the pigment was extracted with hot water and the colorants were known as simple 'extracts of cochineal' [90]. Recently, proteinase enzyme treatment and further purification has provided colorants known as the 'carmines of cochineal'. The word carmine may be used as a general term for this family of anthraquinone pigments but is more usually considered to be the aluminium or magnesium lake of carminic acid on aluminium hydroxide, and contains about 50% carminic acid. These preparations are likely to be the commercial source of cochineal for the future since synthesis is unlikely at this time in view of the expense of toxicity testing. Even assuming that the synthetic pigments were 'nature-alike', it is likely that considerable testing for toxicity would be required.

The structure for carminic acid, one of the major colorants in cochineal, is shown in Figure 7.8. The pigment in Kermes is kermesic acid, the aglycone of carminic acid. It also occurs as an isomer, ceroalbolinic acid. The lac colorants are a complex mixture of laccaic acids (Figure 7.9) and the closely related compounds erythrolaccin and deoxyerythrolaccin (Figure 7.8). Solutions of carminic acid at pH 4 show a pale-yellow colour and actually have little intrinsic colour below pH 7, but they do complex with metals to produce

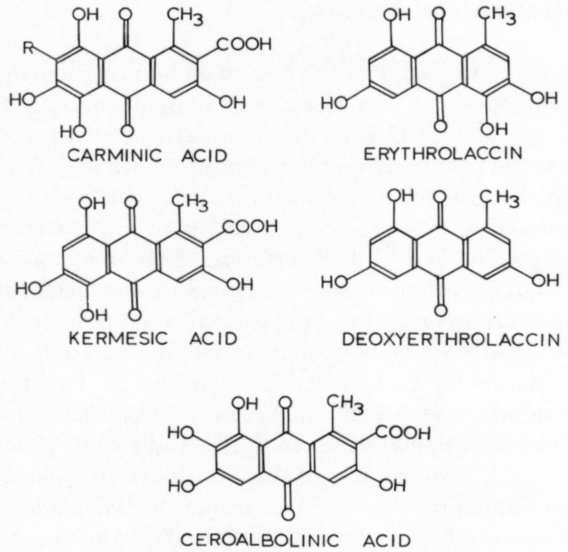

CARMINIC ACID

ERYTHROLACCIN

KERMESIC ACID

DEOXYERTHROLACCIN

CEROALBOLINIC ACID

Figure 7.8 Anthraquinone pigments of cochineal and kermes.

brilliant red hues. Complexes with tin and aluminium produce the most desirable colours and most commercial preparations involve complexes with aluminium. Floyd [80] has shown that a range of colours 'strawberry' to near 'blackcurrant' can be produced by adjusting the ratio of carminic acid to aluminium.

Carmine can be used in powder form to colour a variety of foods. It can be used in a solution of ammonia to colour foodstuffs, such as baked products, confectionery, syrups, jams and preserves. Food with low pH values may cause precipitation of the colorant but this may be an advantage.

Alkannet (Figure 7.13) is a related pigment extracted with alcohol from the roots of *Alkanna tinctoria* Tausch and *Anchusa tinctoria* Lom from southern Europe. The red pigment is only slightly soluble in water but readily soluble in organic solvents. It has been used to colour confectionery, ice-cream and wines [91].

Cochineal and related compounds have had a real resurgence of interest as food colorants in view of the movement towards the 'naturals'. This has led to more research interest—particularly when its safety was questioned [92, 93], and a series of exhaustive toxicological studies were initiated [94–99]. The consensus by an FAO/WHO Expert Committee on Food Additives was that carmine, at doses employed as food colorants, had no adverse effect. The renewed interest led to reports on methods of analysis such as the spectrophotometric study of carminic acid in solution and its application to

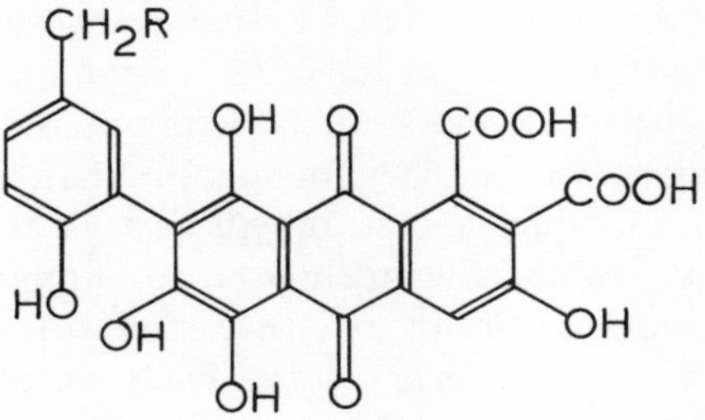

LACCAIC ACIDS

A: R = CH_2–NH–$COCH_3$

B: R = CH_2OH

C: R = $CH(NH_2)COOH$

E: R = CH_2–NH_2

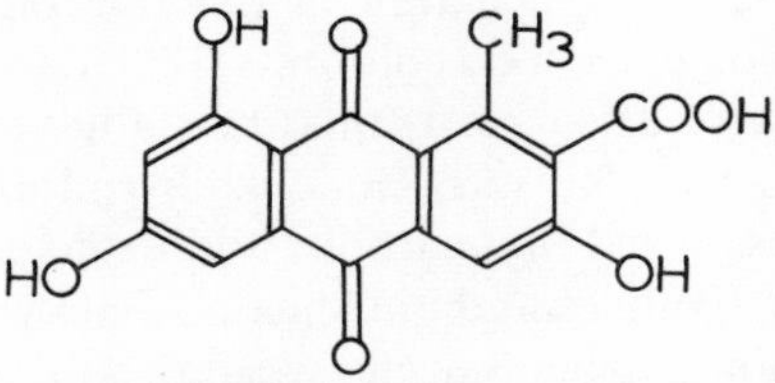

LACCAIC ACID D

Figure 7.9 Anthraquinone pigments in lac.

analysis [100]. Recent interest in the quinone area may have been stimulated by the extensive research on microbiological growth and physiology related to the production of antibiotics, cell culture and general biotechnology. Several new compounds have been reported. Richardson *et al.* [101] reported a new anthraquinone (1,6-dihydroxy-4-methoxy-9,10-anthraquinone) from *Xenorhabdus luminescens*. It occurred together with a hydroxystilbene antibiotic. Mukherjee *et al.* [102] reported a new anthraquinone pigment from *Cassia mimosoides*. The yellow pigment was found in the leaves and stems of the senna plant. Sakuma *et al.* [103] reported two new napthopyrone pigments from the crinoid *Comanthus parvicirrus*. The two new pigments occurred together with comaparvin, 6-methoxycomaparvin, and 6-methoxycomaparvin-5-O-Me-ether. Ali *et al.* [104] reported the presence of deoxyerythrolaccin and a new anthraquinone in *Gladiolus segetum*. Ikenaga *et al.* [105] studied the growth and napthoquinone production in *Lithospermum erythrorhizan*. They were interested in seasonal changes in order to optimize the content of shikonen derivatives for use in oriental medicine. Nedentsov *et al.* [106] studied the biosynthesis of the napthoquinone pigments in the fungus *Fusarium decemcellulase*. They found more than ten pigments including fusarubin, javanicin, nonjavanicin, anhydrojavonicin, anhydrofusarubin, movarubin, and bostrycoidin. Suemitsu *et al.* [107] reported the formation of modified anthraquinones in an extensive study of the physiology and metabolic products of *Alternaria porri*. They named the novel bioanthraquinones, alterporriols. In addition, Billen *et al.* [108] described the stereochemistry of the pre-anthraquinones and the dimeric anthraquinone pigments in a book chapter. Furthermore, Parisot *et al.* [109] reviewed the physiology and genetic control of the napthoquinone pigments produced by the filamentous fungus *Fusarium solani*.

The patent literature from 1969 to 1984 contains twelve patents on napthoquinones and anthroquinones [37]. They are involved mainly with purification by alkali [110], alcohols [111], stabilization with metallic salts [112, 113], sugars [114], and alum [115], and derivatization [116–118]. Two more recent patents involved the application of quinone pigments to cosmetics [119] and the stabilization of lac food colorants [120]. Lac pigments in ethanol, propylene glycol, glycerol or sorbitol could be stabilized by the addition of citric, malic, tartaric, acetic, butyric, lactic, hydrochloric or sulphuric acid. Another patent [121] reported that carmine and its lakes could be stabilized with carboxylic acid derivatives and phosphates. Baldwin [122] patented the use of carmine to colour collagen sausage casings.

The anthraquinone and napthoquinone pigments have a good reputation as food colorants primarily because of their stability and high tinctorial power [89]. They were the chromophore of choice in the 'linked polymer' colorants pioneered by the Dynapol Corporation [37]. Unfortunately this concept is no longer being commercialized but the concept is interesting and may have application for other food additives. Some of the new pigments described in the literature may have potential as food colorants.

7.6 Turmeric

Turmeric, also called Curcuma, is a fluorescent yellow-coloured extract from the rhizomes of several species of the curcuma plant. *Curcuma longa* L. is the usual commercial source. Three pigments occur in every species, curcumin, demethoxycurcumin, and bisdemethoxycurcumin (Figure 7.10) together with compounds that contribute to the flavour. These include turmerone, cineol, zingeroni, and phellandrene. A dark-coloured Turmeric oleoresin containing both pigments and flavour compounds can be prepared by extracting the dried rhizomes of *C. longa* with ethanol and removing the solvent under vacuum. Turmeric colorants of varying degrees of purity can be obtained

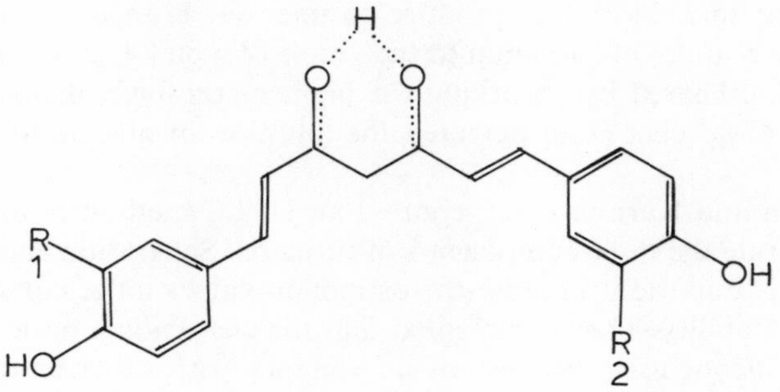

$R_1 = R_2 = OCH_3$ = Curcumin

$R_1 = H$ $R_2 = OCH_3$ = Demethoxycurcumin

$R_1 = R_2 = H$ = Bisdemethoxycurcurmin

Figure 7.10 Pigments of turmeric.

from the oleoresin. Fukazawa [123] patented a process to produce turmeric from *C. domestica* (syn. *C. longa*) by extracting the rhizomes with ether, evaporating the ether, and dissolving the residue in vegetable oil. Verghese and Joy [124] reported an extraction process using ethyl acetate which produced a 2.77% yield. Govindarajan [125] estimated that 160 000 tons were produced annually. The large production has stimulated research on optimization of yield. For example, Tonneson *et al.* [126] studied the variations in turmeric from Nepal. They reported that the major pigment was bisdemethoxycurcumin as opposed to sources from the other major producing areas in which the major pigment was curcumin. Sampathee *et al.* [127] studied the relationship between curing methods and the quality of the final product.

Turmeric oleoresins and colorants are unstable to light and alkaline conditions. Thus several patents are involved with adjustment to acid conditions. Leshnik [128, 129] prepared a stable preparation by spray-drying with citric acid, waxy maize starch, Polysorbate 60, and sodium citrate. The encapsulated product was stable for 16 weeks at 35°C. Schrantz [130] patented a process using gelatin and citric acid. Obata *et al.* [131] prevented colour loss by addition of gentisic acid, gallic acid, tannic acid polyphosphate and citric acid.

Curcumin is insoluble in water but water-soluble complexes can be made by complexing with metals. Maing and Miller [132, 133] employed stannous chloride and zinc chloride to produce an intensely orange colorant using a ratio of 4 to 5 moles of curcumin to each mole of metal. Curcumin colorants can also be prepared by absorbing the pigment on finely divided cellulose [134, 135]. Goldscher [136] patented the addition of glycine to reduce the bitter taste.

Tonneson and Karlsen [137] reported an HPLC method of analysis that would separate the three components of turmeric. Separation and individual estimation is important because the extinction values differ considerably as well as the stability. They commented that the well-known boric acid test is sometimes inconclusive because of interference with other extracted components. With the HPLC method and a UV detector, they could detect greater than 20 ng of pigment. With a fluorescence detector using 420 exciting and 470 emission nm they could detect greater than 2 ng. Rouseff [42] reported an HPLC method that would separate and estimate the three pigments in turmeric and the two in annatto in one analysis. Turmeric and annatto are often found in combination, since apparently the mixture is more stable [44]. *C. domestica* also contains 2-hydroxy methyl anthraquinone [138].

Shah and Netrawadi [139] reported that turmeric extracts did not show mutagenic activity even after activation with mammalian liver microsomal extracts. Nagabhushan *et al.* [140] reported that turmeric extracts and all three pigments reduced the nitrozation of methylurea by sodium nitrite and thus reduced the mutagenicity of *Salmonella typhimurium*. El Gazzar and Marth [141] reported that *Listeria monocytogenes* could not survive in dairy starter cultures when annatto or turmeric colorants were added. These encouraging antibacterial and antimutagenic responses may encourage even wider use of turmeric in foods. Turmeric is already widely used in the food industry in canned products, pickles, soups, mixes and confectionery. Turmeric is the usual yellow colorant in mustard and pickled cauliflower.

7.7 Carthamin

Carthamin, sometimes called carthemone, is a yellow to red preparation extracted from the flowers of *Carthamus tinctorius*. It contains three chalcones [142, 143], the red carthamin, safflor yellow A, safflor yellow B (Figure 7.11) and several precursors. Fresh yellow flower petals contain precarthamin, which oxidizes to form the red colorant carthamin. Carthamin, upon treatment with dilute hydrochloric acid, yields two isomers—red carthamin and yellow isocarthamin. Upon acid hydrolysis, carthamin yields glucose and two aglycones, the flavanones carthamidin and isocarthamidin.

The formation of carthamin involves an oxidizing or enzymatic step. Saito *et al.* [144] and Saito and Fukushima [145], reported the conditions for enzymatic synthesis and oxidation [146, 147]. They reported that exogenous substances such as alcohols, amino acids, amines, carboxylic acids and ethers

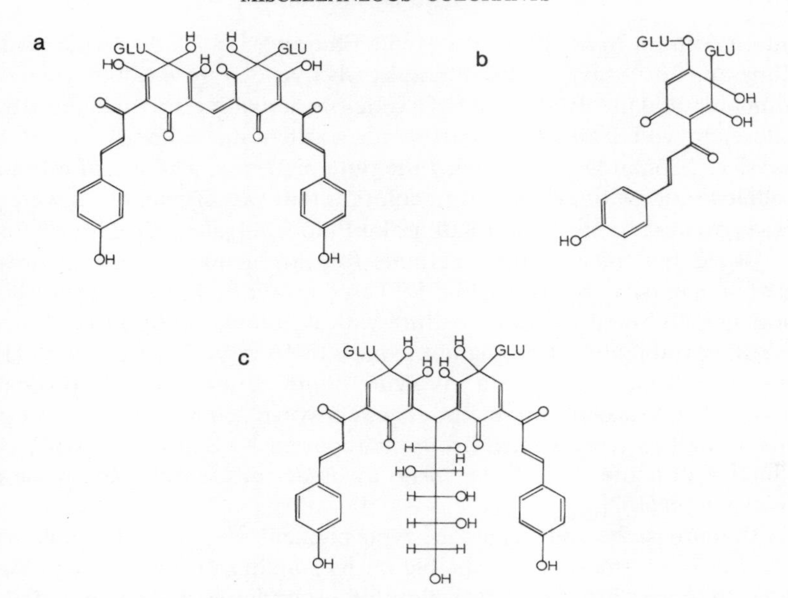

Figure 7.11 Pigments of carthamin: (a) carthamin; (b) safflor yellow A; and (c) safflor yellow B.

produced a bathochromic shift in colour, whereas esters and some fatty acids (except formic and acetic) caused a hypsochromic shift.

The earlier patents on carthamin involved simple aqueous extraction of the petals [148, 149] or crushing the petals prior to extraction to allow oxidation [150]. Both a red and a yellow colorant could be obtained for use with pineapple juice, and yoghurts [151]. Fukushima *et al.* [152] published a method for the purification of the three pigments by adsorption on polystyrene resins. Purification and stabilization has been greatly enhanced by absorption of carthamin on cellulose. Apparently cellulose has great affinity for carthamin and this is known as the 'Saito Effect' [145]. 'The effect is so strong that the carthamin may be retained for more than 1000 years without appreciable change in the red coloration' [153]. No storage data were presented to substantiate this claim. Precarthamins and safflor yellow A + B were not absorbed on a CM Sephadex C-25 column, but carthamin is strongly absorbed and could be eluted with dilute formic or acetic acid or ammonia. Apparently the mechanism of absorption on cellulose powder and cellulose ion exchanges is different. Saito [154] reported that the site of absorption was the primary alcohol hydroxyl on the glucose unit of the macromolecules. He arrived at this conclusion by studying absorption with a series of synthetic glycosyl polymers. The cellulose absorption method was improved and adapted to large-scale production using a methanolic extract and a BD-cellulose column in acidic media [155]. Elution with 3.3 M acetic acid in acetone and successive chromatography also yielded pure safflor yellow B. Another method involved the use of a cellulose derivative, diethylaminoethyl-cellulose. It would absorb precarthamin, carthamin, safflor A and B and successive chromatography yielded the pure pigments. The use of cellulose, or cellulose and chitin, to provide a colorant that was dispersible in water or oil was patented by the Sanyo-Kokusaku Pulp Co. [156, 157]. Three patents were issued for tissue-culture methods for production of carthamin and related pigments. Yomo *et al.* [158, 159] used flower bud cells on Murashiga-Skoog agar followed by liquid culture with cellulose. Addition of chitosan raised the production of carthamin from 5 to 50 mg/l. Daimon *et al.* [160] patented an improved method involving liquid culture and 4% powdered cellulose. The tissue-culture methods may involve production of novel pigments as well as the expected group as reported by Saito *et al.* [161]. The production of pigments of *C. tinctorius* by tissue culture was recently viewed by Wakayama [162].

Carthamin is the only chalcone-type pigment suggested for colouring foods. Little information is available on its stability in formulations but it appears to be very promising. Its yellow to red-orange gives it some flexibility. Its applications were reviewed by Onishi [163].

7.8 Monascus

The genus Monascus includes several fungal species that will grow on various solid substrates, especially steamed rice. The combination of fungus and substrate has been known for many centuries in south and east oriental countries. It was first mentioned in 1590 in a treatise on Chinese medicine [76]. Traditionally in the Orient, Monascus species were grown on rice and the whole mass eaten as such, or dried, powdered and incorporated into food as desired. They were used to colour wine, bean curd, and also as a general food colorant and a medicine [164]. They are currently being produced commercially in Japan, Taiwan and China.

The structure of the Monascus pigments is shown in Figure 7.12. Monascin and ankaflavin are yellow, rubropunctatin and monascorubin are red, whereas rubropunctamine and monascorubramine are purple [165, 166]. The yellow and red pigments are considered to be normal secondary metabolites of the fungal growth whereas the two purple pigments may be produced by chemical or enzymatic modification of the yellow and red pigments. The red pigments readily react with compounds containing amino groups via a ring opening and a Schiff rearrangement to form water-soluble compounds (Figure 7.13). The Monascus pigments have been reacted with amino sugars, polyamino acids, amino alcohols, ethanol, chitin amines or hexamines [167–169]; proteins, peptides and amino acids [170, 171]; bovine serum albumin

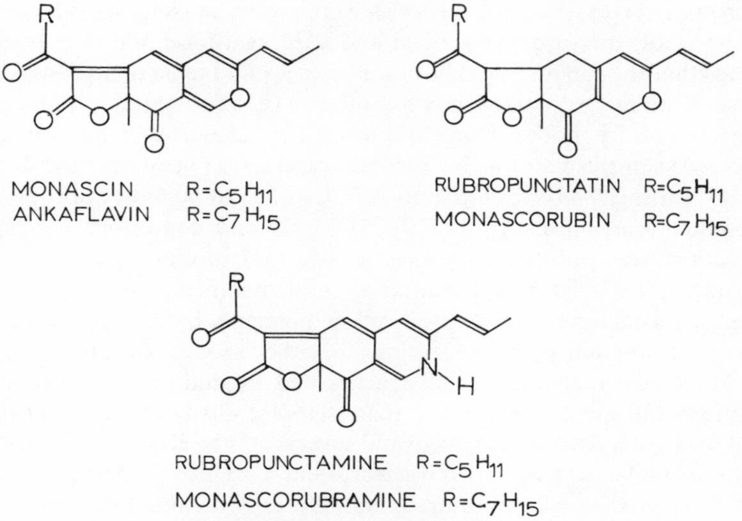

MONASCIN $R=C_5H_{11}$
ANKAFLAVIN $R=C_7H_{15}$

RUBROPUNCTATIN $R=C_5H_{11}$
MONASCORUBIN $R=C_7H_{15}$

RUBROPUNCTAMINE $R=C_5H_{11}$
MONASCORUBRAMINE $R=C_7H_{15}$

Figure 7.12 Pigments of monascus.

[172]; casein [173, 174]; gluten [175]; selected amino acids [172]; RNA [175]; nucleic acids [171]; sugar amino-acid browning-reaction products [175]; aminoacetic acid and aminobenzoic acid [77]. Sweeney *et al.* [176] prepared the N-glucosyl derivatives of rubropunctamine and monascorubramine and investigated the protective effect of these compounds by 1,4,6-trihydroxy-naphthalene. Obviously, several Monascus derivatives are possible, and these apparently show greater water solubility, thermostability and photostability.

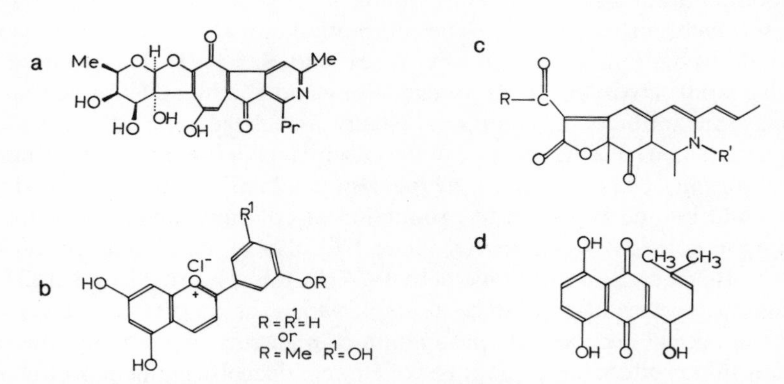

Figure 7.13 Miscellaneous pigments: (a) rubrolone suggested by Iacobucci and Sweeney (1981); (b) is deoxyanthocyanidin also suggested by Iacobucci and Sweeney (1979); (c) is a red pigment suggested by Moll and Farr (1976) where R represents an aliphatic radical and R^1 represents a radical of a compound of the formula HN-R which represents an amino sugar, a polymer of an amino sugar or an amino alcohol; and (d) is alkannet.

For many years, Monascus species were grown on solid cereal substrates and the whole mixture was ground and used as a food additive. It became obvious that the fungus could be cultured in liquid media or semi-solid state, so several studies were published to optimize the production of pigment in a variety of culture modes. Han [177] listed 125 references to growth studies. He screened thirteen strains for pigment production in submerged flask culture and chose *Monascus purpureus* ATCC 16365 to be most appropriate for solid-state fermentation [178, 179]. The optimum conditions for pigment production were polished long rice at 50% moisture and pH 5.0 to 6.0 at 30°C and RH-97-100. Supplementation with zinc increased the red-pigment content by a factor of 2.5 and the yellow pigments by 2.0. Optimization of pigment production by other strains and other modes will obviously differ [37]. Much recent research has been devoted to studies on general culture conditions and substrates. Lin and Iizuka [164] studied rice meal, wheat bran, wheat meal, bread meal, corn meal and several others. Kim *et al.* [180] studied the use of molasses and corn. Shepherd and Carels [181] developed a two-stage process that was optimized first for growth and then for pigment production. Others studied the growth process to optimize the type and amount of pigment. For example, growth of *M. anka* on bread produced only yellow pigments [182], whereas culture on a solid medium gave superpigment production [183, 184]. Lin [185] and Su *et al.* [186] also studied strain selection for pigment production. Recent research has included higher pigment production by selection of mutants induced by UV light or N-methyl-N-nitronitrosoguanidine [187] or produced by protoplast fusion [188].

Monascus species produce other compounds such as ethanol, enzymes, coenzymes, monascolins, which modify fat metabolism and inhibit cholesterol biosynthesis [189, 190], antibiotics, antihypertensives and flocculants. For use as food colorants, the above compounds have to be removed or preferably not produced in the process. The antibiotic content, in particular, would be undesirable in a food colorant. Wong and Bau [191] first reported the antibacterial activity of *M. purpureus*. Wong and Koehler [76] found that the increase in antibiotic content was usually associated with the increase in pigment content but that acetate in the culture media inhibited the formation of the antibiotic and enhanced the pigment production. They suggested that this could be one approach to production of colorant preparations free of antibiotic activity. Wong and Koehler [76] also reported that cultures of ATCC 16365 had no antibiotic activity. This was confirmed by Han [177]. Monascus species also produce ethanol, which, as an alternate metabolic pathway, could decrease the production of pigments. Apparently, the conditions that produce large quantities of ethanol decrease pigment production; however, ethanol production in small quantities apparently stimulates pigment production [177].

The Monascus pigments are readily soluble in ethanol and only slightly soluble in water [192]. Kumasaki *et al.* [193] reported that monascorubrin was soluble in ether, methanol, ethanol, benzene, chloroform, acetic acid and acetone but insoluble in water and petroleum ether. Ethanolic solutions of monascorubrin are orange at pH 3.0 to 4.0, red at 5.0 to 6.0 and purplish red at 7.0 to 9.0. Su and Huang [192] reported that the pigment from *M. anka* faded under prolonged exposure to strong light and the pigment in 70% ethanol was more photostable than solutions in water. They also reported that the pigment in 70% ethanol was very heat stable at temperatures up to 100°C. It also showed excellent heat stability at neutral or alkaline conditions. Gan [194] used UV and visible spectroscopy and NMR to study the stability of the yellow Monascus pigments. They showed good stability at 100°C and much less at 120°C. There is little detailed data on stability of Monascus pigments in food formulations.

Colorants from microbial species offer considerable advantages since they can be produced in any quantity and are not subject to the vagaries of nature. The range of colours from yellow to purplish red combined with their stability in neutral media is an advantage. Sweeney *et al.* [176] suggested that they could be used in many applications such as processed meats, marine products, jam, ice-cream and tomato ketchup. Their stability in ethanol is a real advantage for colouring alcoholic beverages. Hiroi [195] reported their application to saki and koji. Suzuki [196) described their application to soy sauce and kamaboko. There is considerable commercial interest in the Monascus group as shown by thirty-eight patents issued in recent years [37].

7.9 Miscellaneous colorants

Several colorants from unusual sources have appeared in the patent literature. For example, Iacobucci and Sweeney [197] reported the isolation and properties of rubrolone (Figure 7.13), a red pigment produced by the bacteria *Streptomyces echinoruber*. This colorant can be stabilized from exposure to sunlight by addition of quercetin-5-sulphonate-5-sulphonate and is suggested for use in dry beverage mixes for solution in water. Moll and Farr [167, 168] reported a series of red colorants (Figure 7.13) produced by reacting monascus with chitosan or monascorubin with the hydrogen chloride salt of glucosamine. The corresponding water-soluble compounds are substituted furoisoquinoline derivatives. A whole series of compounds can be formed as described previously. Iacobucci and Sweeney [198] reported the synthesis of a series of 3-deoxyanthocyanidins by reduction of acetylated flavanones with sodium borohydride to the corresponding flavan followed by oxidation with chloranil or bromanil. Yeowell and Swearingen [199] described the preparations of a series of colorants synthesized from benzyl alkyl compounds condensed with guanidine to give benzylpyrimidine derivatives. The colorants are suggested for food, textiles and medicines. Watanabe [200] reported the isolation of a red pigment from the mould *Penicillium purpurogenum*. The isolation of berberin, an isoquinoline type pigment, was reported by a Japanese company Kakko Honsha [201]. Berberin can be extracted from the bark of *Phellodendron amurense*, the roots of *Coptis japonica* or the xylem of *Berberis thunbergii*. It also occurs in the leaves of *P. amurense*. The yellow pigment is stable to heat and light. Koda [202] patented a potential blue food colorant from the callus tissue of the plant *Clerodendrum trichotomum*. Several potential colorants have also been reported in the general literature. Hamburger *et al.* [203] reported a novel purple pigment from the tree *Dalbergia candenatensis*, which occurred together with a mixture of orange and red pigments from the heartwood. Achenbach [204] described a new class of pigments of the flexirubin type in the bacteria *Flexibacter elegans*. Sato and Hasegawa [205] reported the blue pigment cyanohermidin in the roots of the plant *Mercurialis leiocarpa*. It resembled allogachrome, a dyestuff used for silk and cotton in ancient Japan. Drewes *et al.* [206], studying the medicinal plants of South Africa, reported four new orange chalcone-type pigments in *Brackenridgea zanguebarica*. De Rosa and De Stefano [207] reported a new guaiane sesquiterpene from the fungus *Lactarius sanguifluus*. Ueda *et al.* [208] studied the formation of a purple pigment from cereals by fermentation. Jaegers and co-workers [209] clarified the chemical structure of leucomelone, and protoleucomelone in the fungus *Boletopsis leucomelaena* and the acetylated derivatives of cycloleucomelone, cyclovariegatin and atromentin in *Anthracophylum* species. The orange pigments in the bacteria *Streptomyces* spp. C-72 were studied by Rachev *et al.* [210] and the blue pigments in the bacteria *Streptomyces lavendula* were described by Mikama *et al.* [211]. Oreshina *et al.* [212] studied the biosynthesis of the pigments in strains of the mould *Actinomycetes* (*Streptoverticillium*). Piskunkova *et al.* [213] and Torapova *et al.* [214] reported the biosynthesis of the pigments in the fungus *Hypomyces rosellus*. Brucker *et al.* [215] reported a broad spectrum of metabolites including the pigments bikaverins and carotenoids in the mould *Fusarium moniliforme*. In the mould *Aspergillus oryzae*, Manonmani and Shreektoniah [215] reported on orange red anthraquinone and Dahiya [216] reported vismiaquinone and a flavone. Gill *et al.* [217] reported a series of red and purple anthraquinones from the fungus *Dermocybe austroveneta*; these included austrocorticin, austrocorticinic acid, austrocorticone, and a series of related pigments [218]. They further reported the dermocanarins, which are unique macrocylic lactones from *Dermocybe canarea* [219]. Suemitsu *et al.* [107, 220, 221] reported on HPLC analysis for macrosporin, altersolanol A and alterporrioles A, B + C in the mould *Alternaria porri*. These are anthraquinone-type pigments. Kondrateva and Moon [222] reported a red pigment in the bacteria *Arthrobacter*. Etoh *et al.* [223] reported the structures of the rhodonocardins produced by a *Nocardia* species of bacteria. Gessner *et al.* [224] reported the synthesis of flazin and perlolyrine found in the β-carboline group of pigments in sake and soy sauce. Ito patented the use of flavins [119] and quinones [225] for use in cosmetics. Sato *et al.* described a blue pigment in the mould *Streptomyces*. Anderson and Chung [226] reported the conversion of versiconal acetate to versiconal and versicolorin C in extracts from the mould *Aspergillus parasiticus*. Nakamura and Homa [227] described prostaglandin-like precursors of a red pigment formed during the autoxidation of methyl arachidonate. Ozawa *et al.* [228] reported the isolation and identification of a novel yellow β-carboline pigment in salted radish roots (*Raphanus sativus* L.). Yang *et al.* [229] patented the extraction method of a yellow pigment from the roots of *Glycyrrhiza*. Cai *et al.* [230] patented the extraction for an edible polyphenol colorant from bean seedcoats (*Phaseolus* spp.). Both a brownish red and a cherry-red colorant could be produced. Aarnio and Agathos [231] reported that variants of the mould *Tolypocladium inflatum* produced white, red, orange and brown colours. This organism is used in production of the immunosuppressive agent cyclosporin and apparently the level of pigment is correlated to the level of cyclosporin produced. It may be possible to prepare a colorant as a by-product of cyclosporin production. Many of the above pigments were found in micro-organisms and may reflect the interest in metabolites that may have physiological effects. Nevertheless, some of them may have potential as colorants, but they would require toxicological clearance.

7.10 Future prospects

The availability of food colorants historically has depended on three sources: specialized plant and animal sources, by-products, and chemical synthesis. Today, a fourth should be added, that of tissue or cell culture possibly with DNA transfer biotechnology.

Specialized plant and animal sources have been known for many centuries. Examples are annatto, safflower, saffron, and turmeric from plants and cochineal from insects. By-product sources are oenocyanins from wine grape skins and carotenoids from many sources such as carrots and palm oil. Chemical synthesis has provided the carotenoids, canthaxanthin, β-carotene, β-8-apo-carotenal, and ethyl β-8-apo-carotenoate. Another example would be the now discontinued linked polymer concept with anthraquinone chromophores. The cell and tissue-culture concepts have been of much research interest lately, possibly because of the pharmaceutical and fermentation industries. The extensive literature on the optimization of growth of the fungus *Monascus* for production of the monascin colorants is one example. Nochida *et al.* [232] patented the manufacture of the colorant canthaxanthin from the *Rhodococcus* species of bacteria. Kumar and Lonsane [233] described the production of a novel red pigment that is associated with the production of gibberellic acid from the fungus *Gibberella fujikuroi*. The new pigment is not the normal indole alkaloids, bikaverin, norbikaverin or *o*-dimethylanhydrofusarubin that are ordinarily associated with *in vivo* growth of *Gibberella*.

Sasaki *et al.* [234] patented the production of pigments from the bacteria *Serratia* SSP-1 (*Enterobacter*). Ishiguro *et al.* [235] patented the production of a heat and light stable blue purple pigment from cultures of the bacterium *Janthinobacterium lividium*. Cioppa *et al.* [236] described the production of melanin by a tyrosinase gene cloned into *E. coli* bacteria. Melanin is of considerable interest as a general brown colorant, particularly in the pharmaceutical industry for skin 'tanning' formulations.

Several reports have appeared on the tissue-culture applications of plants for colorant and flavour applications, possibly due to the background of information available from plant propagation and plant-culture developments. Nozue *et al.* [36] patented the production of the stable acylated anthocyanins from the callus tissue of red sweet potatoes (*Ipomoea batatas*). Koda [237] patented the production of the acylated anthocyanins from red cabbage (*Brassica*) by tissue culture. Other patents involving anthocyanins are: callus culture of the beefsteak plant (*Perilla*) by Ota [238] and Ota *et al.* [239, 240], a red plant (*Basella rubra*) by Fukuzaki *et al.* [241], the red sepals of *Hibiscus sabdariffa* by Oohashi *et al.* [242], and the Evening Primrose *Oenothera* (Yamada and Fugita [243]). Li and Zhu [244] reported the formation of anthocyanins in Panax ginseng. Two reports involved the production of leucoanthocyanins by fermentation from wheat germ (Ueda [245]) and barley (Ohba *et al.* [246]). Kargi and Friedel [247] reported the indole alkaloids and related compounds produced in cell culture of the plant *Catharanthus roseus*. Yomo *et al.* [158, 159] described the production of carthamin and Saito *et al.* [248] reported a novel red pigment from cell suspension cultures of *Carthamus tinctorius*. Three reports involved the production of crocin from the stigmas of *Crocus sativus* [249–251]. Saffron colorants are an obvious choice for cell culture because of their high price. Kilby and Hunter [252] described the production of betanin from beetroot (*Beta vulgaris*) by cell culture. The production of useful chemicals by biotechnology has been reviewed by Trividi [253], Ilker [254], Hosono [255] Whitaker and Evans [256], Fontanel and Tabata [257], Bomar and Knopfel [258] and Knorr *et al.* [259].

Plant products have been used in the pharmaceutical, cosmetic and food industries for many years. The choice of supply by chemical synthesis or plant biosynthesis depends on many factors but three are predominant. First, some chemicals are difficult to synthesize chemically and possibly could be obtained more easily by biosynthesis. One example in the colorant area is the ternitins. These compounds, which are reported (Terahara *et al.* [3]) to be the most stable anthocyanins reported to date, have a complex acylated structure that might be difficult to synthesize. Second, others contain too many components. For example, jasmine with over 300 components and a limited market is a poor candidate for chemical synthesis. Similarly, coffee flavour has over 1000 components and may be a good candidate for tissue culture. Grapes are reported to have as many as 30 anthocyanins (Singleton and Esau [13]), whereas blueberries are reported to have 15 (Francis *et al.* [260]). However, the number of anthocyanins in a given plant is unlikely to be a deterrent for chemical synthesis because not all components would be necessary for a good colorant. A third important aspect is the ability to select strains or conditions that allow much greater concentrations of the desired compounds to be produced. For example, Fontanel and Tabata [257] reported a 1000-fold increase in the production of berberine by *Thalictrum minus* when tissue culture was used as opposed to the whole plant. Ilker [254] reported that shikonen from *Lithospermum* roots showed a 845-fold increase. With increases like this, cell culture approaches may be very attractive. Some of the colorants may be good candidates for this emerging area of biotechnology.

References

1. Yoshitama, K. *Phytochem.* **16** (1977) 1857.
2. Francis, F.J. *Crit. Rev. Food Sci. Nutr.* **28** (1989) 273.
3. Terahara, N., Saito, N., Honda, T., Toki, K. and Osajima, Y. *Tetrahedron Lett.* **31** (1990) 2921.
4. Goto, T. *Proc. 5th Asian Symp. Med. Plants, Spices*, Seoul, Korea (1984) p.593.
5. Stirton, J.Z. and Harborne, J.B. *Biochem. Syst. Ecol.* **8** (1980) 285.
6. Idaka, E., Ogawa, T. and Kondo, T. *Agricol. Biol. Chem.* **51** (1987) 2215.
7. Shi, Z., Lin, M. and Francis, F.J. *J. Food Sci.* (1991) in press.
8. Francis, F.J. Analysis of Anthocyanins, in *Anthocyanins as Food Colours* (P. Markakis, Ed.), Academic Press, New York (1982) p. 181.
9. Brouillard, R. Flavonoids and Flower Culture. in *The Flavonoids: Advances in Research Since 1980*. Ch. 16 (J.B. Harborne, Ed.) Chapman and Hall, London (1988) pp.525.
10. Asen, S., Stewart, R.N. and Norris, K.H. *Phytochem.* **16** (1977) 1118.
11. Hoshimo, T., Matsumoto, U. and Goto, T. *Phytochem.* **19** (1980) 663.
12a. Francis, F.J. and Harborne, J.B. *J. Food Sci.* **31** (1966) 524.
12b. Francis, F.J., Draetta, I., Baldini, V. and Iaderoze, M. *J. Amer. Soc. Hortic. Sci.* **107** (1982) 789.
13. Singleton, V.L. and Esau, P. *Phenolic Substances in Grapes and Wine and Their Significance*, Academic Press, New York (1969).
14a. Bassa, I.A. and Francis, F.J. *J. Food Sci.* **52** (1987) 1753.
14b. Shi, Z., Bassa, I.A., Gabriel, S.L. and Francis, F.J. *J. Food. Sci.* (1991) in press.
15. Teh, L.S. and Francis, F.J. *J. Food Sci.* **53** (1988) 1580.
16. Brouillard, R. *Phytochem.* **20** (1981) 143.
17. Saito, N. Personal communication (1990).
18. Goto, T., Imagawa, H., Kondo, T. and Miura, I. *Heterocycles* **17** (1982) 355.
19. Shi, Z., Lin, M. and Francis, F.J. *J. Food Sci.* (1991) in press.
20. Min, L., Shi, Z. and Francis, F.J. *J. Food Sci.* (1991) in press.
21. Asen, S., Stewart, R.N. and Norris, K.H. *US Patent No.* 4 172 902 (1979).
22. K.K Seiwa Kazuku Co. *Japan Patent No.* 82 058 871 (1982).
23. Sanei Chemical Co. *Japan Patent No.* 80 025 460 (1980).
24. Sanei Chemical Co. *Japan Patent No.* 59 223 756 (1984).
25. Idaka, H. *Japan Patent No.* 87 209 173 (1987).
26. Idaka, H. *Japan Patent No.* 63 113 078 (1988).
27. Idaka, H. *Japan Patent No.* 63 278 971 (1988).
28. Mitsuta, H., Wada, S. and Nishino, Y. *Japan Patent No.* 86 224 979 (1986).
29. Obata, M., Koda, T. and Yasuda, A. *Japan Patent No.* 87 03 765 (1987).
30. Obata, M., Koda, T. and Yasuda, A. *Japan Patent No.* 87 03 775 (1987).
31. Yasuda, A., Koda, T. and Obata, M. *Japan Patent No.* 61 282 032 (1986).
32. Katake, K. and Yasuda, A. *Japan Patent No.* 87 037 39 (1987).
33. Katake, K. and Yasuda, A. *Japan Patent No.* 87 037 41 (1987).

34. Idaka, H. *Japan Patent No.* 89 224 389 (1988).
35. Idaka, H. *Japan Patent No.* 63 110 259 (1988).
36. Nozue, M., Kikuma, M., Miyamoto, Y., Fukuzaki, E., Motsumura, T. and Hashimoto, Y. *Japan Patent No.* 88 233 993 (1988).
37. Francis, F.J. *Handbook of Food Colorant Patents, Food and Nutrition Press*, Westport, CT (1986).
38. Murai, K. and Wilkins, D. *Food Technol.* **44** (1990) 131.
39. LaBell, F. *Food Processing* **51** (1990) 69.
40. Rozinkov, P.F., Gavilov, F.M. and Klochkova, N.V. *USSR Patent No.* 343 987 (1972).
41. Guimares, I.S., Barbosa, A.L.S. and Nassariani, G. *Rev. Bras. Eng. Quin* **12** (1989) 22.
42. Rouseff, R.L. *J. Food Sci.* **53** (1988) 1823.
43. Reith, J.F. and Gielan, J.W. *J. Food Sci.* **36** (1971) 861.
44. Todd, P.H. *US Patent No.* 3 162 538 (1964).
45. Najar, S.V., Bobbio, F.O. and Bobbio, P.A. *Food Chem.* **29** (1988) 283.
46. Heltiarachchy, N.S. and Muffett, D.J. *European Patent EP* 200 043 (1986).
47. Ford, M.A. and Mellor, C. *GB Patent No.* 2 190 922 (1987).
48. Berset, C. and Marty, C. *Lebensmittel-Wissenschaft und -Technologie* **19** (1986) 126.
49. Ford, M.A. and Draisey, A.H. *GB Patent No.* 2 188 823 (1987).
50. Ikawa, Y. and Kagemoto, A. *Japan Patent No.* 75 104 231 (1975).
51. Mori, K., Tonari, K. and Nishiura, Y. *Japan Patent No.* 75 113 533 (1975).
52. Harrison, R.M. *GB Patent No.* 2 186 175 (1987).
53. Alonso, G.L., Varon, R., Gomez, R., Navarro, F. and Salinas, M.R. *J. Food Sci.* **55** (1990) 595.
54. Corradi, C. and Micheli, G. *Boll. Chim. Farm.* **118** (1979) 553.
55. Solinas, M. and Cichelle, A. *Industries Alimentari* **27** (1988) 634.
56. Klaui, H., Manz, U., Rigassi, N., Ryser, G. and Schweiter, U. *German Patent No.* 2 053 381 (1971).
57. Klaui, H., Manz, N.R., Gottleib, R. and Schweiter, R. *US Patent No.* 39 223 68 (1975).
58. Garrido, J.L., Diez de Bethancourt, C. and Revilla, E. *Anales de Bromatologia* **39** (1987) 69.
59. Umetani, Y., Fukui, H. and Tobata, M. *Yakugoku Zarshi* **100** (1980) 920.
60. Yoshizumi, S., Okuyoma, H. and Toyoma, R. *Shouhuhin Kogyo* **23** (1980) 41.
61. Gunatilaka, A.A.L. and Sirimanni, S.R. *J. Chem. Res. Synop.* **7** (1979) 21.
62. Gunatilaka, A.A.L., Sirimone, S.R., Sothersworan, S. and Sreyani, H.T.B. *Phytochem.* **21** (1982) 805.
63. Ishara, N. and Arichi, S. *Wakanayaku Shinpojumu* **14** (1981) 45.
64. Uesato, S., Ueda, S., Kobayashi, K. and Inouye, H. *Proc. Int. Cong. Plant Tissue Cell Cult.* 5th edn. (A. Fujiwara, Ed.) (1982).
65. Uesato, S., Ueda, S. and Kobayashi, K. *Chem. Pharm. Bull.* **31** (1983) 4185.
66. Uesato, S., Uedo, S., Kobayashi, K., Miyauchi, M. and Inouye, H. *Tetrahedron Lett.* **25** (1984) 573.
67. Terao, T., Ohashi, H. and Mizukomi, H. *Plant Sci. Lett.*, **33** (1984) 47.
68. Mizukomi, H., Terao, T., Miura, H. and Ohaski, H. *Phytochem.* **22** (1983) 679.
69. Tobata, M., Umetani, Y., Shima, K. and Tanaka, S. *Plant Cell Tissue Organ Culture* **33** (1984) 3.
70. Yoshizumi, S., Okuyoma, H. and Toyoma, R. *Fureguransu Janaru* **8** (1980) 104.
71. Aburoda, M. *Gendai Toyo Izaku.* **4** (1982) 42.
72. Kuwano, S. *Gendai Toyo Igaku.* **4** (1983) 55.
73. Joshi, K.C., Singh, P. and Pardosoni, R.T. *J. Ind. Chem. Soc.* **56** (1979) 327.
74. Yan, Y. *Yaowa Fenxi Sazhi* **4** (1984) 227.
75. Inouye, H. *Gendai Toyo Izaku.* **4** (1983) 48.
76. Wong, H.C. and Koehler, P.E. *J. Food Sci.* **46** (1981) 956.
77. Wong, H.C. and Koehler, P.E. *J. Food Sci.* **48** (1983) 1200.
78. Hagikara, K., Yamauchi, K. and Kuwano, S., Mukogawa Joshi Daigaku Kiyo, *Yakugaku Hen.* **29** (1981) 2.
79. Noda, N., Yomada, S., Hayakawa, J. and Uno, K. *J. Hygienic. Chem.* **29** (1983) 7.
80. Floyd, A.G. *Food Chem.* **5** (1980) 19.
80. Ishiguro, K., Yamaki, M., Takagi, S. *J. Nat. Prod.* **46** (1983) 532.
81. Komura, K., Okayama, H., Toyoma, R., Sawada, Y. and Ichinose, S. *Japan Patent No.* 76 006 230 (1976).

82. Hasegawa, K. *Japan Patent No.* 79 152 026 (1979).
83. Koga, K., Fujikawa, S. and Fukui, Y. *European Patent No. EP* 0 251 063 9 (1988).
84. Koga, K., Fujikawa, S. and Fukui, Y. *US Patent No.* 4 878 921 (1989).
85. Tomara, I., Komai, T. and Hotae, T. *Japan Patent No.* 61 296 070 (1986).
86. Fujikawa, S., Fukui, Y., Koga, K. and Kumada, J. *J. Ferment. Technol.* **65** (1987) 419.
87. Fujikawa, S., Nakamura, S., Koga, K. and Kumada, J. *J. Ferment. Technol.* **65** (1987) 711.
88. Toyama, R. and Moritome, N. *New Food Indust.* **30** (1988) 13.
89. Riboh, M. *Food Eng.* **49** (1977) 66.
90. Donkin, R.A. *Trans. Amer. Philosoph. Soc.* **67** (1977) Part 5.
91. Coulson, J. *Developments in Food Colours* (J. Walford, Ed.), Applied Science, London (1980) p.189.
92. Gutteridge, J.M.C. and Quinlin, G.J. *Food Add. and Contam.* **3** (1986) 289.
93. Yamada, J., Tomita, Y. and Fukude, K. *Agricol. Biol. Chem.* **52** (1988) 2893.
94. Ford, G.P., Gopal, T., Grant, D., Gaunt, I.F., Evans, J.G. and Butler, W.H. *Food and Chem. Toxicol.* **25** (1987) 897.
95. Ford, G.P., Stevenson, B.I. and Evans, J.G. *Food and Chem. Toxicol.* **25** (1987) 919.
96. Grant, D. and Gaunt, I.F. *Food and Chem. Toxicol.* **25** (1987) 903.
97. Grant, D., Gaunt, I.F. and Carpanini, F.M.B. *Food and Chem. Toxicol.* **25** (1987) 913.
98. Phillips, J.C., Bex, C., Walters, D.G. and Gaunt, I.F. *Food and Chem. Toxicol.* **25** (1987) 927.
99. Ghosh, A. and Chakravarti, I. *J. Food Sci. Tech., India* **25** (1988) 94.
100. Schwing-Weill, M.J. *Analusis* **14** (1986) 290.
101. Richardson, W.H., Schmidt, T.M. and Nealson, K.H. *Appl. Environ. Microbiol.* **54** (1988) 1602.
102. Mukherjee, K.S., Bhattachargee, P., Mukherjee, R.K. and Ghosh, P.K. *J. Indian Chem. Soc.* **63** (1986) 619.
103. Sakuma, Y., Tanaka, J. and Hiza, T. *Aust. J. Chem.* **40** (1987) 1613.
104. Ali, A.A., Abdalla, O.M. and Steglick, W. *Phytochem.* **28** (1988) 281.
105. Ikenaga, T., Kibuta, S., Mimura, K. and Ohashi, H. *Shoyakugaku Zasshi* **40** (1986) 397.
106. Nedentsov, A.G., Baskunov, B.P. and Akimento, V.K. *Biokhimiya (Moscow)* **53** (1988) 413.
107. Suemitsu, R., Horiuchi, K., Ohnishi, K. and Yanagowase, S. *J. Chromat.* **454** (1988) 406.
108. Billen, G., Karl, U., Scholl, T. and Stroech, K.D. *Natural Products Chemistry* (A. Rahmon and P.W. Le Quesne, Eds), Springer Verlag, Berlin **3** (1988) pp.305.
109. Parisot, D., Devys, M. and Barbier, M. *Microbios* **64** (1990) 31.
110. Kasumov, M.A., Amirova, G.S., Alekperov, U.K. and Musaev, V.R. *USSR Patent No.* 988 845 (1983).
111. Tokano, M. and Konishi, T. *Japan Patent No.* 76 124 120 (1976).
112. Baldwin, W.L. *US Patent No.* 4 769 246 (1988).
112. Sanei Kaguku Kogyo Co. *Japan Patent No.* 4 838 154 (1973).
113. Daisho Co., Ltd. *Japan Patent No.* 59 120 075 (1984).
114. Shiaki, Y. *Japan Patent No.* 71 002 386 (1971).
115. Sanei Chemical Co. *Japan Patent No.* 81 139 561 (1981).
116. Eguchi, A., Suzuki, Y., Oonuki, T., Hirata, T., Tosaka, O. and Yokokawa, Y. *Japan Patent No.* 76 050 938 (1976).
117. Eguchi, A., Suzuki, Y., Onuma, T., Yokokawa, Y., Hirata, T. and Tosaka, O. *Japan Patent No.* 52 084 220 (1977).
118. Rikio, G. and Takanoshi, M. *Japan Patent No.* 77 108 061 (1977).
119. Ito, K. *Japan Patent No.* 61 225 257 (1986).
120. Asahi, Y. and Domae, A. *Japan Patent No.* 62 151 467 (1987).
121. Crompton and Knowles Co. *Japan Patent No.* 62 131 071 (1987).
123. Fukazawa, T. *Japan Patent No.* 70 024 672 (1970).
124. Verghese, J. and Joy, M.T. *Flavour Fragrance J.* **4** (1989) 31.
125. Govindarajan, V.S. *Crit. Rev. in Food Science and Nutrition*, CRC Press, Boca Raton, Florida (1980) 199.
126. Tonneson, H.H., Karlsen, J.K., Adhikary, S.R. and Pandey, R. *Zeitschrift fur Lebensmittel-untersuchung und Forschung* **189** (1989) 116.
127. Sampathee, S.R., Krishnamurthy, N., Sowbhagya, H.B. and Shankaranarayana, M.L. *J. Food Sci. Technol. (India)* **25** (1988) 152.

128. Leshnik, R.R. *US Patent No.* 4307 117 (1981).
129. Leshnik, R.R. *European Patent No.* 37 204 (1982).
130. Schrantz, J.L. *US Patent No.* 4 368 208 (1983).
131. Obata, S., Ohaski, S. and Hanatoku, M. *Japan Patent No.* 73 014 941 (1973).
132. Maing, Y. and Miller, I. *European Patent No.* 25 637 (1981).
133. Maing, Y. and Miller, I. *US Patent No.* 4 263 333 (1981).
134. Shinagawa, K., Nagai, S. and Kokagoshi, T. *Japan Patent No.* 79 163 866 (1979).
135. Sanyo-Kokusaku Pulp Co. *Japan Patent No.* 56 019 220 (1981).
136. Goldscher, K. *US Patent No.* 4 163 803 (1979).
137. Tonneson, H.H. and Karlsen, J.K. *J. Chromat.* **259** (1983) 367.
138. Ogbeide, O.N., Eduaveguavoen, O.I. and Parvez, M. *Pakistan J. Sci.* **37** (1985) 15.
139. Shah, R.G. and Netrawadi, M.S. *Bull. Envir. Contam. Toxicol.* **40** (1988) 350.
140. Nagabhushan, M., Nair, U.J., Amonkar, A.J., D'Souza, A.V. and Bhide, S.V. *Mut. Res.* **202** (1988) 163.
141. El Gazzar, F.E. and Marth, E.H. *Lebensmittel-Wissenschaft und Technologie* **22** (1989) 406.
142. Takahashi, Y., Saito, K., Yanagiya, M., Ikura, M., Hikichi, K., Matsumoto, T. and Wada, M. *Tetrahed. Lett.* **25** (1984) 2471.
143. An, X., Li, Y., Chen, J., Li, F., Fang, S. and Chen, Y. *Zhongcaoyao* **21** (1990) 188.
144. Saito, K., Takahashi, Y. and Wada, M. *Biochem. Biophys. Acta* **756** (1983) 217.
145. Saito, K. and Fukushima, A. *Acta. Soc. Bot. Pol.* **55** (1986) 639.
146. Saito, K. and Fukushima, A. *Food Chem.* **26** (1987) 125.
147. Saito, K. and Takahashi, Y. *Acta. Soc. Bot. Pol.* **54** (1985) 231.
148. Tokai University, *Japan Patent No.* 16 346 (1970).
149. Kokusaku, Pulp Co. *Japan Patent No.* 57 025 576 (1982).
150. Wada, M. *Japan Patent No.* 70 016 346 (1970).
151. Umeda, A. and Inoue, K. *Japan Patent No.* 70 040 266 (1970).
152. Fukushima, A., Hase, H. and Saito, K. *Acta Soc. Bot. Pol.* **56** (1987) 485.
153. Saito, K. and Fukushima, A. *Food Chem.* **29** (1988) 161.
154. Saito, K. *Food Chem.* **36** (1990) 243.
155. Saito, K. and Fukushima, A. *Food Chem.* **32** (1989) 297.
156. Sanyo-Kokusaku Pulp Co. *Japan Patent No.* 80 102 659 (1980).
157. Sanyo-Kokusaku Pulp Co. *Japan Patent No.* 80 102 661 (1980).
158. Yomo, H., Miyano, S. and Sekino, Y. *Japan Patent No.* 63 253 386 (1987).
159. Yomo, H., Miyano, S. and Sekino, Y. *Japan Patent No.* 62 257 390 (1987).
160. Daimon, E., Wakayama, Y., Sekino, Y. and Yomo, H. *Japan Patent No.* 01 20 092 (1989).
161. Saito, Y. and Hasegawa, M. *Kobunkazai no Kagagu* **31** (1988) 18.
162. Wakayama, S. *New Food Industry* **30** (1988) 7.
163. Onishi, K. *New Food Industries* **29** (1987) 22.
164. Lin, C.F. and Iizuka, H. *Appl. Envir. Microbiol.* **43** (1982) 671.
165. Fielding, B.C., Holker, J.S.E., Jones, D.F., Powel, A.D.G., Richmond, K.W., Robertson, A. and Whalley, W.B. *J. Chem. Soc.* (1961) 4579.
166. Manchard, P.S. and Whalley, W.B. *Phytochem.* **12** (1973) 2531.
167. Moll, H.R. and Farr, D.H. *US Patent No.* 3 993 789 (1976).
168. Moll, H.R. and Farr, D.H. *German Patent No.* 2 461 642 (1976).
169. Moll, H.R. and Farr, D.H. *Switzerland Patent No.* 606 433 (1978).
170. Yamaguchi, Y., Ito, H., Watanabe, S., Yoshida, T. and Kumatsu, A. *Japan Patent No.* 3 765 906 (1973).
171. Kawabato, S. and Sato, K. *Japan Patent No.* 76 099 519 (1976).
172. Broder, C.U. and Koehler, P.E. *J. Food Sci.* **45** (1980) 567.
173. Nakagawa, N., Watenabe, S. and Kobayoshi, J. *Japan Patent No.* 76 091 937 (1976).
174. Nakagawa, N., Watenabe, S. and Kobayoshi, J. *Japan Patent No.* 76 091 939 (1976).
175. Nakagawa, N., Watenabe, S. and Kobayoshi, J. *Japan Patent No.* 76 091 938 (1976).
175. Tsunenaga, T. and Onoe, A. *Japan Patent No.* 75 036 519 (1975).
175. Yoshimura, M., Yamanaka, S. and Hirose, Y. *Japan Patent No.* 76 130 428 (1976).
176. Sweeney, J.G., Estrada-Valdez, M.C., Iacobucci, G.A., Sato, H. and Sakamura, S. *J. Agricol. Food Chem.* **29** (1981) 1189.

177. Han, O.H. *Optimization of Monascus Pigment Production in Solid-State Fermentation.* PhD Thesis. Univ. Massachusetts, Amherst (1990).
178. Han, O.H. and Mudgett, R.E. *Enzyme Microb. Technol.* (1991) in press.
179. Han, O.H. and Mudgett, R.E. *Biotechnol. Bioengin.* (1991) in press.
180. Kim, C., Rhee, S. and Kim, I. *Korean J. Food Sci. and Technol.* **9** (1977) 277.
181. Shepherd, D. and Carels, M.S.C. *Swiss Patent No.* 606 421 (1978).
182. Tadao, H., Suzuki, T., Tsukioka, M. and Takahashi, T. *Japan Patent No.* 81 006 263 (1981).
183. Hiroi, T., Shima, T., Suzuki, T., Tsukioka, M. and Takahashi, T. *Japan Patent No.* 75 025 766 (1975).
184. Hiroi, T., Shima, T., Suzuki, T. and Tsukioka, M. *Agricol. Biol. Chem.* **43** (1979) 1975.
185. Lin, C.F. *J. Ferm. Tech.* **53** (1973) 407.
186. Su, Y.C., Dhen, W.L. and Lee, Y.H. *Res. Rep. Coll. Agri. Nat. Taiwan Univ.* **14** (1973) 41.
187. Ryu, B.H., Lee, B.H., Park, B.G., Kim, H.S., Kim, D.S. and Lee, J.H. *Korean J. Food Sci. Technol.* **21** (1989) 31.
188. Ryu, B.H., Lee, B.H., Park, B.G., Kim, H.S., Kim, D.S. and Roh, M.H. *Korean J. Food Sci. Technol.* **21** (1989) 37.
189. Fink-Gremmels, J. and Leistner, L. *Mitteilungsblatt der Bundesonstalt fur Fleischforschung, Kulmback* **101** (1988) 8073.
190. Fink-Gremmels, J. and Leistner, L. *Fleischwirtshaft* **69** (1989) 115.
191. Wong, H.C. and Bau, Y.S. *Wert. Plant Physiol.* **60** (1977) 578.
192. Su, Y.C. and Huang, J.H. *Proc. Nat. Sci. Council* **4** (1980) 201.
193. Kumasaki, S., Nakanishi, K., Nishikawa, E. and Ohashi, M. *Tetrahed.* **18** (1962) 1171.
194. Gan, C. *Kexueban* **4** (1988) 63.
195. Hiroi, T. *New Food Indust.* **30** (1988) 1.
196. Suzuki, H. *New Food Indust.* **30** (1988) 20.
197. Iacobucci, G.A. and Sweeney, J.G. *US Patent No.* 4 285 985 (1981).
198. Iacobucci, G.A. and Sweeney, J.G. *British Patent No.* 1 552 402 (1979).
199. Yeowell, D.A. and Swearingen, R.A. *German Patent No.* 2 725 992 (1977).
200. Watanabe, H. *Japan Patent No.* 74 093 587 (1974).
201. Kakko Honsha, K.K. *Japan Patent No.* 80 155 056 (1980).
202. Koda, T. *Japan Patent No.* 62 83 892 (1987).
203. Hamburger, M., Cordell, G.A., Ruangrungsi, N. and Tantivatana, P. *J. Org. Chem.* **53** (1988) 4161.
204. Achenbach, H. *Prog. Chem. Org. Nat. Prod.* **52** (1988) 73.
205. Sato, Y. and Hasegawa, M. *Kobunkagai no Kagagu* **31** (1986) 18.
206. Drewes, S.E., Hudson, N.A., Bates, R.B., Linz, G.S. *J. Chem. Soc.* (Perkin Trans.) **12** (1987) 2809.
207. De Rosa, S. and De Stefano, S. *Phytochem.* **26** (1987) 2007.
208. Ueda, S., Koba, Y. and Ohba, R. *Hakko Kogaku Kaishi* **65** (1987) 237.
209. Jaegers, E., Hillen-Maske, E., Schmidt, H., Steglich, W. and Horan, E. *Z. Naturforsch B: Chem. Soc.* **42** (1987) 1354.
210. Rachev, R., Sakanekova, M., Darakchieva, M., Tsbetkova, R., Gesheva, R., Puntsag, T. and Tsetseg, B. *Acta Microbiol. Bulg.* **21** (1987) 57.
211. Mikama, Y., Yazawa, K., Yokoyama, K., Takahashi, K. and Arai, T. *Symp. Biol. Hung.* **32** (1986) 297.
212. Oreshina, M.G., Penzikova, G.A. and Bartoshevich, Y. *Antibiot. Khimioter* **33** (1988) 323.
213. Piskunkova, N.F., Maksimov, V.N., Toropova, E.G., Egorov, N.S. and Mardamshina, A.D. *Biol. Nauki (Moscow)* **2** (1988) 84.
214. Torapova, E.G., Mardomshina, A.D. and Piskunkova, N.F. *Antibiot. Khimioter* **33** (1988) 96.
215. Bruker, B., Blechsmidt, D. and Schubert, B. *Zentralblatt fur Mikrobiologie* **144** (1989) 3.
215. Manonmani, H.K. and Shreektoniah, K.R. *J. Food Sci. Tech. India* **21** (1984) 195.
216. Dahiya, J.S. *Ind. J. Microbiol.* **27** (1987) 12.
217. Gill, M., Gimenez, A. and McKenzie, R.W. *J. Nat. Prod.* **51** (1988) 1255.
218. Gill, M. and Gimenez, A. *J. Chem. Soc.* (Perkin Trans.) **4** (1990) 1159.

219. Gill, M. and Gimenez, A. *Tetrahed. Lett.* **31** (1990) 3503.
220. Suemitsu, R., Ohnishi, K., Yanagawase, S., Yanamoto, K. and Yamada, Y. *Phytochem.* **28** (1989) 1621.
221. Suemitsu, R., Sakurai, Y., Nakachi, K., Miyoshi, I., Kubota, M. and Ohnishi, K. *Agricol. Biol. Chem.* **53** (1989) 1301.
222. Kondrateva, L.M. and Moon, T.H. *Microbiologiya* **55** (1987) 1010.
223. Etoh, H., Iguchi, M., Nagasawa, T., Tani, Y., Yanada, H. and Fukomi, H. *Agricol. Biol. Chem.* **51** (1987) 1819.
224. Gessner, W.P., Brossi, A., Bembenek, M.E. and Abell, C.W. *Archiv. der Pharmazie* **321** (1988) 95.
225. Ito, K. *Japan Patent No.* 61 225 259 (1986).
226. Anderson, J.A. and Chung, C.H. *Mycopathologia* **110** (1990) 31.
226. Sato, K., Nihira, T., Sakuda, S., Yanagimoto, M. and Yamada, Y. *J. Ferment. Bioengin.* **68** (1989) 170.
227. Nakamura, T. and Homa, Y. *J. Agricol. Food Chem.* **36** (1988) 15.
228. Ozawa, Y., Kawakishi, S., Uda, Y. and Maeda, Y. *Agricol. Biol. Chem.* **54** (1990) 1241.
229. Yang, B., Chang, X. and Pan, X. *Chinese Patent No.* 1 033 459 (1989).
230. Cai, C., Dong, H., Dong, X. and Dong, T. *Chinese Patent No.* 1 031 549 (1989).
231. Aarnio, T. and Agathos, S. *Appl. Microb. Biotechnol.* **33** (1990) 435.
232. Nochida, K., Uejima, T., Ochiai, K. and Kawamoto, I. *Japan Patent No.* 02 138 996 (1990).
233. Kumar, P.K.R. and Lonsane, P.K. *Appl. Microbiol. Biotech.* **28** (1988) 537.
234. Sasaki, M., Bandai, S. and Kamiyama, H. *Japan Patent No.* 63 255 293 (1988).
235. Ishiguro, Y., Morita, K. and Ito, Y. *Japan Patent No.* 63 245 666 (1988).
236. Cioppa, G., Garger, S.J., Sverlow, G.G., Turpen, T.H. and Grill, L.K. *Biotechnol.* **8** (1990) 634.
237. Koda, T. *Japan Patent No.* 62 181 796 (1987).
238. Ota, S. *Gekkan Fudo Kemikaree* **4** (1988) 55.
239. Ota, S., Tamaoki, Y. and Niwada, S. *Japan Patent No.* 63 156 865 (1988).
240. Ota, S., Tamaoki, Y., Niwada, S. and Kawachi, M. *Japan Patent No.* 64 02 593 (1989).
241. Fukuzaki, E., Miyamoto, Y., Kikuma, M., Matsumura, T. and Hashimoto, Y. *Japan Patent No.* 63 109 786 (1988).
242. Oohashi, Y., Mizokami, H., Tomita, K., Hiraoka, N. and Fugimoto, K. *Japan Patent No.* 01 30 594 (1989).
243. Yamada, O. and Fugita, T. *Japan Patent No.* 62 210 992 (1987).
244. Li, S. and Zhu, W. *Zhiwu Xuebao* **32** (1990) 103.
245. Ueda, S. *Japan Patent No.* 63 230 774 (1988).
246. Ohba, R., Koba, Y. and Ueda, S. *Hakko Kogoku Kaishi* **65** (1987) 507.
247. Kargi, F. and Freidel, I. *Biotech. Lett.* **10** (1988) 409.
248. Saito, K., Daimon, E., Kusaka, K., Wakayama, S. and Sekino, Y. *Z. Naturforsch C. Biosci.* **43** (1988) 862.
249. Himeno, H. and Sano, K. *Agricol. and Biol. Chem.* **51** (1987) 2395.
250. Fakhrai, F. and Evans, P.K. *J. Exper. Bot.* **41** (1990) 47.
251. Sarama, K.S., Maesato, K., Hara, T. and Sonoda, Y. *J. Exper. Bot.* **41** (1990) 745.
252. Kilby, N.J. and Hunter, C.S. *Appl. Microbiol. Biotechnol.* **33** (1990) 448.
253. Trividi, N. *Biotechnology of Food Processing* (S. Harlander and T.P. LaBuza, Eds), Noyes Data Corp., Park Ridge, New Jersey (1986) pp.15–132.
254. Ilker, R. *Food Technol.* **41** (1987) 70.
255. Hosono, T. *Up-to-date Food Proc.* **22** (1987) 41.
256. Whitaker, R.J. and Evans, D.A. *Chemtech.* **17** (1987) 674.
257. Fontanel, A. and Tabata, M. *Nestlé Research News* (1987) 93.
258. Bomar, M.T. and Knopfel, S.A. Biosynthesis of Food Colorants by Microorganisms (T. Kutsch, Ed.), in *Research in the Service of Nutrition: A Tribute*. German Federation Research Inst. for Nutrition, Prague (1989) pp.253.
259. Knorr, D., Beaumont, M.D., Castor, C.S., Dornenburg, H., Gross, B., Pandya, Y. and Romagnoli, L.G. *Food Technol.* **44** (1990) 71.
260. Francis, F.J., Harborne, J.B. and Barker, W.G. *J. Food Sci.* **31** (1966) 583.

Index